Floods in a Changing Climate

Hydrologic Modeling

Hydrologic modeling of floods enables more accurate assessment of climate change impacts on flood magnitudes and frequencies. This book synthesizes various modeling methodologies available to aid planning and operational decision-making, with emphasis on methodologies applicable in data-scarce regions, such as developing countries. Topics covered include: physical processes that transform precipitation into flood runoff, flood routing, assessment of likely changes in flood frequencies and magnitudes under climate change scenarios, and use of remote sensing, GIS, and DEM technologies in modeling of floods to aid decision-making. Problems included in each chapter, and supported by links to available online data sets and modeling tools accessible at www.cambridge.org/mujumdar, engage the reader with practical applications of the models.

This is an important resource for academic researchers in the fields of hydrology, climate change, and environmental science and hazards, and will also be invaluable to professionals and policy-makers working in hazard mitigation, remote sensing, and hydrologic engineering.

This volume is the second in a collection of four books within the International Hydrology Series on flood disaster management theory and practice within the context of anthropogenic climate change. The other books are:

1 – Floods in a Changing Climate: Extreme Precipitation *by Ramesh Teegavarapu*
2 – Floods in a Changing Climate: Inundation Modelling *by Giuliano Di Baldassarre*
3 – Floods in a Changing Climate: Risk Management *by Slodoban Simonović*

P. P. Mujumdar is a Professor in the Department of Civil Engineering at the Indian Institute of Science (IISc), Bangalore, India, and holds an Associate Faculty position in the Divecha Center for Climate Change at IISc Bangalore. He is also Chairman of the Water Resources Management Committee of the International Association for Hydro-Environment Research (IAHR). Professor Mujumdar's area of specialization is water resources with a focus on climate change impacts that includes downscaling and uncertainty modeling. He also works as a professional consultant across areas including floodplain management, reservoir operation, urban storm water drainage, lift irrigation, and hydropower development. He is a Distinguished Visiting Fellow of the Royal Academy of Engineering, UK, and is currently a member of the Editorial Board of the journal *Advances in Water Resources*.

D. Nagesh Kumar is a Professor in the Department of Civil Engineering at the Indian Institute of Science (IISc), Bangalore, India, and holds an Associate Faculty position in the Center for Earth Sciences. His primary research areas include applications of remote sensing and GIS in hydrologic modeling, soft computing, hydrologic teleconnections, and water resources systems. Professor Kumar also works in professional consultancy in areas such as river basin planning and management, flood routing, downscaling for climate change projections, and reservoir operation. He is currently an Associate Editor of the *Journal of Hydrologic Engineering*.

INTERNATIONAL HYDROLOGY SERIES

The **International Hydrological Programme** (IHP) was established by the United Nations Educational, Scientific and Cultural Organization (UNESCO) in 1975 as the successor to the International Hydrological Decade. The long-term goal of the IHP is to advance our understanding of processes occurring in the water cycle and to integrate this knowledge into water resources management. The IHP is the only UN science and educational programme in the field of water resources, and one of its outputs has been a steady stream of technical and information documents aimed at water specialists and decision-makers.

The **International Hydrology Series** has been developed by the IHP in collaboration with Cambridge University Press as a major collection of research monographs, synthesis volumes, and graduate texts on the subject of water. Authoritative and international in scope, the various books within the series all contribute to the aims of the IHP in improving scientific and technical knowledge of fresh-water processes, in providing research know-how and in stimulating the responsible management of water resources.

TITLES IN PRINT IN THIS SERIES

M. Bonell, M. M. Hufschmidt and J. S Gladwell *Hydrology and Water Management in the Humid Tropics: Hydrological Research Issues and Strategies for Water Management*
Z. W. Kundzewicz *New Uncertainty Concepts in Hydrology and Water Resources*
R. A. Feddes *Space and Time Scale Variability and Interdependencies in Hydrological Processes*
J. Gibert, J. Mathieu and F. Fournier *Groundwater/Surface Water Ecotones: Biological and Hydrological Interactions and Management Options*
G. Dagan and S Neuman *Subsurface Flow and Transport: A Stochastic Approach*
J. C. van Dam *Impacts of Climate Change and Climate Variability on Hydrological Regimes*
D. P. Loucks and J. S. Gladwell *Sustainability Criteria for Water Resource Systems*
J. J. Bogardi and Z. W. Kundzewicz *Risk, Reliability, Uncertainty, and Robustness of Water Resource Systems*
G. Kaser and H. Osmaston *Tropical Glaciers*
I. A. Shiklomanov and J. C. Rodda *World Water Resources at the Beginning of the Twenty-First Century*
A. S. Issar *Climate Changes during the Holocene and their Impact on Hydrological Systems*
M. Bonell and L. A. Bruijnzeel *Forests, Water and People in the Humid Tropics: Past, Present and Future Hydrological Research for Integrated Land and Water Management*
F. Ghassemi and I. White *Inter-Basin Water Transfer: Case Studies from Australia, United States, Canada, China and India*
K. D. W. Nandalal and J. J. Bogardi *Dynamic Programming Based Operation of Reservoirs: Applicability and Limits*
H.S. Wheater, S. Sorooshian and K.D. Sharma *Hydrological Modelling in Arid and Semi-Arid Areas*
J. Delli Priscoli and A. T. Wolf *Managing and Transforming Water Conflicts*
H. S. Wheater, S. A. Mathias and X. Li *Groundwater Modelling in Arid and Semi-Arid Areas*
L. A. Bruijnzeel, F. N. Scatena and L. S. Hamilton *Tropical Montane Cloud Forests*
S. Mithen and E. Black *Water, Life and Civilization: Climate, Environment and Society in the Jordan Valley*
K.A. Daniell *Co-Engineering and Participatory Water Management*
R. Teegavarapu *Floods in a Changing Climate: Extreme Precipitation*
P. P. Mujumdar and D. Nagesh Kumar *Floods in a Changing Climate: Hydrologic Modeling*
G. Di Baldassarre *Floods in a Changing Climate: Inundation Modelling*
S. Simonović *Floods in a Changing Climate: Risk Management*

Floods in a Changing Climate

Hydrologic Modeling

P. P. Mujumdar and D. Nagesh Kumar

Indian Institute of Science, Bangalore

CAMBRIDGE
UNIVERSITY PRESS

University Printing House, Cambridge CB2 8BS, United Kingdom

One Liberty Plaza, 20th Floor, New York, NY 10006, USA

477 Williamstown Road, Port Melbourne, VIC 3207, Australia

4843/24, 2nd Floor, Ansari Road, Daryaganj, Delhi - 110002, India

79 Anson Road, #06-04/06, Singapore 079906

Cambridge University Press is part of the University of Cambridge.

It furthers the University's mission by disseminating knowledge in the pursuit of education, learning and research at the highest international levels of excellence.

www.cambridge.org
Information on this title: www.cambridge.org/9781108447027

First published 2012
First paperback edition 2017

A catalogue record for this publication is available from the British Library

Library of Congress Cataloging in Publication data
Mujumdar, P. P., 1958–
Floods in a changing climate. Hydrologic modeling / P. P. Mujumdar, D. Nagesh Kumar.
pages cm. – (International hydrology series)
Includes bibliographical references and index.
ISBN 978-1-107-01876-1
1. Hydrologic models – Remote sensing. 2. Flood forecasting. 3. Flood damage prevention. 4. Climatic changes – Environmental aspects. I. Nagesh Kumar, D. (Dasika), 1963– II. Title.
GB656.2.H9M85 2012
511.48′9011 – dc23 2012015666

ISBN 978-1-107-01876-1 Hardback
ISBN 978-1-108-44702-7 Paperback

Contents

Color plates appear between pages 18 and 19.

Foreword

How much *higher* does the flood wall need to be built? How much *larger* does the reservoir spillway need to be? How much *more* capacity is needed in the urban drainage system? These are the type of questions that floods are confronting, given expected changes in rainfall intensity, timing and volume of peak flows, and sea levels under climate change.

There are no simple answers. In fact, climate change is just one driver of flood risk amongst many others. Rapid urbanization, development in floodplains, removal of natural vegetation, artificial drainage, and so forth, can all compound flooding. Therefore, it is helpful to regard the climatic hazard as a multiplier of flood risk that may already be increasing because of other socio-economic trends and/or human changes to the landscape. It then follows that the adaptation responses will involve cooperation amongst many disciplines: the hydrologist to quantify the impact of environmental changes on river flows; the economist to weigh up the costs and benefits of different options; the geographer to see the broader socio-physical context of the scheme and non-structural choices; the engineer to implement the design and devise operational rules; and the politician to persuade taxpayers to pay for the project!

The unavoidable truth is that structural measures for reducing flood risk need hard numbers to dimension the flood wall, spillway, and drainage system. This is challenging enough under present climate conditions because we seldom have long enough hydrometeorological records (if any); the records we have may be corrupted by other long-term changes in the river basin; and there is plenty of subjectivity in our methods for estimating extreme flow statistics. To add climate change to this situation is to present a technical challenge steeped in uncertainty. Meanwhile, the engineer still needs numbers to get the project underway.

The scientific community offers the practitioner a bewildering set of methodologies. For example, there are many different ways of conjuring future rainfall intensity: regional climate downscaling (using physical or statistical models); Monte Carlo and re-sampling techniques for constructing large synthetic rainfall series based on available data; or just specifying a plausible range of values for sensitivity testing. Indeed, given the large and irreducible uncertainties in climate model predictions, a strong case can often be made for simpler methods. Depending on the risk–reward, a climate change safety margin may be set without even referring to climate models. In situations where the structure has a short (less than 20 years) life-span, the concern will be climate variability not climate change. In other heavily discounted cases, future climate changes may be irrelevant to the viability of a scheme and shorter-term goals may be paramount.

Despite the large uncertainty in future flood risk, hydrologists still have plenty to offer adaptation planners. More accurate, timely, and clearly articulated flood warnings will continue to improve the safety of vulnerable populations living with unavoidable flooding. Simulations of flood depths and areas will help planners to zone floodplains and regulate development in hazardous places. Field monitoring and modeling will contribute to improved understanding of the physical processes that generate extreme floods in different environments. Remotely sensed data and seasonal forecasts can inform the operation of infrastructure to optimize benefits across multiple objectives (such as flood control, hydropower production, and water supply). Such activities are not always perceived as adaptation, but they do contribute to flood risk management just as much as the wall, spillway, and drainage system.

The authors of this book are to be commended for guiding practitioners, step-by-step, through some of the analytical tools and techniques for evaluating flood risk under non-stationary climate conditions. The true benefit will come from recognizing the part played by these approaches within a larger armoury of strategies for managing flood hazards over coming decades and beyond.

Robert L. Wilby
Professor of Hydroclimatic Modelling,
Department of Geography,
Loughborough University, UK

Preface

This book has been developed under the coordination of Professor Siegfried Demuth, International Hydrology Programme, Chief of the Hydrological Systems and Global Change Section and Scientist in Charge of the International Flood Initiative, and Dr. Biljana Radojevic of the Division of Water Sciences, UNESCO.

The book presents methodologies for hydrologic modeling of floods and for assessing climate change impacts on hydrology, with specific focus on impacts on flood magnitudes and frequencies. The following topics are covered with a view to training the reader in the use of hydrologic models for assessing climate change impacts: (i) review of basic hydrology that includes physical processes transforming precipitation into flood runoff, intensity–duration–frequency relationships, and flood routing; (ii) recent modeling tools of artificial neural networks and fuzzy inference systems for river discharge forecasting and flood routing; (iii) use of climate projections provided by the general circulation models (GCMs) for assessing likely changes in flood frequencies; and (iv) use of remote sensing, global information system (GIS) and digital elevation model (DEM) technologies in modeling of floods to aid decision-making.

The novelty of the book lies in synthesizing various methodologies available to help in planning and operational decision-making, in the face of climate change. Exercise problems discussed in the text and those presented at the end of the chapters are intended to train a reader on applications in practical settings. Two case studies are discussed in detail in Chapter 6 to help the reader understand the use of the methodologies for long-term planning, in real situations. The book is intended to serve as a reference document to practicing hydrologists, especially those with responsibilities of assisting in decision-making with respect to floods, graduate students, professionals, researchers, and members of governmental and non-governmental agencies interested in impacts of changing climate on floods.

Bringing together the basic material available in classical textbooks on hydrology and the advanced material available only in the recently published research papers in a coherent form in a book is a challenge. The authors were confronted with the difficulty of presenting fundamental topics, such as estimation of flood runoff, hydrologic flood routing, and frequency analysis, that are available in standard text books along with advanced topics of statistical downscaling of GCM simulations, fuzzy inference systems, and use of satellite remote sensing in hydrologic modeling. The authors believe that such a mix of topics will be of immense use to a new entrant to the field of hydrologic modeling for climate change impact assessment.

The book comprises six chapters. Chapter 1 provides an introduction to climate change impacts on hydrology, and sets the tone for the remainder of the book. Chapter 2 provides the necessary background on basic hydrology and use of hydrologic models for planning and management of floods. Chapter 3 presents methodologies commonly employed to obtain projections of flood magnitudes and frequencies under future climate change, with GCMs and hydrologic models. A review of techniques for downscaling large-scale atmospheric variables to station-scale hydrologic variables is also provided in this chapter. Chapter 4 introduces remote sensing (RS), covering the topics of spectral reflectance curves, RS platforms, digital images, image processing including rectification, enhancement, and information extraction. The role of RS in hydrologic modeling is discussed through examples. Use of GIS and DEMs in hydrologic modeling is discussed in Chapter 5. Synthesis of climate change impacts, uncertainties, RS, GIS, DEM, and hydrologic models is demonstrated through two case studies in Chapter 6. Future perspectives on hydrologic modeling under climate change are presented in Chapter 6.

The topic of climate change impact assessment is of recent origin, and the authors had to rely on contemporary research, especially in the chapters dealing with climate change impact assessment and use of remote sensing and GIS for hydrologic modeling. The authors have used material from their own recently published research papers extensively in the book.

The contribution to Chapter 3 by Dr. Deepashree Raje, a postdoctoral research associate at IISc Bangalore when the first draft of the book was prepared, was significant and valuable. Considerable help was provided by Ms. J. Indu, research student, in the preparation of Chapter 5. The authors are grateful to their students, Ms. Arpita Mondal, Mr. Ujjwal Saha, Ms. Divya Bhatt, Mr. Laxmi Raju, Dr. Reshmidevi, Dr. Dhanya, and Ms. Sonali Pattanayak for their help in providing solutions to the numerical examples and in preparation of figures. Ms. Chandra Rupa, a Project Associate at IISc, helped the authors in the preparation of numerical examples and in formatting the figures, tables, and text. Her assistance is gratefully acknowledged.

Glossary

Climate change Climate change refers to a change in the state of the climate that can be identified by changes in the mean and/or the variability of its properties, and that persists for an extended period, typically decades or longer. Climate change may be due to natural internal processes or external forcings, or to persistent anthropogenic changes in the composition of the atmosphere or in land use (IPCC, 2007).

Climate model A numerical representation of the climate system based on the physical, chemical, and biological properties of its components, their interactions and feedback processes, and accounting for all or some of its known properties (IPCC, 2007).

Climate projection A projection of the response of the climate system to emission or concentration scenarios of greenhouse gases and aerosols, or radiative forcing scenarios, often based upon simulations by climate models. Climate projections are distinguished from climate predictions in order to emphasize that climate projections depend upon the emission/concentration/ radiative forcing scenarios used, which are based on assumptions concerning, for example, future socio-economic and technological developments that may or may not be realized and are therefore subject to substantial uncertainty (IPCC, 2007).

Climate scenario A plausible and often simplified representation of the future climate, based on an internally consistent set of climatological relationships, that has been constructed for explicit use in investigating the potential consequences of anthropogenic climate change, often serving as input to impact models (IPCC, 2007).

Climate variability Climate variability refers to variations in the mean state and other statistics (such as standard deviations, the occurrence of extremes, etc.) of the climate on all spatial and temporal scales beyond that of individual weather events. Variability may be due to natural internal processes within the climate system (internal variability), or to variations in natural or anthropogenic external forcing (external variability) (IPCC, 2007).

Digital elevation models (DEMs) DEMs are data files that contain the elevation of the terrain over a specified area, usually at a fixed grid interval over the surface of the Earth. The intervals between each of the grid points are always referenced to some geographical coordinate system such as latitude/longitude or Universal Transverse Mercator (UTM) coordinates. When the grid points are located closer together, more detailed information is available in the DEM. The details of the peaks and valleys in the terrain are better modeled with a closer grid spacing than when the grid intervals are very large.

Downscaling Downscaling is a method that derives local- to regional-scale (10 to 100 km) information from larger-scale models or data analyses. Two main methods are distinguished: dynamical downscaling and empirical/statistical downscaling. The dynamical method uses the output of regional climate models, global models with variable spatial resolution, or high-resolution global models. The empirical/statistical methods develop statistical relationships that link the large-scale atmospheric variables with local/regional climate variables. In all cases, the quality of the downscaled product depends on the quality of the driving model (IPCC, 2007).

Emissions scenario A scenario is a plausible representation of the future development of emissions of substances that are potentially radiatively active (e.g., greenhouse gases, aerosols), based on a coherent and internally consistent set of assumptions about driving forces (such as demographic and socio-economic development, technological change) and their key relationships. Concentration scenarios, derived from emissions scenarios, are used as input to a climate model to compute climate projections (IPCC, 2007).

Geographic information systems (GIS) GIS is a system of hardware and software used for storage, retrieval, mapping, and analysis of geographic data. Spatial features are stored in a latitude/longitude or UTM coordinate system, which references a particular place on the Earth. Descriptive attributes in tabular form are associated with spatial features. Spatial data and associated attributes in the same coordinate system can then be layered together for mapping and analysis. GIS can be used for scientific investigations, resource management, and development planning.

Remote sensing Remote sensing (RS) is the art and science of obtaining information about an object or feature without physically coming in contact with that object or feature. It is the process of inferring surface parameters from measurements of the electromagnetic radiation (EMR) from the Earth's surface. This EMR can be either reflected or emitted radiation from the Earth's surface. In other words, RS is detecting and measuring electromagnetic energy emanating or reflected from distant objects made of various materials, to identify and categorize these objects by class or type, substance, and spatial distribution (American Society of Photogrammetry, 1975).

Troposphere The lower part of the terrestrial atmosphere, extending from the Earth's surface up to a height varying from about 9 km at the poles to about 17 km at the equator, in which the temperature decreases fairly uniformly with height.

Abbreviations

AMC	Antecedent Moisture Content
ANN	Artificial Neural Network
AOGCM	Atmospheric Oceanic General Circulation Model
ARI	Average Recurrence Interval
AVSWAT	ArcView Soil and Water Assessment Tool
BP	Back Propagation
CCA	Canonical Correlation Analysis
CGCM	Coupled General Circulation Model
CN	Curve Number
CP	Circulation Pattern
CRF	Conditional Random Field
CWG	Conditional Weather Generator
DEM	Digital Elevation Model
DHM	Distributed Hydrologic Model
DN	Digital Number
EMR	ElectroMagnetic Radiation
ENSO	El Niño–Southern Oscillation
FCC	False Color Composite
FDFRM	Fuzzy Dynamic Flood Routing Model
GCM	General Circulation Model (Global Climate Model)
GEV	Generalized Extreme Value
GHG	GreenHouse Gas
GIS	Geographic Information System
HEC-HMS	Hydrologic Engineering Center – Hydrologic Modeling System
HEC-RAS	Hydrologic Engineering Center – River Analysis System
HRU	Hydrologic Response Unit
IDF	Intensity–Duration–Frequency
IMD	India Meteorological Department
IPCC	Intergovernmental Panel on Climate Change
IR	Infrared
ISRO	Indian Space Research Organisation
KNN	K-Nearest Neighbor
LSAV	Large-Scale Atmospheric Variable
LSM	Land Surface Model
MATLAB	MATrix LABoratory
MF	Membership Function
MHM	Macroscale Hydrologic Model
MSLP	Mean Sea Level Pressure
NARR	North American Regional Reanalysis
NCAR	National Center for Atmospheric Research
NCEP	National Center for Environmental Prediction
NHMM	Non-homogeneous Hidden Markov Model
NIR	Near Infrared
NRCS	Natural Resources Conservation Service (formerly known as Soil Conservation Service)
NRMSE	Normalized Root Mean Square Error
PCA	Principal Component Analysis
PDO	Pacific Decadal Oscillation
RBF	Radial Basis Function
RCM	Regional Climate Model
RMSE	Root Mean Square Error
RS	Remote Sensing
SCS	Soil Conservation Service (now known as Natural Resources Conservation Service)
SDSM	Statistical DownScaling Model
SMOS	Soil Moisture and Ocean Salinity
SPI	Standardized Precipitation Index
SRES	Special Report on Emissions Scenarios
SRTM	Shuttle Radar Topography Mission
SSFI	Standardized Stream Flow Index
SVL	Soil–Vegetation–Land
SVM	Support Vector Machine
SWAT	Soil and Water Assessment Tool
SWMM	Storm Water Management Model
TCC	True Color Composite
TRMM	Tropical Rainfall Measuring Mission
UNESCO	United Nations Education, Social and Cultural Organisation
USACE	United States Army Corps of Engineers
USEPA	United States Environmental Protection Agency
USGS	United States Geological Survey
VCS	Variable Convergence Score
VIC	Variable Infiltration Capacity
VPMS	Variable Parameter Muskingum Stage hydrograph
WG	Weather Generator

1 Introduction

Adequate scientific evidence (e.g., IPCC, 2007) exists now to show that the global climate is changing. The three prominent signals of climate change, namely, increase in global average temperature, rise in sea levels, and change in precipitation patterns, convert into signals of regional-scale hydrologic change in terms of modifications in extremes of floods and droughts, water availability, water demand, water quality, salinity intrusion in coastal aquifers, groundwater recharge, and other related phenomena. Increase in atmospheric temperature, for example, is likely to have a direct impact on river runoff in snow-fed rivers and on the evaporative demands of crops and vegetation, apart from its indirect impacts on all other processes of interest in hydrology. Similarly, a change in the regional precipitation pattern may have a direct impact on magnitude and frequency of floods and droughts and water availability. Changes in precipitation patterns and frequencies of extreme precipitation events, along with changes in soil moisture and evapotranspiration, will affect runoff and river discharges at various time scales from sub-daily peak flows to annual variations. At sub-daily and daily time scales, flooding events are likely to cause enormous socio-economic and environmental damage, which necessitates the use of robust and accurate techniques for projections of flood frequencies and magnitudes under climate change, and development of flood protection measures to adapt to the likely changes. Knowledge about hydrologic modeling and use of GCMs is critical in planning and operation for flood management under climate change. The main objective of this book is to provide a basic background on hydrologic modeling, impact assessment methods, uncertainties in impacts, and use of satellite products for enhancing the capability of the models.

This chapter provides the background and a brief introduction to the topics covered in the book.

1.1 HYDROLOGIC MODELS

Hydrologic models are concerned with simulating natural processes related to movement of water, such as the flow of water in a stream, evaporation and evapotranspiration, groundwater recharge, soil moisture, sediment transport, chemical transport, growth of microorganisms in water bodies, etc. The hydrologic processes that occur in nature are *distributed*, in the sense that the time and space derivatives of the processes are both important. The hydrologic models are classified as distributed models or lumped models depending on whether the models consider the space derivatives (distributed) or not (lumped). The semi-distributed models account for spatial variations in some processes while ignoring them in others. On any time scale, the models may be discrete or continuous in time. Flood management requires models for two consecutive phases: planning and operation, which demand different kinds of models. Plate (2009) provides a classification of hydrologic models specifically for flood management. The different levels of hydrologic models considered by Plate (2009) are: (i) the data level, consisting of the GIS and data banks at different time scales, such as seasonal and event scales; (ii) the model level, consisting of (a) a basic hydrologic model incorporating the topography, digital terrain models, channel networks, sub-catchments, and long-term water and material balance and (b) a transport model operating at seasonal and event scales; (iii) the output level, which provides outputs in terms of maps and tables for use in decision-making; and (iv) the decision level, which uses information provided by the output level, for arriving at management decisions. Hydrologic models for floods function on the basis of partitioning the rainfall into various components. Several hydrologic models are available to estimate the peak flood discharge, flood hydrograph at specified locations in a catchment, and for flood routing. The models differ essentially with respect to the methods used for estimating the various hydrologic components and assumptions made, and with respect to how they account for the distributed processes on spatial scales. Hydrologic models typically operate at a river basin or a watershed scale. They play a significant role in providing an understanding of a range of problems dealing with water resources and hydrologic extremes at river basin and watershed scales. These problems could be, for example, the magnitude and duration of flood discharges for specified intensities and durations of rainfall, movement of a flood wave along a river extent of water backing up due to an obstruction to flow caused by a dam and other structure, sediment deposition and bank erosion, and so on. The inputs required by hydrologic models depend on the purpose for which the model is built. A river flow simulation model, for example, will need inputs such as precipitation, catchment characteristics such as the soil type, slope of the catchment, type of vegetation, type of land use, temperature, solar radiation groundwater contribution, etc. The typical output from such a model includes the river flow at a location during a period (such as a day, a week, or a month), and soil moisture and evapotranspiration during the period.

A number of hydrologic models, with user-friendly interfaces, are freely available today for useful applications. These include the hydrologic models developed by the US Army Corps of Engineers (e.g., the HEC-HMS and the HEC-RAS), the SWMM developed by the US Environmental Protection Agency (USEPA), the AVSWAT (Neitsch *et al.*, 2000, 2001), and the VIC (Liang *et al.*, 1994; Gao *et al.*, 2010) models. Such models, along with the global data sets, are extremely useful in assessing the possible impacts of climate change on regional hydrology, especially in developing countries with limited resources, data, and capacity. Recently developed empirical models, such as those based on artificial neural networks and fuzzy logic, are useful in real-time flood forecasting.

1.2 REMOTE SENSING FOR HYDROLOGIC MODELING

Remote sensing (RS) is the art and science of obtaining information about an object or feature without physically coming into contact with that object or feature. It is the process of inferring surface parameters from measurements of the reflected electromagnetic radiation (EMR) from the Earth's surface. Satellite RS provides the essential inputs required for more effectively modeling different components of the hydrologic cycle. Remote sensing provides a means of observing hydrologic states or fluxes over large areas with a synoptic view. Remote sensing applications in hydrology have primarily focused on developing approaches for estimating hydrometeorological states and fluxes. The primary set of variables include land use, land cover, land surface temperature, near-surface soil moisture, snow cover/water equivalent, water quality, landscape roughness, and vegetation cover. The hydrometeorological fluxes of interest are evaporation and plant transpiration (or evapotranspiration) and snowmelt runoff.

The spectral reflectance curve of a particular feature is a plot of the spectral reflectance of that feature versus wavelength. Spectral reflectance curves of different features on the Earth's surface, such as vegetation, soil, and water, help in identifying various spectral bands that are useful in estimating different land surface features. For example, for vegetation monitoring, the red and near-infrared (NIR) bands of EMR are useful.

The concepts of color composites and false color composites are essential to visualize and interpret the spectral reflectance information available beyond the visible region, such as in the NIR and thermal-IR regions of EMR. It is important to understand various characteristics of satellite RS systems, such as spatial resolution, spectral resolution, radiometric resolution, and temporal resolution, to interpret digital images. Digital image processing techniques are continually evolving for rectification, enhancement, and information extraction of satellite images. Information extraction from microwave and hyperspectral images requires special image processing techniques.

Satellite RS has played a significant role in identification of potential flood zones, flood hazard estimation, flood inundation mapping and mitigation. The distributed hydrologic models may be rendered more effective with use of data from satellite RS. In the recent past, specialized RS satellites and space missions have been launched to obtain detailed information on soil moisture and hydrometeorological fluxes. They include the Soil Moisture and Ocean Salinity (SMOS) Satellite, the Global Precipitation Measurement (GPM) Mission, the Tropical Rainfall Measuring Mission (TRMM), and Megha-Tropiques. Data available from these and other similar satellite missions will facilitate better modeling of various components of the hydrologic cycle and better estimation of hydrometeorological states and fluxes.

Many regions on the globe, such as oceans, deserts, polar regions, and uninhabited regions, suffer from lack of ground-based measurements of hydrometeorological variables. Prediction in ungauged basins is a major challenge for the hydrologist. Satellite RS plays a vital role in providing the essential inputs required in addressing this challenge.

1.3 GIS AND DEM FOR HYDROLOGIC MODELING

A geographic information system (GIS) is an excellent tool for stacking, analyzing, and retrieving large numbers of non-spatial and geo-spatial databases including RS images. It is essential to understand representations of spatial features and various formats in which spatial databases are stored on GIS. Distance-based proximity tools in GIS are useful for analyzing and interpreting multiple spatial databases. Web-based GIS facilitate integration of various spatial and non-spatial databases available at different locations for hydrologic modeling and dissemination of the results and decisions to remote locations in real time.

Digital elevation models (DEMs) are powerful tools to analyze the digital elevation data available at regular grid spacing. DEMs provide essential inputs about the topographical features of a river basin, such as slope/aspect, flow direction, flow pathways, flow accumulation, stream network, catchment area, upstream contributing area for each grid cell, etc. The D8 algorithm is very useful in interpreting digital elevation data to extract the above-mentioned features. Satellite and space shuttle missions are launched specifically to provide digital elevation data over the globe using radar interferometry and light detection and ranging (LIDAR). These missions include the Shuttle Radar Topography Mission (SRTM), the Advanced Spaceborne Thermal Emission and Reflection Radiometer (ASTER) onboard the Terra satellite and LIDAR sensors.

Integration of RS, GIS, and DEM into a distributed hydrologic model will significantly improve the modeling of various components of the hydrologic cycle. Such integration has played a vital role in near-real-time monitoring of flood events, and their damage estimation and mitigation.

1.4 ASSESSMENT OF CLIMATE CHANGE IMPACTS

It is important to distinguish between climate change and climate variability. *Climate change* refers to a change in the state of the climate that persists for an extended period, typically decades or longer, as distinct from *climate variability*, which refers to variations in the mean state and other climate statistics, on all space and time scales. The year to year variations in the rainfall at a location, for example, indicate climate variability, whereas a change in the long-term mean rainfall over a few decades is a signal of climate change. In this book, we are concerned with assessment of the hydrologic impacts of *climate change*.

Climate change is generally expected to increase the intensity (flood discharges) and duration of floods. However, there will be a large variation in how the hydrology of different regions responds to signals of climate change. Regional assessment of the impacts of climate change is therefore important. A commonly adopted methodology for assessing the regional hydrologic impacts of climate change is to use the climate projections provided by the GCMs for specified emissions scenarios in conjunction with the process-based hydrologic models to generate the corresponding hydrologic projections. The scaling problem arising because of the large spatial scales at which the GCMs operate compared to those required in most distributed hydrologic models is commonly addressed by downscaling the GCM simulations to hydrologic scales. This commonly used procedure of impact assessment is burdened with a large amount of uncertainty due to the choice of GCMs and emissions scenarios, small samples of historical data against which the models are calibrated, downscaling methods used, and several other sources. Development of procedures and methodologies to address such uncertainties is a current area of research. Vulnerability assessment, adaptation to climate change, and policy responses all depend on the projected impacts, with quantification of the associated uncertainty.

General circulation models, also commonly known as global climate models, are the most credible tools available today for projecting the future climate. The GCMs operate on a global scale. They are used for weather forecasting, understanding climate, and projecting climate change. They use quantitative methods to simulate the interactions of the atmosphere, oceans, land surface, and ice. The most frequently used models in the study of climate change are the ones relating air temperature to emissions of carbon dioxide. These models predict an upward trend in the surface temperature, on a global scale. A GCM uses a large number of mathematical equations to describe physical, chemical, and biological processes such as wind, vapor movement, atmospheric circulation, ocean currents, and plant growth. A GCM relates the interactions among the various processes. For example, it relates how the wind patterns affect the transport of atmospheric moisture from one region to another, how ocean currents affect the amount of heat in the atmosphere, and how plant growth affects the amount of carbon dioxide in the atmosphere and so on. The models help in providing an understanding of how the climate works and how the climate is changing. A typical climate model projection used in the impact studies is that of global temperatures over the next century. GCMs project an increasing trend in the global average temperature over the next century, with some estimates even showing an increase of more than 4 °C with respect to the temperature during 1980–99 (e.g., see IPCC, 2007). Such projections of temperature and other climate variables provided by GCMs are used to obtain projections of other variables of interest (but which are not well simulated by GCMs), such as precipitation and evapotranspiration, in the impact studies.

GCMs are more skillful in simulating the free troposphere climate than the surface climate. Variables such as wind, temperature, and air pressure can be predicted quite well, whereas precipitation and cloudiness are less well predicted. Other variables of key importance in the hydrologic cycle, such as runoff, soil moisture, and evapotranspiration are not well simulated by GCMs. Runoff predictions in GCMs are over-simplified and there is no lateral transfer of water within the land phase between grid cells (Xu, 1999a; Fowler *et al.*, 2007). The GCM simulation of rainfall has been found to be especially poor. The ability of GCMs to predict spatial and temporal distributions of climatic variables declines from global to regional to local catchment scales, and from annual to monthly to daily amounts. This limitation becomes particularly pronounced in assessing likely impacts of climate change on flood frequencies and magnitudes of flood peak flows. Flood peak flows in a catchment are generated by high-intensity storms of durations typically ranging from a few hours to a few days. At these time scales the simulations provided by GCMs are almost of no direct consequence. Stochastic disaggregation techniques have been used to disaggregate the longer time simulations provided by the GCMs to the shorter time events necessary in flood hydrology studies. The spatial scale mismatch between the scales of GCM simulations (with grid size of the order of tens of thousands of square kilometers) and those typically required for hydrologic modeling (with spatial scales of the order of a few hundred square kilometers and less) is classically addressed by spatial downscaling.

The impacts of climate change on floods are essentially assessed in the planning context by addressing the likely changes in the frequencies of given magnitudes of flood discharges. Flood frequencies are believed to be increasing due to climate change. High-resolution regional climate projections are necessary for assessing, with reasonable confidence, such impacts on flood frequencies. To overcome the limitations due to the coarse resolution of most existing GCMs, approaches such as stochastic weather generators and delta change methods are employed to examine the likely change in flood frequency due to climate change. Quantification of uncertainties in the projected impacts is particularly critical in the context of flood management, due to the huge economic implications of the adaptation measures.

1.5 ORGANIZATION OF THE BOOK

This book presents methodologies for hydrologic modeling of floods and for assessing climate change impacts on flood magnitudes and frequencies. The following topics are covered with a view to training the reader in the use of hydrologic models with climate change scenarios: (i) physical processes that transform precipitation into flood runoff; (ii) flood routing; (iii) assessing likely changes in flood frequencies and magnitudes under climate change scenarios and quantifying uncertainties; and (iv) use of RS, GIS, and DEM technologies in modeling of floods to aid decision-making. This chapter, Chapter 1, sets the scene and provides an introduction to climate change impacts on hydrology.

The objective of Chapter 2 is to provide the necessary background on hydrologic models for use in planning and operations related to floods. The chapter includes a brief review of hydrologic models and presents the use of empirical models for flood forecasting and flood routing.

Chapter 3 presents methodologies commonly employed to obtain projections of floods under future climate change, with GCMs and hydrologic models. The use of climate change scenarios developed by the Intergovernmental Panel on Climate Change (IPCC) and issues of spatio-temporal scale mismatches between GCMs and hydrologic models are explained in this chapter. A review of techniques for downscaling large-scale atmospheric variables to station-scale hydrologic variables is also provided. Applications of macroscale hydrologic models and hypothetical scenarios for generating future projections as an alternative to downscaling are discussed.

Chapter 4 introduces RS, covering the topics of spectral reflectance curves, RS platforms, digital images, and image processing including rectification, enhancement, and information extraction. The role of RS in hydrologic modeling is elaborated. Recent satellite missions for precipitation and soil moisture estimation are elaborated. Image processing techniques are demonstrated using MATLAB.

Chapter 5 presents GIS and DEMs for hydrologic modeling. Representation of spatial objects and different data formats in GIS is discussed. Types of DEMs, sources of DEM data, and extraction of drainage pattern and sub-watersheds using the D8 algorithm are explained and illustrated. The roles of RS and GIS in flood zone and flood inundation mapping are explained in the chapter. Web-based GIS and its role in integrating various spatial and non-spatial databases available at different locations for hydrologic modeling are also discussed.

Synthesis of climate change impacts, uncertainties, RS, GIS, DEM, and hydrologic models is explained and demonstrated through two case studies in Chapter 6. Future perspectives on hydrologic modeling under climate change are presented in this chapter.

2 Hydrologic modeling for floods

2.1 INTRODUCTION

The objective of this chapter is to provide the necessary background on hydrologic models for use in planning and operations related to floods. Partitioning of precipitation into different components is discussed briefly first. These components include the channel flow, overland flow, unsaturated flow, groundwater flow, soil moisture storage, surface storage, infiltration, interception, and evapotranspiration. The concept of excess rainfall and direct runoff is introduced next. The commonly used SCS curve number method and the rational method to estimate the flood discharge in a watershed are introduced. Hydrologic and kinematic flood routing, empirical models of artificial neural networks, and fuzzy inference systems for forecasting river discharges and flood routing are discussed. A focus of the chapter is on the modeling approach to be adopted in data-scarce regions, especially in countries where many river basins are poorly gauged, and data on river discharges, soil types, land use patterns, and catchment characteristics are not readily available. Information on global data sets that may be useful in such situations is provided. A review of commonly used hydrologic models in decision-making for flood modeling is given, along with a list of references where model applications are discussed. On going through this chapter, it is expected that the reader will become well versed in procedures for estimating flood discharges and in the use of hydrologic models for estimating and predicting flood discharges for use in decision-making.

2.1.1 Partitioning of rainfall

Hydrologic models for floods function on the basis of partitioning the rainfall into various components. Several hydrologic models are available to estimate the flood hydrograph at specified locations in a catchment and for flood routing. The models differ essentially with respect to the methods used for estimating the various hydrologic components and assumptions made, and with respect to how they account for the distributed processes in spatial scales. In the context of floods, estimating the flood runoff volume and hydrograph resulting from a given storm is a critical exercise, and therefore generation of flood runoff from a storm is first discussed. Figure 2.1 shows the various processes that take place once the precipitation occurs. As the precipitation falls, part of it is *intercepted* by vegetation and other surfaces and this part will not be available for runoff immediately during a storm. Once the precipitation reaches the land surface, part of it may *infiltrate* into the soil. Part of the rainfall is also trapped by surface depressions including lakes, swamps, and smaller depressions down to the size of small grain size cavities. A small amount is also lost as *evaporation* from bare surfaces and, from vegetation, as *evapotranspiration*. The infiltrated water may join the stream (channel) as interflow or may add to the aquifer recharge and deep groundwater storage. The direct runoff, or rainfall excess, is that part of the rainfall from which all losses have been removed and which eventually becomes flood runoff. As seen from Figure 2.1, the direct runoff hydrograph consists of contributions from the channel input to the streamflow through various routes of overland flow, interflow (throughflow), and groundwater flow.

2.1.2 Overland flow

As the rainfall intensity increases, and exceeds the infiltration capacity of the soil, water starts running off in the form of a thin sheet on the land. This type of flow is called Hortonian overland flow. The runoff rate in a Hortonian overland flow may be simply estimated by $(I - f)$, where I is the rainfall intensity in cm/hr and f is the infiltration capacity of the soil, also in cm/hr. When the rainfall intensity is less than the infiltration capacity of the soil, all of the rainfall is absorbed by the soil as infiltration. Hortonion overland flow is the most commonly occurring overland flow. The sheet of overland flow is quite thin before it joins a channel, to become channel flow. The *detention storage* – the storage that is held by the sheet flow corresponding to the depth of overland flow – contributes continuously to the channel flow, whereas part of the *retention storage*, held by surface depressions, is released slowly to the streams in the form of subsurface flow or is lost as evaporation. Other parts of the retention storage may add to the infiltration and subsequently recharge the groundwater.

Hortonian overland flow occurs when the soil is saturated from above by precipitation. Saturated overland flow, on the other hand, occurs when the soil is saturated from below – most commonly

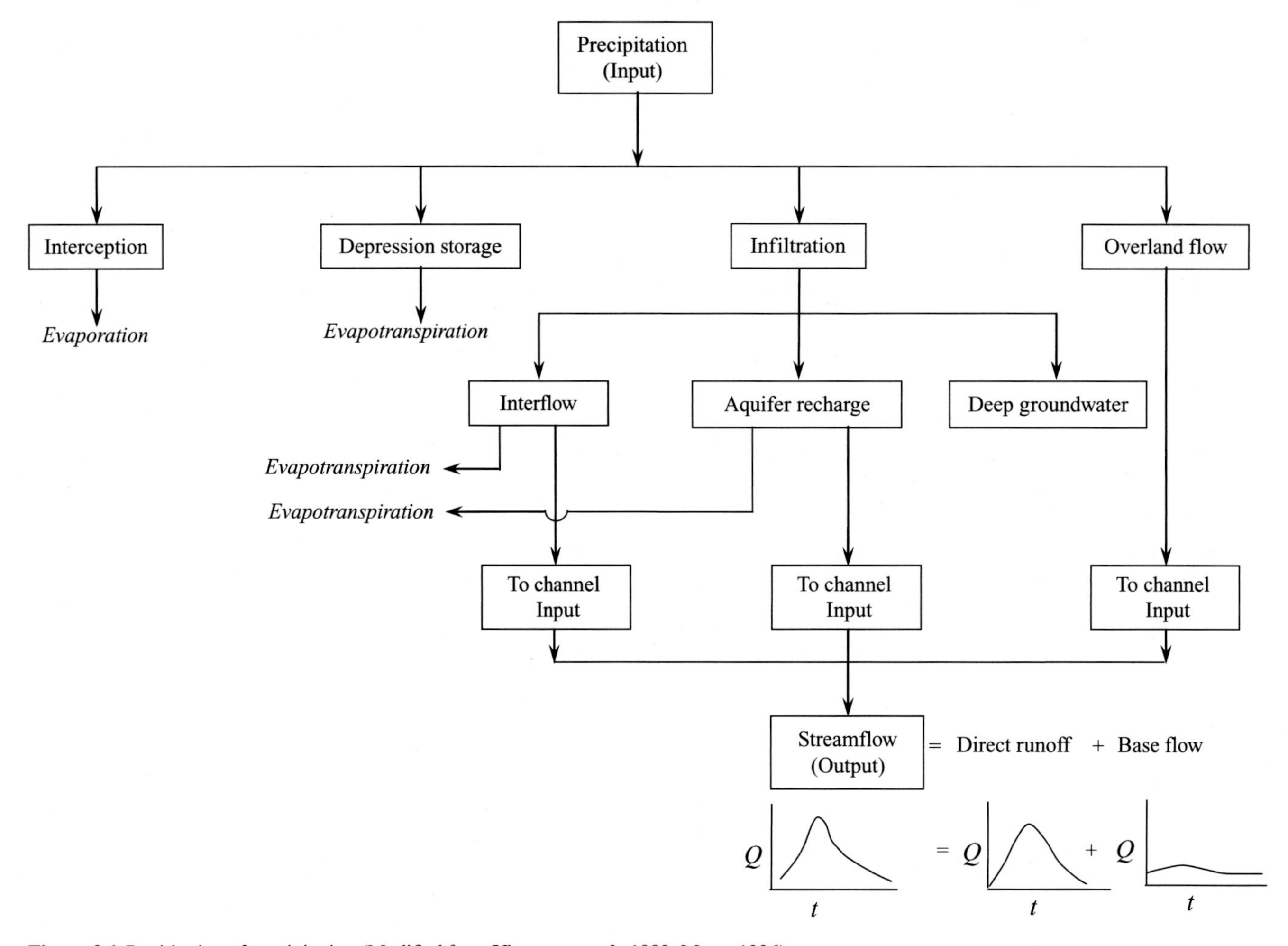

Figure 2.1 Partitioning of precipitation (Modified from Viessman *et al.*, 1989; Mays, 1996).

because of subsurface flow. Saturation overland flow occurs commonly at valleys and near river banks. *Throughflow* occurs through macropores in the soil such as cracks, animal holes, and roots. Throughflow reaches the stream channel relatively quickly.

All three types of overland flows – Hortonian overland flow, saturated overland flow, and throughflow – may occur simultaneously during a storm. It is also possible that only a part of a drainage basin – and not the entire basin – may be contributing to the flood runoff at a location. This part of the drainage area, called the *source area*, may be different for different storms in the drainage basin and may also change within the same storm as the storm evolves.

2.1.3 Excess rainfall and direct runoff

Excess rainfall is that part of the rainfall that directly contributes to the runoff – it is neither retained in storage nor is lost as infiltration, interception, and evapotranspiration. Direct runoff is caused by excess rainfall after it travels over the surface as Hortonian overland flow. In flood studies, obtaining the direct runoff hydrograph (DRH) from an observed total runoff hydrograph is an important step. Procedures for obtaining a DRH from a total runoff hydrograph and methods of estimation of various losses or abstractions are available in standard textbooks (e.g., Chow *et al.*, 1988; Singh, 1992), and are not discussed here.

Two procedures – the SCS curve number method and the rational formula – for estimation of flood runoff are discussed here. These methods may be used for estimating flood runoff and flood peaks for hydrologic designs, even with limited data. Many commonly used hydrologic models employ these methods for flood runoff estimation.

2.2 ESTIMATION OF FLOOD PEAK DISCHARGE

2.2.1 Soil Conservation Service curve number method

Estimating flood runoff from a given storm involves estimating losses from the rainfall. The Soil Conservation Service (SCS) – now called the Natural Resources Conservation Service (NRCS) –

curve number method (Soil Conservation Service, 1969) is the most commonly used and simple method for practical applications. It is based on accounting for infiltration losses from rainfall depending on the antecedent moisture content (AMC) and the soil type. The rainfall is assumed to occur uniformly over the entire watershed, during the storm. The fundamental basis for the SCS curve number method is that the runoff starts after initial losses due to abstractions, I_a, are accounted for. These losses consist of interceptions due to vegetation and built area that prevent the rainfall from reaching the ground immediately after it occurs, surface storage consisting of water bodies such as lakes, ponds, and depressions, and infiltration. An assumption in developing the curve numbers is that the ratio of actual retention of rainfall in the watershed to potential retention, S, in the watershed is equal to the ratio of direct runoff to rainfall minus the initial abstractions, I_a (before commencement of the runoff).

The parameter S depends on the catchment characteristics of soil, vegetation, and land constituting the soil–vegetation–land (SVL) complex (Singh, 1992), and the AMC. With a parameter, CN, to represent the relative measure of water retention on the watershed by a given SVL complex, the potential retention in a watershed with a given SVL complex is calculated as

$$S = \frac{25400}{\text{CN}} - 254 \text{ in millimeters} \tag{2.1}$$

The parameter CN is called the curve number; it takes values between 0 and 100. The value of CN depends on the soil type and the AMC in the watershed. CN has no physical meaning. The equation for runoff (rainfall excess) is given as

$$P_e = \frac{(P - 0.2S)^2}{P + 0.8S} \tag{2.2}$$

The following points must be kept in mind while using the SCS curve number method:

1. CN is a parameter that ranges from 0 to 100. A value of 100 indicates that all of the rainfall is converted into runoff and that there are no losses. Completely impervious and water surfaces are examples of this. For normal watersheds, CN < 100. A value of CN close to 0 indicates that almost all of the rainfall is accounted for losses, indicating highly dry conditions and therefore negligible runoff results.
2. CN has no physical meaning. Since it is the only parameter used to compute the runoff from rainfall, it accounts for the combined effects of soil type, AMCs, and vegetation type on the runoff.
3. The soil group is assumed to be uniform throughout the watershed. The rainfall is assumed to be uniformly distributed over the watershed.
4. When a watershed consists of different soil types, AMCs and vegetation types, a composite CN may be determined for the watershed, as an area-weighted CN.
5. The SCS method, when used to estimate runoff from rainfall that has actually occurred, may produce poor results and is rather heavily dependent on the AMCs assumed for the watershed. It is more useful for estimating design flood runoff resulting from a design storm (see for example, Maidment, 1993).
6. The SCS method may over-predict the volume of runoff in a watershed (Maidment, 1993).
7. The method is generally used for non-urban catchments.

Soils are classified into four groups, A, B, C, and D, based on their runoff potential, with soil group A comprising soils having the lowest runoff potential and soil group D having the highest runoff potential. The AMC accounts for the moisture content in the soil preceding the storm for which the runoff is to be computed. The AMC of the watershed is classified into three groups, I, II, and III, based on the rainfall in the previous 5 days and based on whether it is a growing season or a dormant season.

The SCS curve numbers (now referred to as the NRCS curve numbers) are available in standard textbooks (e.g., Chow *et al.*, 1988; Singh, 1992) and on the Web (e.g., http://emrl.byu.edu/gsda/data_tips/tip_landuse_cntable.html, accessed December 17, 2011).

2.2.1.1 FLOOD HYDROGRAPH FROM THE SCS METHOD

For most hydrologic designs for floods, the peak flood discharge rather than the total flood runoff is of interest. The following procedure may be adopted for constructing the design flood hydrograph, once the flood runoff has been estimated (Maidment, 1993). The time to peak, Tp, is estimated by

$$Tp = 0.5D + 0.6t_c \tag{2.3}$$

where, Tp is the time to peak (hours), D is duration of the rainfall excess in hours, and t_c is the time of concentration in hours. The time of concentration is the time it takes from the beginning of the storm for the entire watershed to contribute to the runoff, and is given by the time it takes for the rain to reach the mouth of the watershed from the remotest part of the watershed. Empirical expressions commonly used to estimate the time of concentration for a watershed are available in standard textbooks (e.g., Chow *et al.*, 1988) The most commonly used is the Kirpich formula (Kirpich, 1940), given below:

$$t_c = 0.0078L^{0.77}S^{-0.385}$$

L = length of channel/ditch from head water to outlet, ft

S = average watershed slope, ft/ft (2.4)

Equating the total runoff volume, V_Q, computed from the SCS method to the area of the triangular direct runoff hydrograph

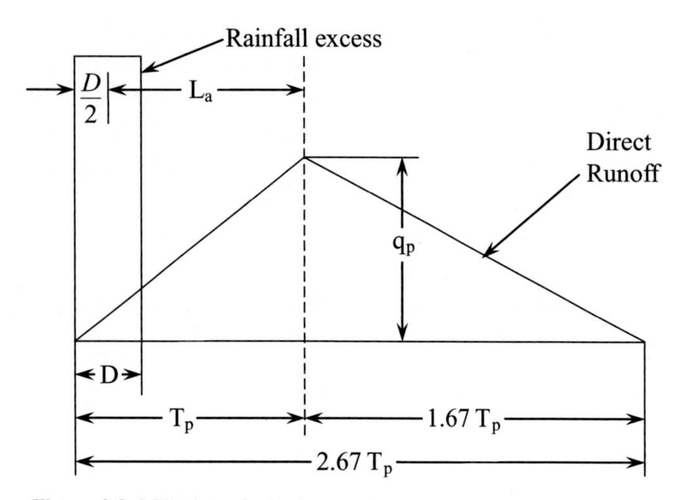

Figure 2.2 SCS triangular hydrograph.

shown in Figure 2.2, the peak discharge, q_p, can be expressed as

$$q_p = \frac{0.208AV_Q}{0.5D + 0.6t_c} \quad \text{SI units} \tag{2.5}$$

where A is the watershed area in square kilometers, V_Q is runoff in mm, D and t_c are in hours. In Figure 2.2, L_a is the basin lag, which is the time from the centroid of the excess rainfall hyetograph to the centroid of the hydrograph, which in the figure also coincides with the time to peak.

Example 2.1

Compute runoff from 100 mm rainfall from 250 km^2 watershed with the data in Table 2.1. Assume the AMC III condition. Consider the time of concentration of 3 hr and rainfall duration of 3 hr.

Solution

The curve numbers are obtained for the AMC II condition, from Chow *et al.* (1988); the computed weighted curve numbers are presented in Table 2.2.

Table 2.1 *Land use conditions of a watershed*

Land use	Area (%)	Soil group
Residential (30% impervious)	40	B
Forest: Poor cover	20	C
Open space:		
Fair grass	15	D
Good grass	15	D
Impervious parking space, schools, shopping complex, etc.	10	B

Thus,

$$\text{Weighted CN(II)} = \frac{3860 + 1540 + 2460}{100} = 78.6$$

This is for the AMC II condition.

For the AMC III condition,

$$\text{CN(III)} = \frac{23\text{CN(II)}}{10 + 0.13\text{CN(II)}} = \frac{23 \times 78.6}{10 + 0.13 \times 78.6} = 89.41$$

$$S = \frac{25400}{89.41} - 254 = 30.08 \text{ mm}$$

Hence the rainfall excess

$$P_e = \frac{(P - 0.2S)^2}{P + 0.8S} = \frac{(100 - 0.2 \times 30.08)^2}{100 + 0.8 \times 30.08} = 71.19 \text{ mm}$$

The runoff hydrograph is formulated using the time of concentration of 3 hr and the duration of 3 hr.

Table 2.2 *Computation of weighted curve number for Example 2.1, for the AMC II condition*

	Hydrologic soil group								
	B			C			D		
Land use	%	CN	Product	%	CN	Product	%	CN	Product
Residential (30% impervious)	40	72	2880						
Forest: Poor cover				20	77	1540			
Open space:									
Fair condition							15	80	1200
Good condition							15	84	1260
Impervious parking space, schools, shopping complex, etc.	10	98	980						
Total			3860			1540			2460

The components of the SCS triangular hydrograph shown in Figure 2.2 are obtained as follows:

The peak discharge,

$$q_p = \frac{0.208AV_Q}{0.5D + 0.6t_c}$$

where A: area of the catchment = 250 km^2,
P_e: rainfall excess = 71.19 mm,
D: duration of rainfall = 3 hr,
t_c: time of concentration of the catchment = 3 hr.

$$q_p = \frac{0.208 \times 250 \times 71.19}{0.5 \times 3 + 0.6 \times 3}$$
$$= 1121.78 \text{ m}^3\text{/s}$$

Time to peak from the start of rainfall, $Tp = 0.5^*D + 0.6^*t_c$
$= 0.5^*3 + 0.6^*3$
$= 3.3$ hr

Total time base of the hydrograph $= 2.67^*Tp$
$= 8.81$ hr

2.2.2 Rational method

The rational method is most commonly used in urban flood designs but is also sometimes used for non-urban catchments. The rational formula is

$$Q_p = CIA \quad (2.6a)$$

where Q_p is the peak discharge, I is the rainfall intensity, A is the area of the watershed, and C is a dimensionless runoff coefficient. Originally developed in FPS units, the formula may be used in SI units with appropriate units for Q_p, I, and A. For example, if Q_p is in m^3/s, I is in mm/hr, and A is in square kilometers, the rational formula may be written as

$$Q_p = 0.278CIA \quad (2.6b)$$

The rational formula (Equation 2.6a) is developed based on the assumption that the intensity of rainfall, I, is constant over the duration and that the peak flow occurs once the entire watershed starts contributing simultaneously to flow at the outlet of the watershed. This duration is the same as the time of concentration defined earlier (Section 2.2.1).

The rational method is most commonly used for hydrologic designs, rather than for estimating peak flows from actual rainfall. The intensity of the rainfall for design purposes is obtained from the intensity–duration–frequency (IDF) relationship for the watershed (Section 2.3). The frequency used to determine the intensity is the same as that required for the design flood. That is, the average recurrence interval (ARI) or the return period chosen for the hydrologic designs is used as the frequency in the IDF relationship. Duration of the design rainfall is generally taken as the time of concentration, t_c, for design purposes.

Table 2.3 *Land use of catchment*

Land use	Area (%)
Residential (30% impervious)	40
Forest: Poor cover	20
Open space:	
Fair grass	15
Good grass	15
Impervious parking space, schools, shopping complex, etc.	10

Table 2.4 *Design rainfall intensity of catchment*

Catchment	Design rainfall intensity (mm/hr)
Residential	150
Forest	75
Open land	100

The coefficient of runoff, C, for a given watershed is a major source of uncertainty in the rational method. The coefficient, as seen from the rational formula, Equation 2.6a, aggregates the effect of soil type, AMC, vegetation, land use, degree of soil compaction, depression storage, catchment slope, rainfall intensity, proximity to water table, and other factors that determine the peak runoff for a given storm in a catchment. Suggested values for the coefficient C are given in Chow *et al.* (1988).

Example 2.2

Compute the peak discharge generated from the following catchment using the rational method. Total area of the catchment is 250 km^2, and the details are as shown in Table 2.3.

Divide this catchment into three sub-catchments and compute the maximum discharge generated from each of the sub-catchments. Take design intensity of rainfall as in Table 2.4.

Solution

The C values for different sub-catchments are calculated from the data given in Chow *et al.* (1988):

Residential, $C = 0.5$

Forest, $C = 0.3$

Open land, $C = 0.2$

The total catchment area is 250 km^2. From the given data, the peak discharge is computed for each sub-catchment, as shown in Table 2.5.

Table 2.5 *Peak discharge calculation for Example 2.2*

Sub-catchment	C value	Area (A)(km^2)	Rainfall intensity (I) (mm/hr)	Peak discharge = $0.278CAI$ (m^3/s)
Residential	0.56	125	150	2919
Forest	0.3	50	75	312.75
Open land	0.2	75	100	417

The following points must be noted with respect to the rational formula:

1. It is assumed in the rational formula that the frequency of the peak discharge is the same as the frequency of the rainfall.
2. The runoff coefficient C is the same for storms of different frequencies.
3. All losses are constant during a storm – the value of C does not change with hydrologic conditions (such as the AMC).
4. As the intensity of the rainfall is assumed to be constant over the duration considered, the rational method is valid for relatively small catchments. Some investigators believe that the maximum area should be about 100 acres (about 40 ha) (Maidment, 1993).
5. The rational method is generally used for urban storm-water drainage designs. It is also used for small non-urban catchments to estimate peak flows. Where the catchment size is large, it is divided into sub-catchments to obtain the peak flows from each sub-catchment and then the resulting hydrographs are routed using flood routing procedures (Section 2.4) to obtain the peak flow at the outlet.
6. To account for a non-linear response of the catchment to increasing intensities of rainfall, the value of C is sometimes assumed to increase as ARI increases. Chow *et al.* (1988) provide a table of runoff coefficients for various return periods (average recurrence intervals).
7. A probabilistic rational method (Maidement, 1993) may be used to obtain the runoff coefficient as a function of the ARI. The rational formula is written, in this case, as

$$Q_p(Y) = C(Y)\, I\,(t_c, Y)\, A \tag{2.7}$$

indicating that the peak discharge is a function of the ARI, the intensity I is a function of the time of concentration and the ARI, and that the coefficient C can be determined as a function of the ARI. The probabilistic method may be employed in watersheds where the frequency analyses of peak flows and the durations of storms are available. However, such information is often not available for most watersheds, and increasing the value of C with ARI using judgment and experience may be necessary.

The rainfall intensity, I, is obtained from the IDF relationships. Derivation of the IDF relationship at a location is discussed in the next section.

2.3 INTENSITY–DURATION–FREQUENCY RELATIONSHIP

Hydrologic designs for floods require the peak flows expected to be experienced. The designs are normally developed for a given return period of a flood event. For example, an embankment along a river may be designed to protect against a flood of return period of 100 years, whereas the urban drainage systems may be typically designed for storms of return periods 2 to 5 years. The return period of an event indicates the ARI of the event.

As discussed previously in Section 2.2, a design rainfall depth or intensity is required for determination of peak flood flows. The design rainfall intensity is obtained from the IDF relationships developed for a given location. The IDF relationships provide the expected rainfall intensity (I) for a given duration (D) of the storm and a specified frequency (F). In this section, determination of IDF relationships for use in hydrologic designs is discussed.

IDF relationships are provided as plots with duration as abscissa and intensity as ordinate and a series of curves, one for each return period. The intensity of rainfall is the rate of precipitation, i.e., depth of precipitation per unit time. This can be either instantaneous intensity or average intensity over the duration of rainfall.

The average intensity is determined as

$$i = \frac{P}{t} \tag{2.8}$$

where P is the rainfall depth and t is the duration of rainfall.

The frequency is expressed in terms of return period (T), which is the average length of time between the rainfall events that equal or exceed the design magnitude. If local rainfall data are available, IDF curves can be developed using frequency analysis. A minimum of 20 years of data is desirable for development of the IDF relationship.

The following steps describe the procedure for developing IDF curves:

Step 1: Preparation of annual maximum rainfall data series
From the available rainfall data, rainfall series for different durations (e.g., 1, 2, 6, 12, and 24 hr) are developed. For each duration, the annual maximum rainfall depths are calculated.

Step 2: Fitting a probability distribution
A suitable probability distribution is fitted to each of the selected duration data series.

Generally used probability distributions are Gumbel's extreme value distribution, normal distribution, log-normal distribution (two parameter), gamma distribution (two parameter), and the log-Pearson type III distribution. The most commonly used distribution is Gumbel's extreme value distribution. The parameters of the distribution are calculated for the selected distribution. Statistical tests such as the Kolmogorov–Smirnov goodness of fit test (e.g., McCuen and Snyder, 1986) may be performed to ensure that the chosen distribution fits the data well.

Step 3: Determining the rainfall depths

The precipitation depth corresponding to a given return period is calculated from the analytical frequency procedures as

$$x_T = \bar{x} + K_T s \quad (2.9)$$

where x_T is the precipitation depth corresponding to the return period T, $\bar{x}$ is mean of the annual maximum values obtained in Step 1, s is the standard deviation of the annual maximum values obtained in Step 1, and K_T is the frequency factor, which depends on the distribution used.

For Gumbel's distribution (Chow *et al.*, 1988),

$$K_T = -\frac{\sqrt{6}}{\pi}\left\{0.5772 + \ln\left[\ln\left(\frac{T}{T-1}\right)\right]\right\} \quad (2.10)$$

For the normal distribution, K_T is the same as the standard normal deviate z and is given by (Chow *et al.*, 1988)

$$K_T = z = w - \frac{2.515517 + 0.802853w + 0.010328w^2}{1 + 1.43788w + 0.189269w^2 + 0.001308w^3} \quad (2.11)$$

where

$$w = \left[\ln\left(\frac{1}{p^2}\right)\right]^{1/2} \qquad 0 \le p \le 0.5 \quad (2.12)$$

In Equation 2.12, p is the excedence probability $P(X \ge x_T)$, where X denotes the random variable, rainfall, and x_T is the magnitude of the rainfall corresponding to the return period T. It may be noted that $p = 1/T$. When $p > 0.5$, $(1 - p)$ is substituted in place of p in Equation 2.12, and the resulting K_T value is used with a negative sign in Equation 2.9.

For the log-Pearson type III distribution, the frequency factor depends on the return period and coefficient of skewness, C_s (Chow *et al.*, 1988). When $C_s = 0$, K_T is equal to the standard normal deviate (Equation 2.11). When $C_s \neq 0$,

$$K_T = z + (z^2 - 1)k + \frac{1}{3}(z^3 - 6z)k^3 - (z^2 - 1)k^3 + zk^4 + \frac{1}{3}k^5 \quad (2.13)$$

where

$$k = \frac{C_s}{6} \quad (2.14)$$

and z is given by Equation 2.11.

Table 2.6 *Series of annual maximum rainfall (mm) for different durations*

Year	15 min	30 min	1 hr	2 hr	6 hr	12 hr	24 hr
1	21.8	34.8	44.5	61.6	104.1	112.3	115.9
2	17	28.5	37	48.2	62.5	69.6	92.7
3	22.5	31.1	41	52.9	81.4	86.9	98.7
4	20	24.1	30	40	53.9	57.8	65.2
5	23.5	33	40.5	53.9	55.5	72.4	89.8
6	26	35.5	52.4	62.4	83.2	93.4	152.5
7	21	40	59.6	94	95.1	95.1	95.3
8	17.9	31.4	22.1	42.9	61.6	64.5	71.7
9	23	32.7	42.2	44.5	47.5	60	61.9
10	21.5	33.5	35.5	36.8	52.1	54.2	57.5
11	29.2	41.5	59.5	117	132.5	135.6	135.6
12	30	30	48.2	57	82	86.8	89.1
13	26	46	41.7	58.6	64.5	65.1	68.5
14	18	30.5	37.3	43.8	50.5	76.2	77.2
15	20	35	37	60.4	70.5	72	75.2
16	30	48	60.2	74.1	76.6	121.9	122.4
17	31	56	65.2	73.7	97.9	103.9	104.2
18	45.1	60.1	47	55.9	64.8	65.6	67.5
19	26	51	148.8	210.8	377.6	432.8	448.7
20	21	31	41.7	47	51.7	53.7	78.1
21	22	36.5	40.9	71.9	79.7	81.5	81.6
22	20.5	29	41.1	49.3	63.6	93.2	147
23	18	30.2	31.4	56.4	76	81.6	83.1
24	15.5	25	34.3	36.7	52.8	68.8	70.5
25	14	15.7	23.2	38.7	41.9	43.4	50.8
26	26	35	44.2	62.2	72	72.2	72.4
27	28.5	42	57	74.8	85.8	86.5	90.4
28	20	30	50	71.1	145.9	182.3	191.3
29	30	49.5	72.1	94.6	111.9	120.5	120.5
30	26	39	59.3	62.9	82.3	90.7	90.9
31	28	44	62.3	78.3	84.3	84.3	97.2
32	20.3	39.6	46.8	70	95.9	95.9	100.8
33	20	37.5	53.2	86.5	106.1	106.2	106.8

The precipitation depth thus calculated from the frequency relationship (Equation 2.9) can be adjusted to match the depths derived from annual maximum series by multiplying the depths by 0.88 for the 2-year return period values, 0.96 for the 5-year return period values, and 0.99 for the 10-year return period values (Chow *et al.*, 1988). No adjustment of the estimates is required for longer return periods (>10-year return period). The corresponding intensities are obtained simply by dividing the depth by the duration.

Example 2.3

Annual maximum rainfall at a station available for 33 years at intervals of 15 minutes is given in Table 2.6. Determine the IDF relationship for these data, using Gumbel's extreme value distribution.

Table 2.7a *Mean and standard deviaton (SD) of the annual maximum rainfall*

	15 min	30 min	1 hr	2 hr	6 hr	12 hr	24 hr
Mean (mm)	23.61	36.56	48.7	66.33	86.77	96.57	105.18
SD (mm)	5.96	9.36	21.52	31.78	57.51	66.25	68.6

Table 2.7b *Frequency factor values for different return periods*

T	K_T
2	−0.164
5	0.719
10	1.305
25	2.044
50	2.592
100	3.137

Gumbell's extreme value distribution for each duration is considered. In some situations, however, different durations may require different distributions to be fitted.

The design rainfall depth for a given return period T is determined using Equations 2.9 and 2.10, $x_T = \bar{x} + K_T s$, where

$$K_T = -\frac{\sqrt{6}}{\pi}\left\{0.5772 + \ln\left[\ln\left(\frac{T}{T-1}\right)\right]\right\}$$

The mean and the standard deviations of the annual maximum series are calculated and presented in Table 2.7a.

The design rainfall depth is calculated using Equation 2.9 and then the intensity in mm/hr is computed by dividing the depth by the duration. This is done for various return periods, T, of 2, 5, 10, 25, 50, and 100 years. The K_T values obtained are shown in Table 2.7b.

One sample calculation is illustrated below.

For 2-hour duration and 5-year return period,

$$\bar{x} = 66.33 \text{ mm}, S = 31.78 \text{ mm}, K_T = 0.719$$

The design rainfall depth for 2-hour duration and 5-year return period is determined by $x_T = \bar{x} + K_T s$

$$x_T = 66.33 + 0.719 * 31.78 = 89.2 \text{ mm}$$

Rainfall intensity for 2-hour duration = 89.2/2 = 44.6 mm/hr.

The rainfall intensities obtained for different return periods and durations, thus obtained are tabulated (Table 2.7c).

The IDF curves are plotted, with these rainfall intensities, and shown in Figure 2.3.

2.4 FLOOD ROUTING

In the previous sections, construction of (design) flood hydrographs was discussed, by estimating the design storm intensity for a given return period from the IDF relationship corresponding to the design duration, the time of concentration, t_c, and the peak flood runoff, q_p. The traverse of a flood in a river stretch is discussed in this section, and methodologies to estimate the hydrographs at various locations in the river stretch are provided.

A flood travels along a river reach as a wave, with velocity and depth continuously changing with time and distance. While it is difficult to forecast with accuracy the time of occurrence and magnitude of floods, it is possible to estimate fairly accurately the movement of the flood wave along a river, once it is known that a flood wave is generated at some upstream location in the river. Such estimation is of immense practical utility, as it can be used in flood early warning systems. In this section, the specific problem of how a flood wave propagates along a channel – a stream, a river, or any open channel – is discussed along with the theoretical framework available for forecasting the propagation of the flood wave.

Imagine a flood wave travelling along a straight reach of a stream, which initially has uniform flow conditions. As the flood wave crosses a section, the velocity (or discharge) and the depth of flow at that section change. Determining the track of these changes in depth and discharge, with time and along the length of the river, is called flood routing.

Table 2.7c *Rainfall intensity for different return periods and duration*

Return period (years)	Rainfall intensity (mm/hr)						
	15 min	30 min	1 hr	2 hr	6 hr	12 hr	24 hr
2	90.54	70.06	45.17	30.55	12.89	7.14	3.91
5	111.62	86.61	64.19	44.60	21.36	12.02	6.44
10	125.57	97.57	76.79	53.90	26.97	15.25	8.11
25	143.21	111.42	92.70	65.65	34.05	19.33	10.22
50	156.29	121.70	104.51	74.36	39.31	22.36	11.79
100	169.27	131.90	116.23	83.02	44.53	25.37	13.35

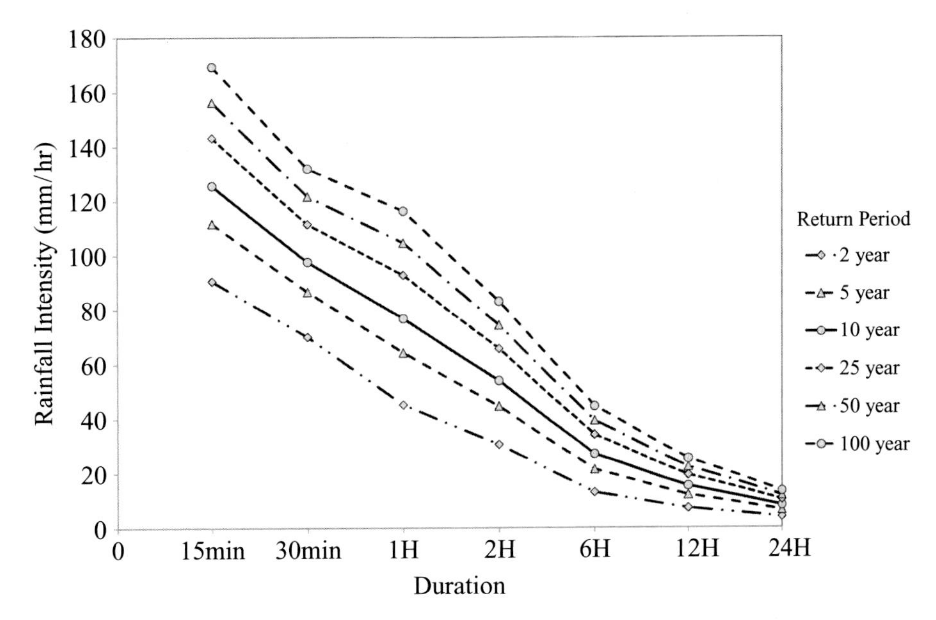

Figure 2.3 IDF relationships for Example 2.3.

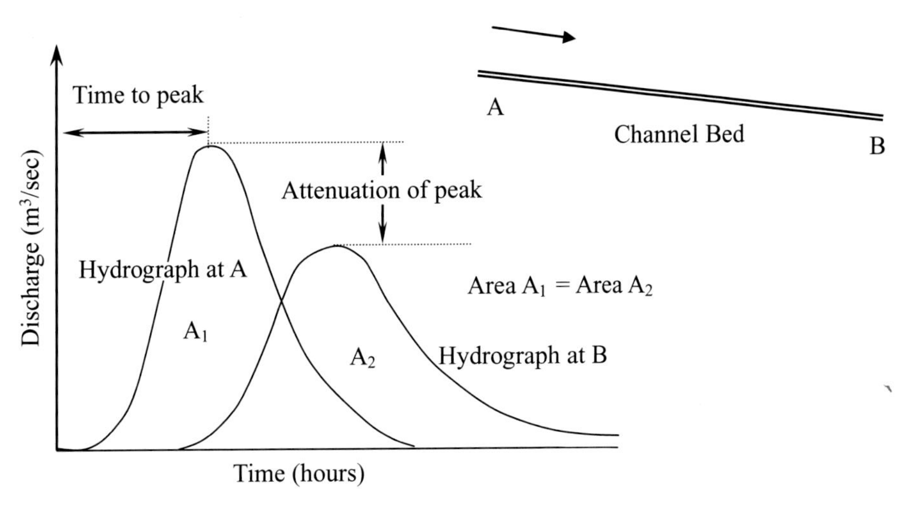

Figure 2.4 Flood hydrographs.

The depth and discharge of the wave front itself change with time. A flood wave is, therefore, represented by a hydrograph, as shown in Figure 2.4. In practical applications, the general interest would be in estimating the discharge and depth of flow at a given location along the stream at a specified time, given the flood hydrograph at an upstream section. The hydraulic method of flood routing uses the Saint-Venant equations discussed in the next section, whereas the hydrologic method of flood routing uses a simple hydrologic continuity in terms of relating the change in storage in the channel length to the difference between inflow and outflow. In this section, only the channel routing methods are covered. The reservoir routing method, which is used in routing the floods through a reservoir, is not discussed here.

2.4.1 Hydraulic routing: the Saint-Venant equations

The flood flow is unsteady – because the flow properties (depth and velocity) change with time – and is gradually varied, because such change with time is gradual. To derive the governing equations for such a wave movement, we use the principles of continuity (conservation of mass) and momentum (essentially, Newton's second law of motion) for the one-dimensional unsteady open channel flow.

In differential form, the governing equations are written as:

Continuity:

$$\frac{\partial Q}{\partial x} + \frac{\partial A}{\partial t} - q = 0 \tag{2.15}$$

Momentum:

$$\left(\frac{1}{A}\right)\frac{\partial Q}{\partial t} + \left(\frac{1}{A}\right)\frac{\partial (Q^2/A)}{\partial x} + g\frac{\partial y}{\partial x} - g(S_0 - S_f) = 0$$

Local acceleration term	Convective acceleration term	Pressure force term	Gravity force term	Friction force term

(2.16)

In these equations, Q is the discharge (m^3/s), A is the area, q is the lateral flow per unit length of the channel (m^3/s per m), x is the distance along the channel, y is the depth of flow, g is the acceleration due to gravity, S_0 is the bed slope of the channel, and S_f is the friction slope. Equations 2.15 and 2.16 are together called the *Saint-Venant equations*, in honor of Saint-Venant who developed these equations in 1871. In the momentum equation, the *local acceleration* term describes the change in momentum due to change in velocity over time, the *convective acceleration* term describes the change in momentum due to change in velocity along the channel, the *pressure force* term denotes a force proportional to the change in water depth along the channel, the *gravity force* term denotes a force proportional to the bed slope, and the *friction force* term denotes a force proportional to the friction slope.

It is not possible to solve Equations 2.15 and 2.16 together analytically, except in some very simplified cases. Numerical solutions are possible, and are used in most practical applications. Depending on the accuracy desired, alternative flood routing equations are generated by using the continuity equation (except the lateral flow term, in some cases) while eliminating some terms of the momentum equation. Based on the terms retained in the momentum equation, the flood wave is called the *kinematic* wave, the *diffusion* wave, and the *dynamic wave*, as shown:

$$\left(\frac{1}{A}\right)\frac{\partial Q}{\partial t} + \left(\frac{1}{A}\right)\frac{\partial (Q^2/A)}{\partial x} + g\frac{\partial y}{\partial x} - g(S_0 - S_f) = 0$$

|— Kinematic wave

|———— Diffusion wave

|———————————— Dynamic wave

(2.16)

The kinematic wave is thus represented by $g(S_0 - S_f) = 0$, or $S_0 = S_f$, the diffusion wave by $g\ \partial y/\partial x - g(S_0 - S_f) = 0$, resulting in $\partial y/\partial x = (S_0 - S_f)$, and the dynamic wave by the complete momentum equation, Equation 2.16.

In most practical applications, the wave resulting either from the simplest form of the momentum equation, i.e., the kinematic wave, or from the complete momentum equation, i.e., the dynamic wave, is used. For the kinematic wave, the acceleration and pressure terms in the momentum equation are neglected, hence the name *kinematic*, referring to the study of motion exclusive of the influence of mass and force. The remaining terms in the momentum equation represent the steady uniform flow. In other words, the flow is considered to be steady for momentum conservation and the effects of unsteadiness are taken into consideration through the continuity equation. Analytical solution is possible for the simple case of the kinematic wave, where the lateral flow is neglected and the wave celerity is constant. However, the *backwater effects* (propagation upstream of the effects of change in depth or flow rate at a point) are not reproduced through a kinematic wave. Such backwater effects are accounted for in flood routing only through the local acceleration, convective acceleration, and the pressure terms, all of which are neglected in the kinematic wave. For more accurate flood routing, numerical solutions of the complete dynamic equation are used.

2.4.2 Numerical solutions

With the availability of high-speed computers, it is presently possible to solve the dynamic wave equations through numerical methods (such as the finite difference method). The numerical methods start with *initial* and *boundary* conditions. At time $t = 0$, the uniform steady flow conditions are specified at all locations. These constitute the initial conditions. At distance $x = 0$, the flood hydrograph is known. This, together with other flow conditions (such as free over-fall at a location, flows from other sources joining the channel, submergence at a junction, etc.) along the length of the channel define the boundary conditions. Solution of the Saint-Venant equations gives the variation of discharge (and depth) with time along the length of the water body (river, stream, or channel), which may be used for real-time flood forecasting.

Following Chow *et al.* (1988), a finite difference scheme is presented here to solve the Saint-Venant equations numerically.

Continuity:

$$\frac{\partial Q}{\partial x} + \frac{\partial A}{\partial t} = q \tag{2.17}$$

Momentum:

$$S_0 = S_f \tag{2.18}$$

The momentum equation can be, in general, expressed in the form

$$A = \alpha Q^{\beta} \tag{2.19}$$

The advantage of expressing the momentum equation in this form is that the variable A can be eliminated in the continuity equation, by differentiating Equation 2.19 with respect to t:

$$\frac{\partial A}{\partial t} = \alpha\beta Q^{\beta-1}\left(\frac{\partial Q}{\partial t}\right) \tag{2.20}$$

Substituting for $\partial A/\partial t$ in Equation 2.17 to give

$$\frac{\partial Q}{\partial x} + \alpha\beta Q^{\beta-1}\left(\frac{\partial Q}{\partial t}\right) = q \tag{2.21}$$

The kinematic wave celerity, c_k, may be expressed as

$$c_k = \frac{dQ}{dA} = \frac{dx}{dt} \tag{2.22a}$$

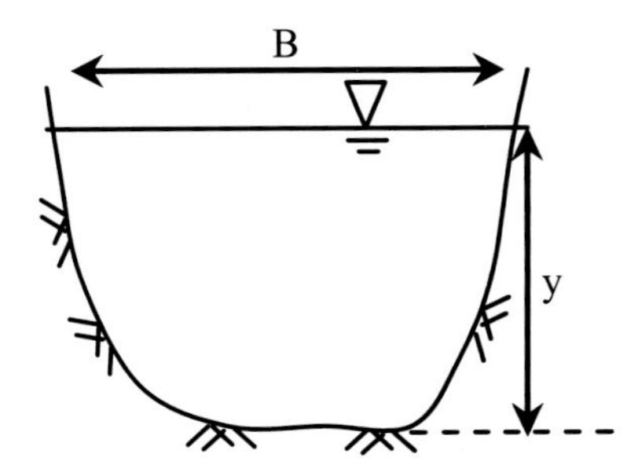

Figure 2.5 Cross-section of a channel.

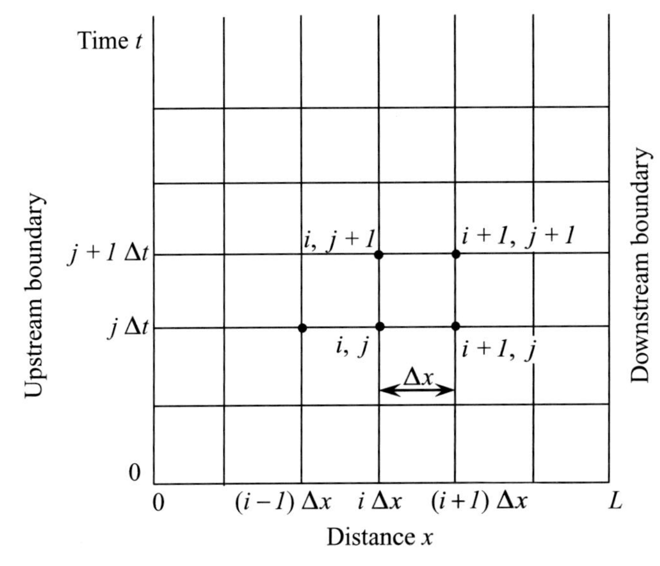

Figure 2.6 Grid on the x–t plane used for numerical solution of the Saint-Venant equations.

with $dA = Bdy$,

$$c_k = \frac{1}{B}\frac{dQ}{dy} \tag{2.22b}$$

B is the top width of flow.

Figure 2.6 shows a general x–t plane of a finite difference scheme. In the explicit method of solution the initial conditions at $x = 0$ and $t = 0$ are both known. For example, the stage or discharge hydrograph at $x = 0$ may be specified as occurring over an initial known stage or discharge at $t = 0$. Starting with $t = 0$, the solution proceeds from one time step to the next, obtaining at each time step the stage (or discharge) value from $x = 0$ (which is known from the initial condition), $x = 1, 2, \ldots, L$. This procedure defines an explicit scheme of solution.

A finite difference form of the derivative of Q_{i+1}^{j+1} may be obtained as follows:

$$\frac{\partial Q}{\partial x} \approx \frac{Q_{i+1}^{j+1} - Q_i^{j+1}}{\Delta x} \tag{2.23}$$

$$\frac{\partial Q}{\partial t} \approx \frac{Q_{i+1}^{j+1} - Q_{i+1}^{j}}{\Delta t} \tag{2.24}$$

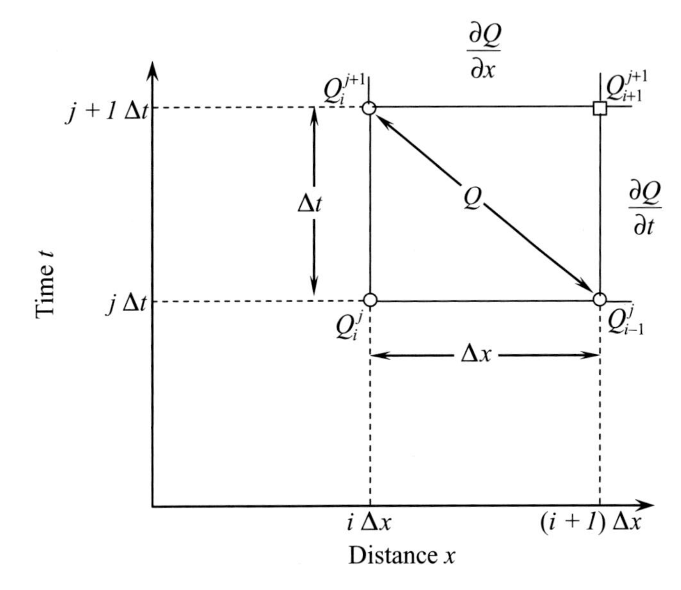

Figure 2.7 Finite difference box for solution of the linear kinematic wave equation showing finite difference equations.

$$Q \approx \frac{Q_{i+1}^{j} + Q_i^{j+1}}{2} \tag{2.25}$$

$$\frac{Q_{i+1}^{j+1} - Q_i^{j+1}}{\Delta x} + \alpha\beta\left(\frac{Q_{i+1}^{j} + Q_i^{j+1}}{2}\right)^{\beta-1}\left(\frac{Q_{i+1}^{j+1} - Q_{i+1}^{j}}{\Delta t}\right) = \frac{q_{i+1}^{j+1} + q_{i+1}^{j}}{2} \tag{2.26}$$

$$Q_{i+1}^{j+1} = \frac{\left[\frac{\Delta t}{\Delta x}Q_i^{j+1} + \alpha\beta Q_{i+1}^{j}\left(\frac{Q_{i+1}^{j}+Q_i^{j+1}}{2}\right)^{\beta-1} + \Delta t\left(\frac{q_{i+1}^{j+1}+q_{i+1}^{j}}{2}\right)\right]}{\left[\frac{\Delta t}{\Delta x} + \alpha\beta\left(\frac{Q_{i+1}^{j}+Q_i^{j+1}}{2}\right)^{\beta-1}\right]} \tag{2.27}$$

It can be shown (Henderson, 1966) that, by choosing Q as the dependent variable rather than A, the relative errors resulting from the numerical computations may be reduced. The coefficients α and β are determined by writing the momentum equation in the form $A = \alpha Q^{\beta}$.

The following steps summarize the solution procedure:

1. Initial values of Q are set at $x = 0$ (for all t) and at $t = 0$ (for all x). These may be specified from the known base flow conditions.
2. The time and space steps, Δt and Δx, are chosen to satisfy the courant condition, $\Delta t \le \Delta x / c_k$, for numerical stability. For a given Δx, the smallest time step, Δt, that satisfies the courant condition is chosen.

Table 2.8 *Inflow hydrograph of channel*

Time (min)	0	12	24	36	48	60	72	84	96	108	120
Inflow rate (m^3/s)	60	60	70	80	90	100	90	80	70	60	60

3. Advancing from the time step j (corresponding to $t = 0$) to the next time step, $j + 1$, Equation 2.27 is applied from $i = 1$ to $i = i_{max}$, one step at a time, to obtain the values of Q at all space steps at the time step $j + 1$, where i_{max} corresponds to last time step at the downstream boundary. This is repeated until the last time step is reached.

Example 2.4

Consider a rectangular channel of 60 m width and 5000 m length, with bed slope of 1% and Manning's $n = 0.052$. The inflow hydrograph at the upstream end for this channel is given in Table 2.8.

Route the inflow hydrograph considering $\Delta x = 1000$ m and $\Delta t = 3$ min and no lateral inflow. The initial condition is a uniform flow of 60 m^3/s along the channel.

Solution

$B = 60$ m, $L = 5000$ m, $S = 0.01$

$\Delta x = 1000$ m, $\Delta t = 3$ min

Manning's equation: $Q = \frac{1}{n}\left(\frac{A}{P}\right)^{2/3} A S_0^{1/2}$

Rearranging the terms, $A = \left(\frac{nP^{2/3}}{S_0^{1/2}}\right)^{3/5} Q^{3/5}$ (which is of the form $A = \alpha Q^{\beta}$)

With the data given, we obtain $\alpha = 3.48$ and $\beta = 0.6$, taking it to be a wide rectangular channel so that $P = B = 60$ m.

Substituting these values in Equation 2.27, the discharge values at various locations along the channel are obtained, tabulated in Table 2.9, and plotted in Figure 2.8. Note that the discharge values at node $i = 1$ (upstream end of the channel) are the hydrograph ordinates, obtained by interpolation from the inflow hydrograph data.

A sample calculation for $j = 6$, $i = 2$ is given below.

$$Q_2^6 = \frac{\left[\frac{\Delta t}{\Delta x} Q_1^6 + \alpha\beta Q_2^5 \left(\frac{Q_2^5 + Q_1^6}{2}\right)^{\beta-1} + \Delta t\left(\frac{q_2^6 + q_2^5}{2}\right)\right]}{\left[\frac{\Delta t}{\Delta x} + \alpha\beta\left(\frac{Q_2^5 + Q_1^6}{2}\right)^{\beta-1}\right]}$$

Substituting the values of $\Delta x = 1000$ m, $\Delta t = 180$ s, $\alpha = 3.48$, $\beta = 0.6$, $Q_1^6 = 62.5$, $Q_2^5 = 60$, q_2^5 and $q_2^6 = 0$ in the above equation,

$$Q_2^6 = \frac{\left[\frac{180}{1000} \times 62.5 + 0.6 \times 3.48 \times 60\left(\frac{60+62.5}{2}\right)^{0.6-1} + 0\right]}{\left[\frac{180}{1000} + 0.6 \times 3.48\left(\frac{60+62.5}{2}\right)^{0.6-1}\right]}$$

$= 60.77$ m^3/s

2.4.3 Hydrologic routing of floods: Muskingum method

2.4.3.1 BASIC EQUATIONS

The Muskingum method of flood routing is a hydrologic routing method. The basis for the equation of continuity in hydrologic routing is that the difference between the inflow and outflow rates is equal to the rate of change of storage (Figure 2.9), i.e.,

$$I - Q = \frac{ds}{dt} \tag{2.28}$$

where I = inflow rate, Q = outflow rate, and S = storage. Considering a small time interval Δt, the difference between the total inflow volume and total outflow volume may be written as

$$\bar{I}\Delta t - \bar{Q}\Delta t = \Delta S \tag{2.29}$$

where $\bar{I}$ = average inflow in time Δt, $\bar{Q}$ = average outflow in time Δt, and ΔS = change in storage. With $\bar{I} = (I_1 + I_2)/2$, $\bar{Q} = (Q_1 + Q_2)/2$, and $\Delta S = S_2 - S_1$, where subscripts 1 and 2 denote the variables at the beginning and end of the time interval,

$$\left(\frac{I_1 + I_2}{2}\right)\Delta t - \left(\frac{Q_1 + Q_2}{2}\right)\Delta t = S_2 - S_1 \tag{2.30}$$

The time interval Δt should be sufficiently small that the inflow and outflow hydrographs may be assumed to be straight lines in that time interval.

The Muskingum equation is written as

$$S = K\left[xI + (1 - x)Q\right] \tag{2.31}$$

In this equation, the parameter x is known as the *weighting factor* and takes a value between 0 and 0.5. A value of $x = 0$ indicates that the storage is a function of discharge only, in which case,

$$S = KQ \tag{2.32}$$

Such storage is known as *linear storage* or *linear reservoir*. When $x = 0.5$ both the inflow and outflow contribute equally in determining the storage. The coefficient K, which has the dimensions of time, is called the *storage-time constant*. It is approximately equal to the time of travel of a flood wave through the channel reach.

When a set of inflow and outflow hydrograph values is available for a given reach, values of S at various time intervals can be determined. By choosing a value of x, values of S at any time t are plotted against the corresponding $[xI + (1 - x)Q]$ values. If the value of x is chosen correctly, a straight-line relationship as in Equation 2.31 will result. If the chosen value is incorrect, the points will yield a looping curve. A trial and error procedure is used to obtain the correct value of x so that the points yield nearly a straight line. The value of K is given by the inverse slope of this straight line. The value of x lies between 0 and 0.3, for natural channels. Within a given reach, the values of x and K are assumed to be constant, in routing. The flow rate Q_2 at the downstream

Table 2.9 *Computation of flow along the channel*

		Distance along channel in m					
		0	1000	2000	3000	4000	5000
Time (min)	Time index j	$i = 1$	2	3	4	5	6
0	1	60	60	60	60	60	60
3	2	60	60	60	60	60	60
6	3	60	60	60	60	60	60
9	4	60	60	60	60	60	60
12	5	60	60	60	60	60	60
15	6	62.5	60.77	60.24	60.07	60.02	60.01
18	7	65	62.09	60.81	60.30	60.11	60.04
21	8	67.5	63.79	61.73	60.74	60.30	60.12
24	9	70	65.75	62.99	61.44	60.65	60.28
27	10	72.5	67.91	64.54	62.40	61.19	60.56
30	11	75	70.19	66.33	63.63	61.95	60.99
33	12	77.5	72.57	68.33	65.11	62.94	61.60
36	13	80	75.01	70.48	66.82	64.16	62.39
39	14	82.5	77.49	72.76	68.72	65.60	63.39
42	15	85	79.99	75.13	70.79	67.25	64.61
45	16	87.5	82.52	77.58	73.00	69.09	66.03
48	17	90	85.05	80.07	75.32	71.11	67.64
51	18	92.5	87.60	82.60	77.73	73.26	69.45
54	19	95	90.14	85.16	80.21	75.55	71.42
57	20	97.5	92.69	87.73	82.74	77.93	73.54
60	21	100	95.24	90.32	85.31	80.39	75.80
63	22	97.5	96.03	92.29	87.70	82.85	78.13
66	23	95	95.67	93.46	89.68	85.16	80.48
69	24	92.5	94.57	93.85	91.11	87.19	82.74
72	25	90	92.99	93.55	91.95	88.82	84.80
75	26	87.5	91.11	92.71	92.22	89.99	86.56
78	27	85	89.03	91.44	91.95	90.66	87.96
81	28	82.5	86.82	89.86	91.23	90.86	88.95
84	29	80	84.53	88.05	90.14	90.61	89.52
87	30	77.5	82.19	86.07	88.75	89.98	89.68
90	31	75	79.81	83.97	87.13	89.01	89.45
93	32	72.5	77.41	81.79	85.33	87.76	88.87
96	33	70	75.00	79.55	83.39	86.28	87.99
99	34	67.5	72.59	77.27	81.36	84.62	86.85
102	35	65	70.17	74.96	79.24	82.83	85.49
105	36	62.5	67.75	72.64	77.08	80.92	83.96
108	37	60	65.33	70.30	74.88	78.92	82.28
111	38	60	63.67	68.20	72.73	76.89	80.49
114	39	60	62.53	66.42	70.71	74.88	78.64
117	40	60	61.75	64.96	68.88	72.95	76.78
120	41	60	61.21	63.79	67.27	71.13	74.94
123	42	60	60.84	62.87	65.89	69.46	73.17
126	43	60	60.58	62.16	64.73	67.96	71.50
129	44	60	60.40	61.62	63.76	66.64	69.94
132	45	60	60.28	61.21	62.97	65.48	68.53
135	46	60	60.19	60.89	62.32	64.50	67.25
138	47	60	60.13	60.66	61.81	63.66	66.12
141	48	60	60.09	60.48	61.40	62.96	65.13

(*cont.*)

Table 2.9 (*cont.*)

		Distance along channel in m					
		0	1000	2000	3000	4000	5000
Time (min)	Time index j	$i = 1$	2	3	4	5	6
144	49	60	60.06	60.36	61.08	62.37	64.27
147	50	60	60.04	60.26	60.83	61.90	63.53
150	51	60	60.03	60.19	60.63	61.50	62.90
153	52	60	60.02	60.14	60.48	61.19	62.37
156	53	60	60.01	60.10	60.36	60.93	61.93
159	54	60	60.01	60.07	60.27	60.73	61.56
162	55	60	60.01	60.05	60.21	60.57	61.25
165	56	60	60.00	60.04	60.15	60.44	61.00
168	57	60	60.00	60.03	60.11	60.34	60.80
171	58	60	60.00	60.02	60.09	60.26	60.63
174	59	60	60.00	60.01	60.06	60.20	60.50
177	60	60	60.00	60.01	60.05	60.15	60.39
180	61	60	60.00	60.01	60.03	60.12	60.31
183	62	60	60.00	60.01	60.03	60.09	60.24
186	63	60	60.00	60.00	60.02	60.07	60.19
189	64	60	60.00	60.00	60.01	60.05	60.15
192	65	60	60.00	60.00	60.01	60.04	60.11
195	66	60	60.00	60.00	60.01	60.03	60.09
198	67	60	60.00	60.00	60.01	60.02	60.07
201	68	60	60.00	60.00	60.00	60.02	60.05
204	69	60	60.00	60.00	60.00	60.01	60.04
207	70	60	60.00	60.00	60.00	60.01	60.03
210	71	60	60.00	60.00	60.00	60.01	60.02
213	72	60	60.00	60.00	60.00	60.01	60.02
216	73	60	60.00	60.00	60.00	60.00	60.01
219	74	60	60.00	60.00	60.00	60.00	60.01
222	75	60	60.00	60.00	60.00	60.00	60.01
225	76	60	60.00	60.00	60.00	60.00	60.01
228	77	60	60.00	60.00	60.00	60.00	60.00
231	78	60	60.00	60.00	60.00	60.00	60.00

point 2 will be

$$Q_2 = C_0 I_2 + C_1 I_1 + C_2 Q_1 \tag{2.33}$$

where

$$C_0 = \frac{-Kx + 0.5\Delta t}{K - Kx + 0.5\Delta t}, \quad C_1 = \frac{Kx + 0.5\Delta t}{K - Kx + 0.5\Delta t}$$

and

$$C_2 = \frac{K - Kx - 0.5\Delta t}{K - Kx + 0.5\Delta t} \tag{2.34}$$

and

$$C_0 + C_1 + C_2 = 1 \tag{2.35}$$

The following example illustrates flood routing with the Muskingum method.

Example 2.5

Route the following hydrograph (Table 2.10) through a river reach for which $K = 7.03$ hr and $x = 0.25$. The initial outflow discharge is 10 m^3/s.

Solution

$\Delta t = 6$ hr is selected to suit the given inflow hydrograph ordinate interval.

The coefficients C_0, C_1, and C_2 are calculated as (refer to Equations 2.34 and 2.35):

$$C_0 = \frac{-7.03 \times 0.25 + 0.5 \times 6}{7.03 - 7.03 \times 0.25 + 0.5 \times 6} = 0.150196 = 0.15$$

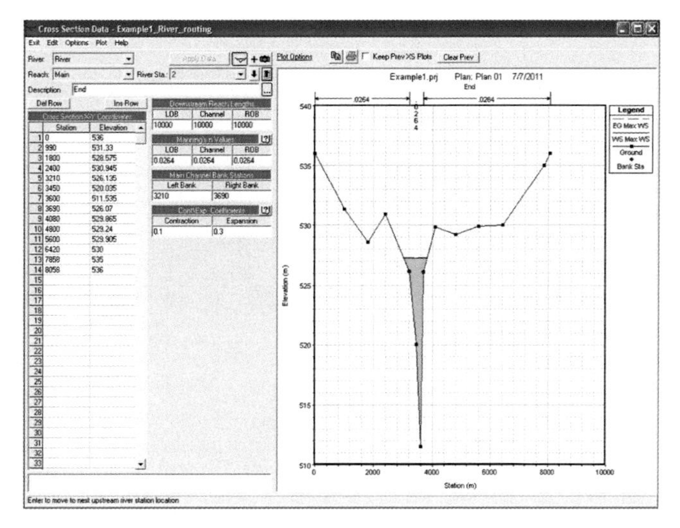

Figure 2.14 Cross section data editor.

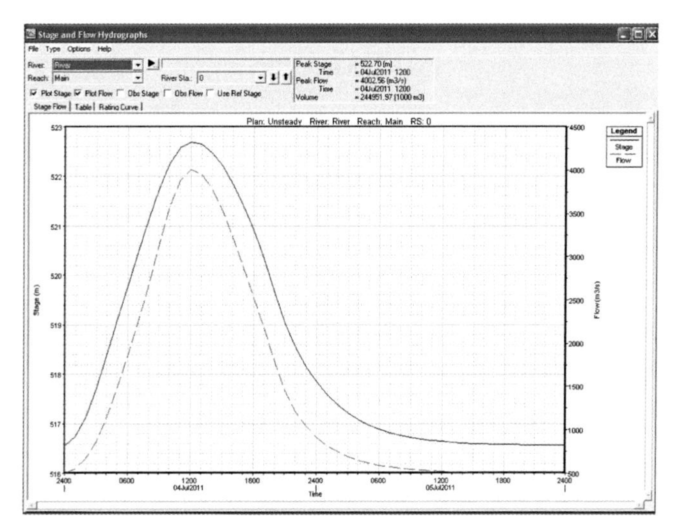

Figure 2.16 Outflow hydrographs at downstream boundary.

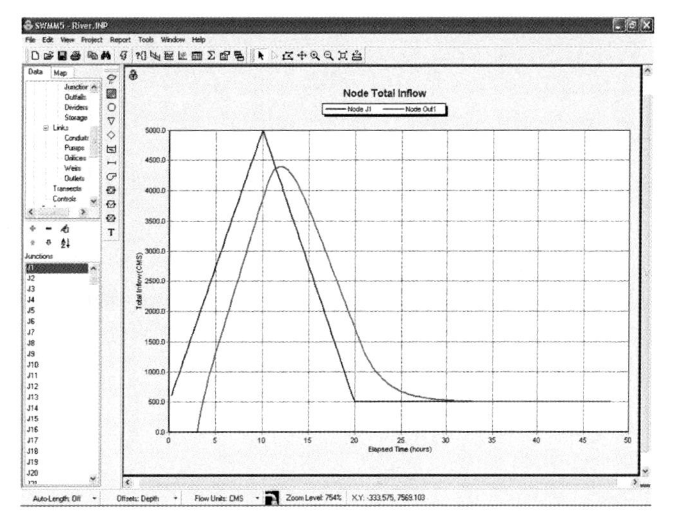

Figure 2.17 Inflow and outflow hydrographs.

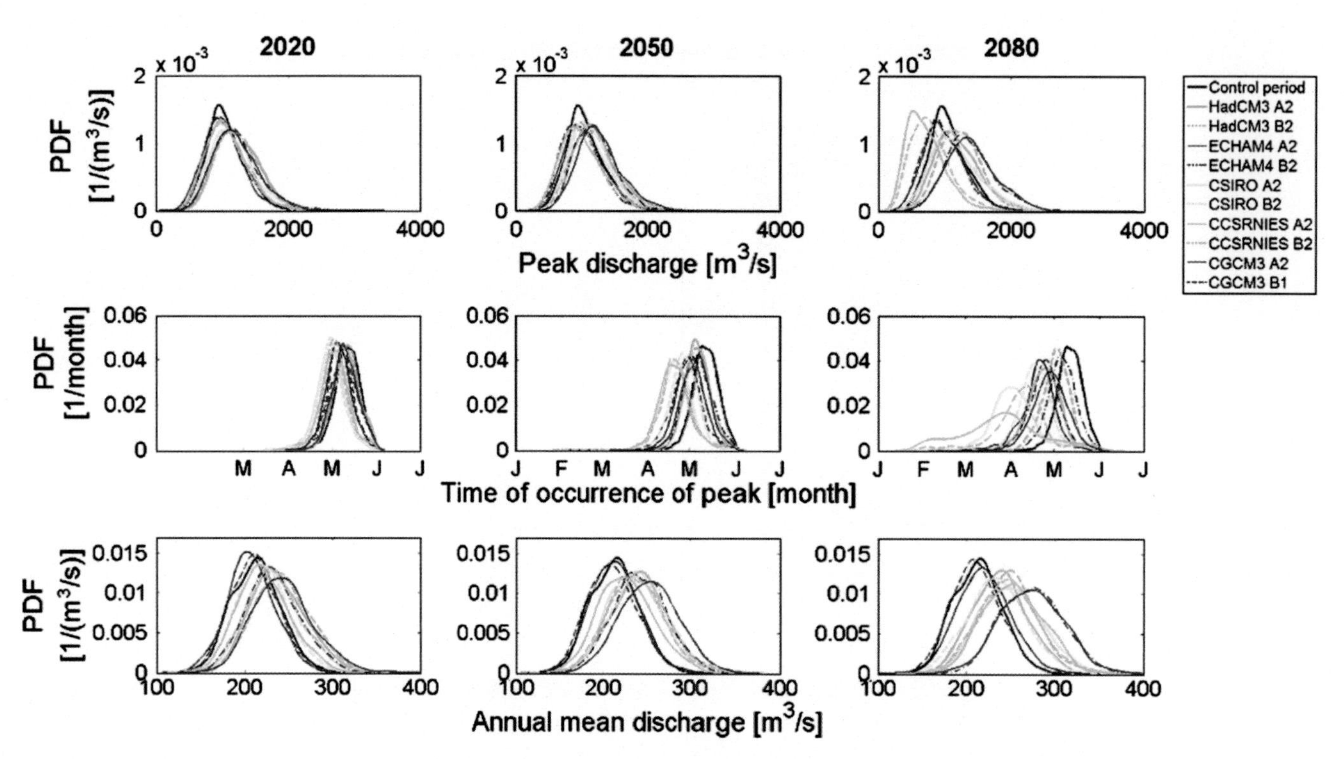

Figure 3.20 Probability density functions of peak discharge (top), time of occurrence of peak (middle), and annual mean discharge (bottom) from the GCM-scenario ensemble for a Nordic watershed for the 2020, 2050, and 2080 time horizons (*Source*: Minville *et al.*, 2008).

Figure 4.5 Standard false color composite (FCC) prepared using IRS–1C LISS III data.

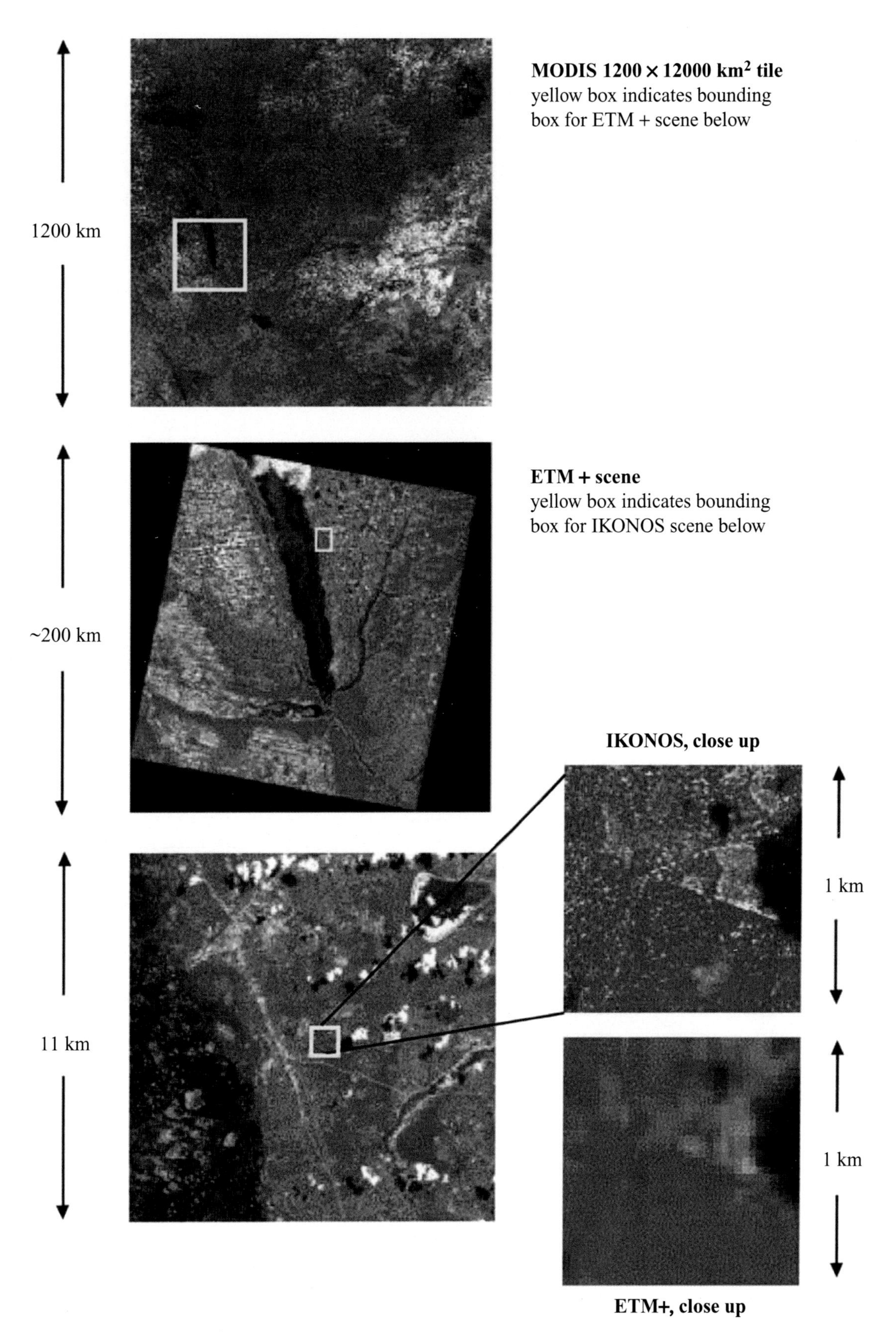

Figure 4.6 Remote sensing images in wide ranges of spatial, radiometric, spectral, and temporal resolutions (*Source*: Morisette *et al.*, 2002).

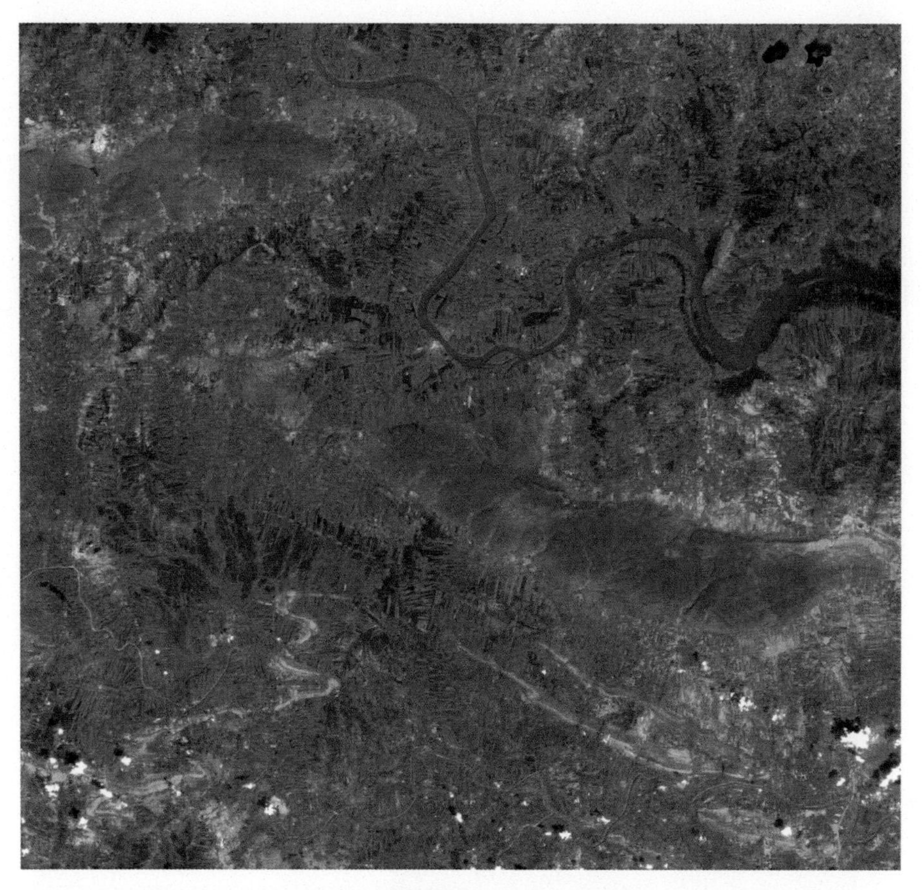

Figure 4.11 Synergetic image prepared using IRS–1C LISS III and PAN data.

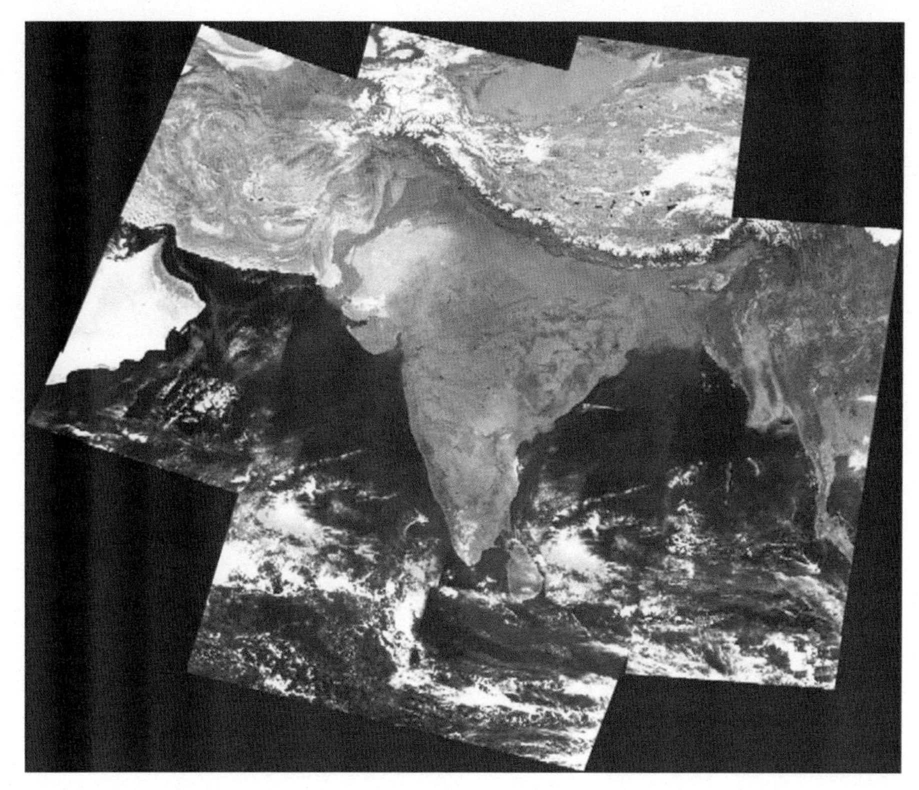

Figure 4.12 Digital mosaic of images obtained using the Ocean Color Monitor (OCM) sensor onboard the IRS-P4 (Oceansat) satellite (*Source*: http://nrsc.gov.in).

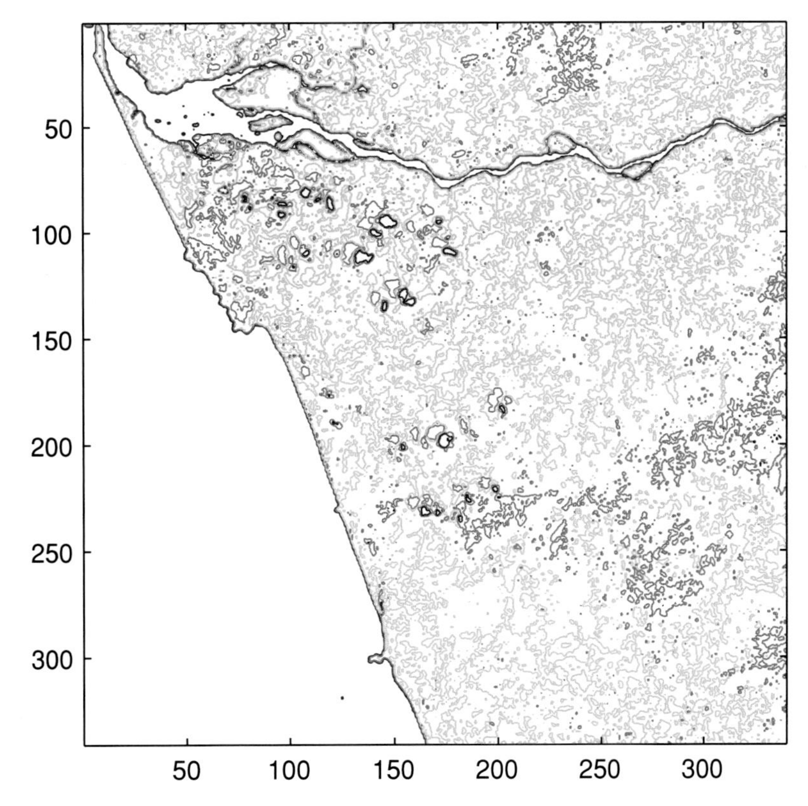

Figure 4.23 Contour plot of an image.

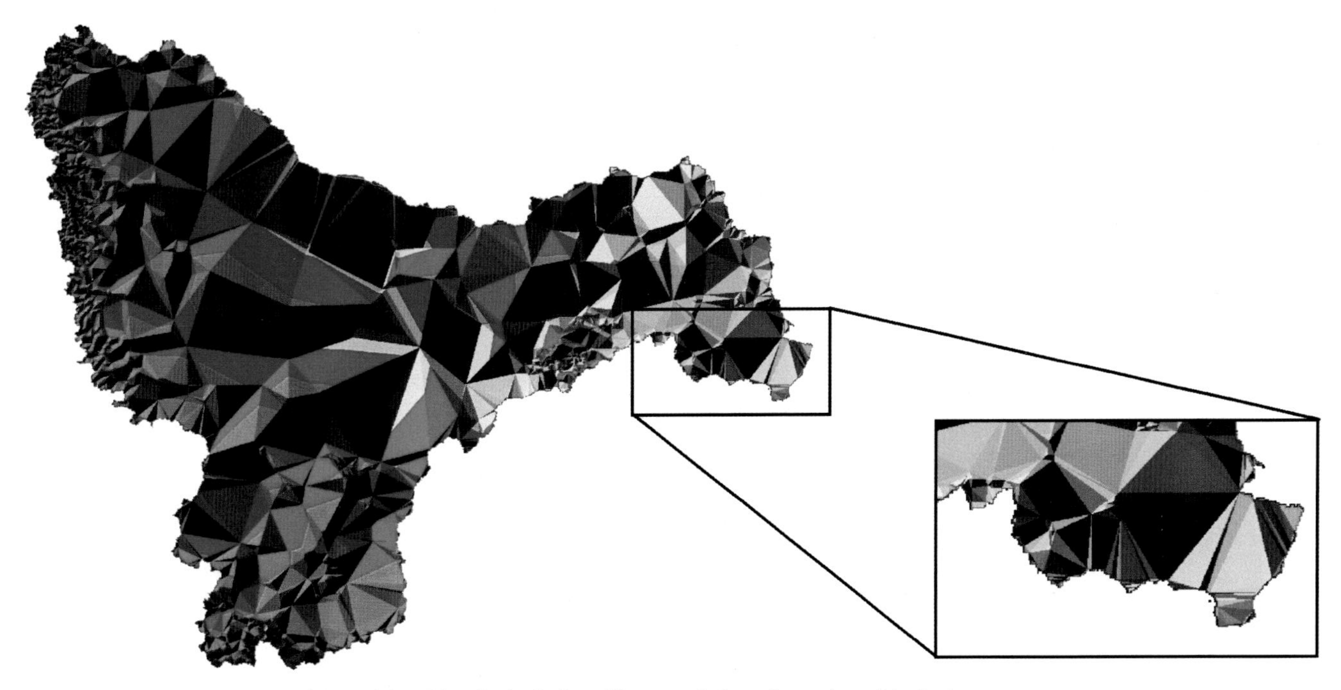

Figure 5.3 TIN representation of the Krishna River basin, India, with zoomed view of a portion of the basin.

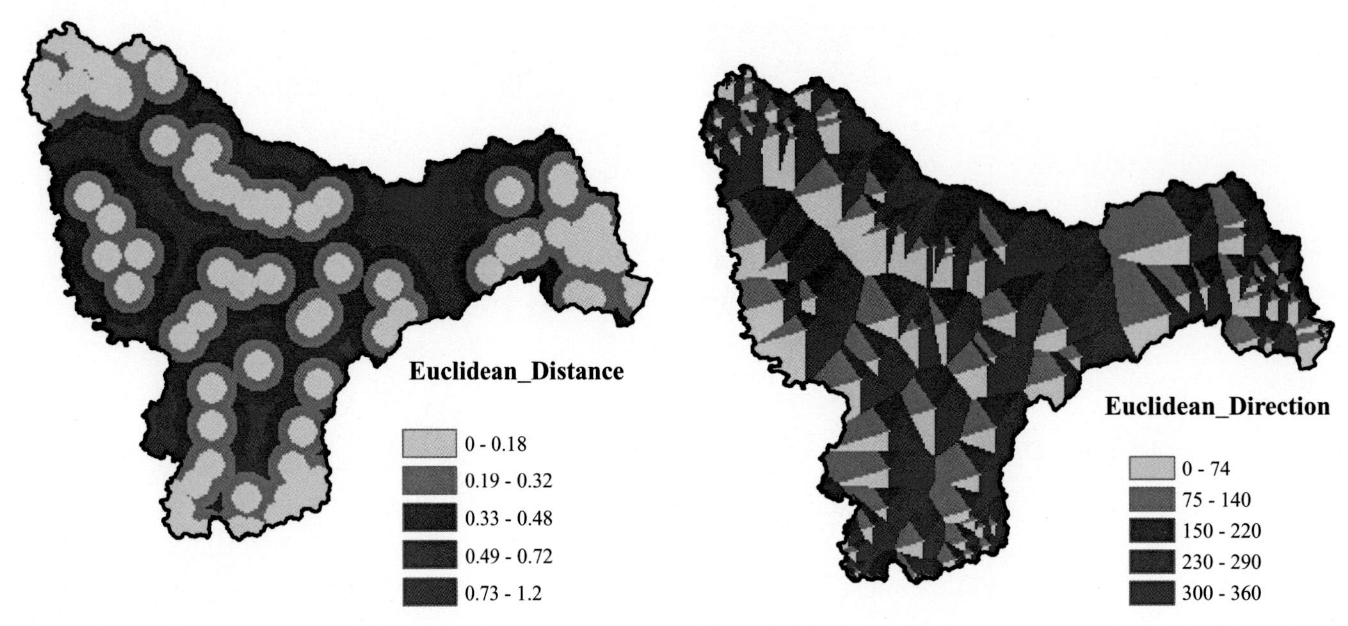

Figure 5.5 Euclidean distance image created around randomly generated point feature class.

Figure 5.6 Euclidean direction image created for the set of data points used in Figure 5.5.

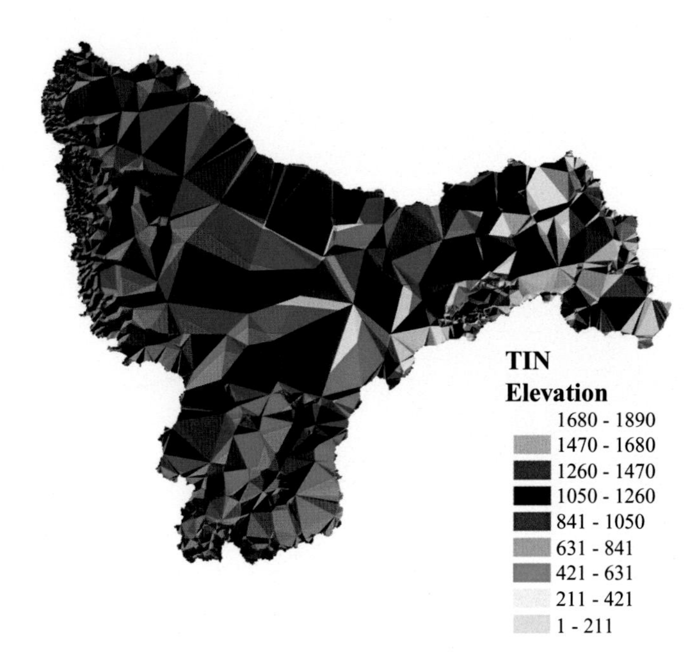

Figure 5.7 TIN for the Krishna basin created from USGS DEM data.

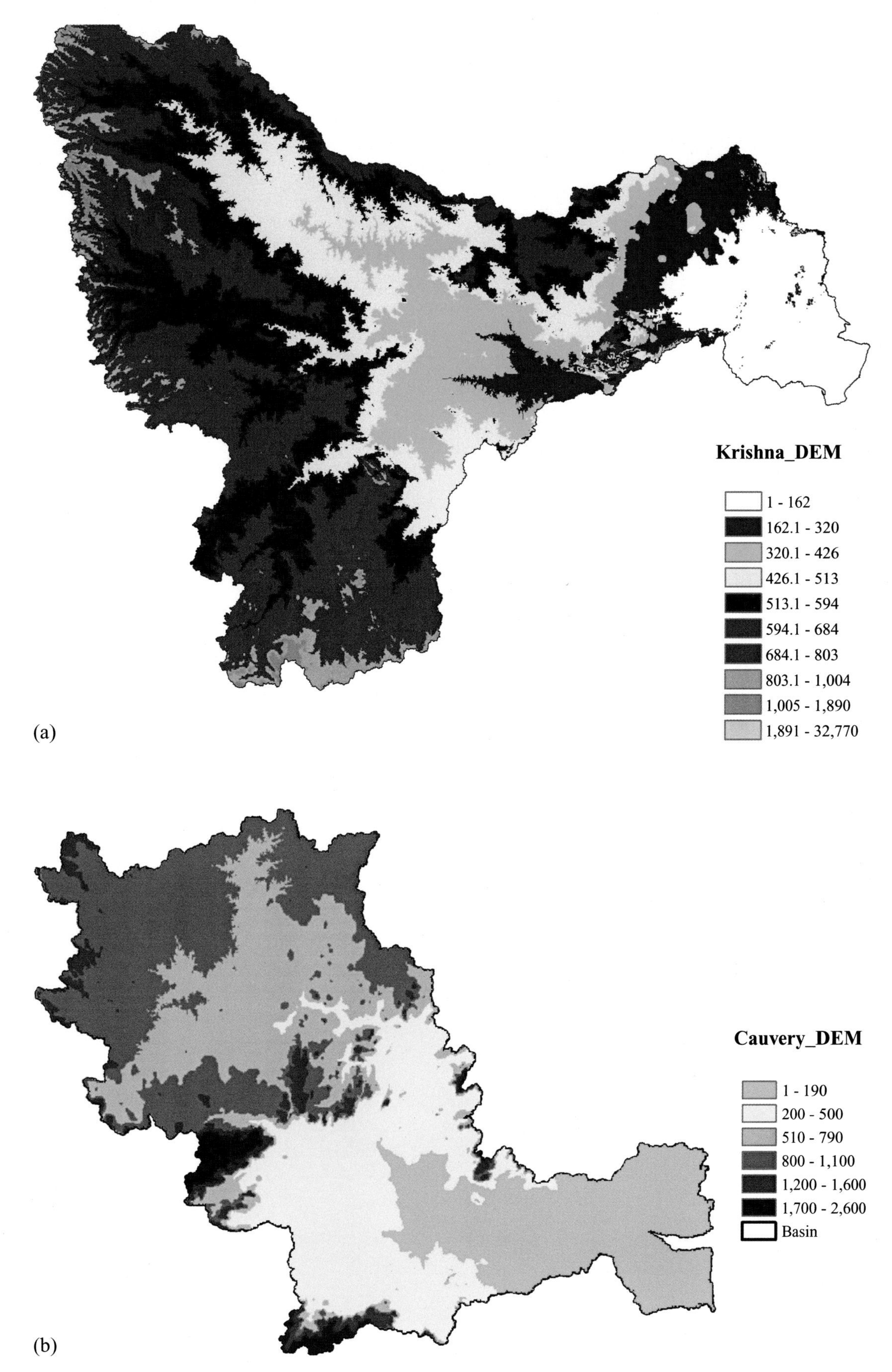

Figure 5.18 SRTM data for (a) Krishna basin and (b) Cauvery basin.

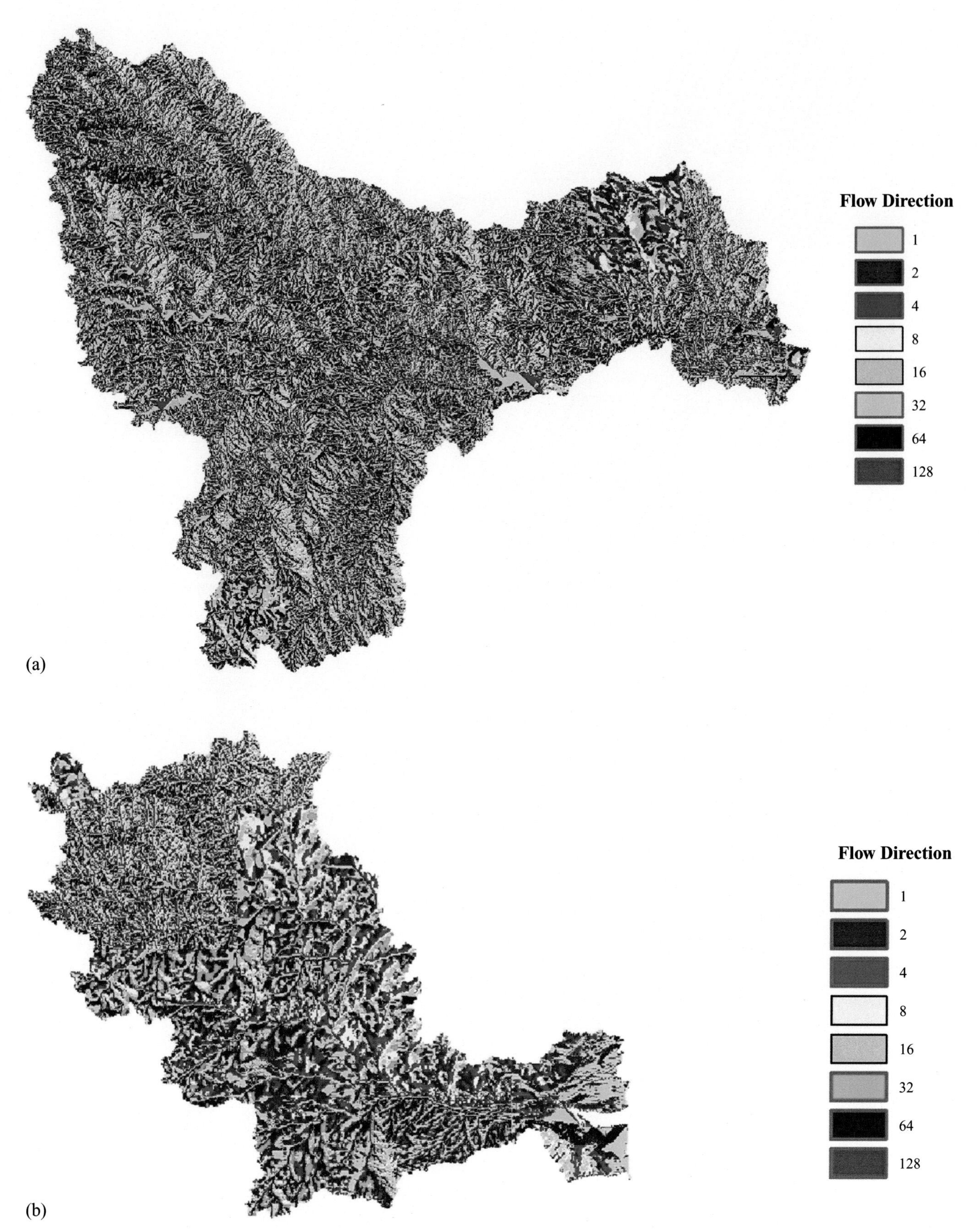

Figure 5.19 Flow direction image for (a) Krishna basin and (b) Cauvery basin.

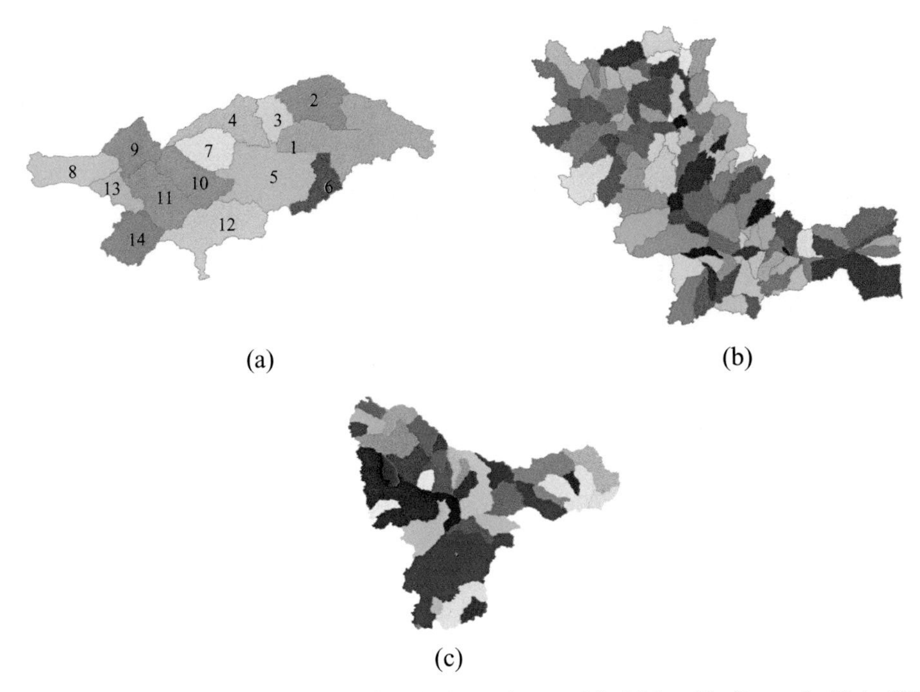

Figure 5.23 (a) Sub-basins formed by AVSWAT for the outlets fixed in the catchment of the Malaprabha Reservoir. (b) ArcGIS-generated sub-watersheds for the Cauvery basin. (c) ArcGIS-generated sub-watersheds for the Krishna basin overlain with stream networks.

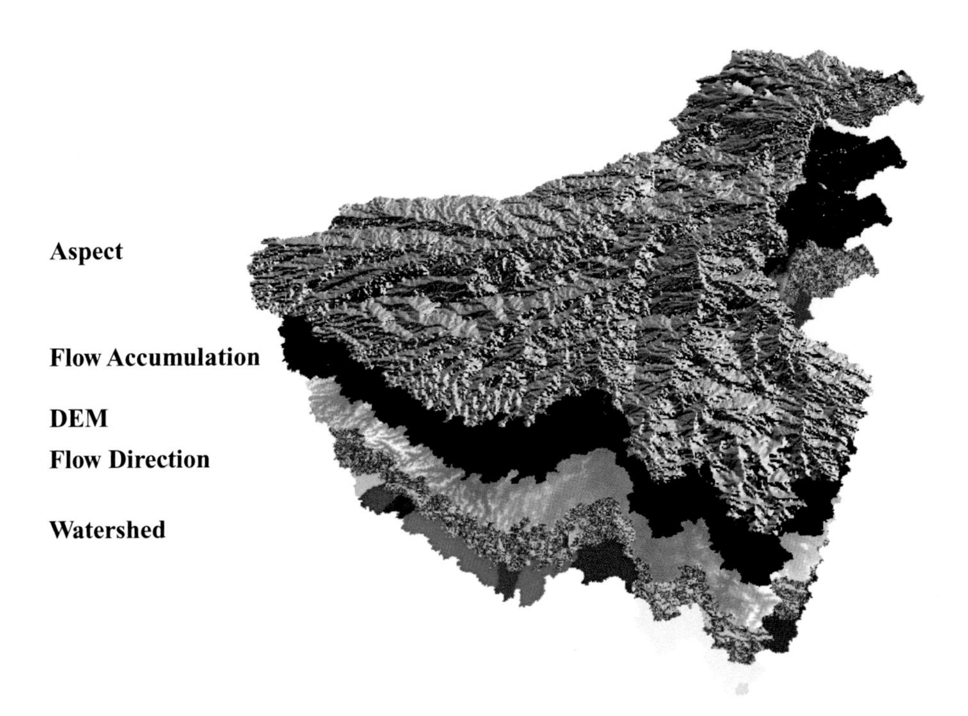

Figure 5.25 Integration of layers in a GIS environment.

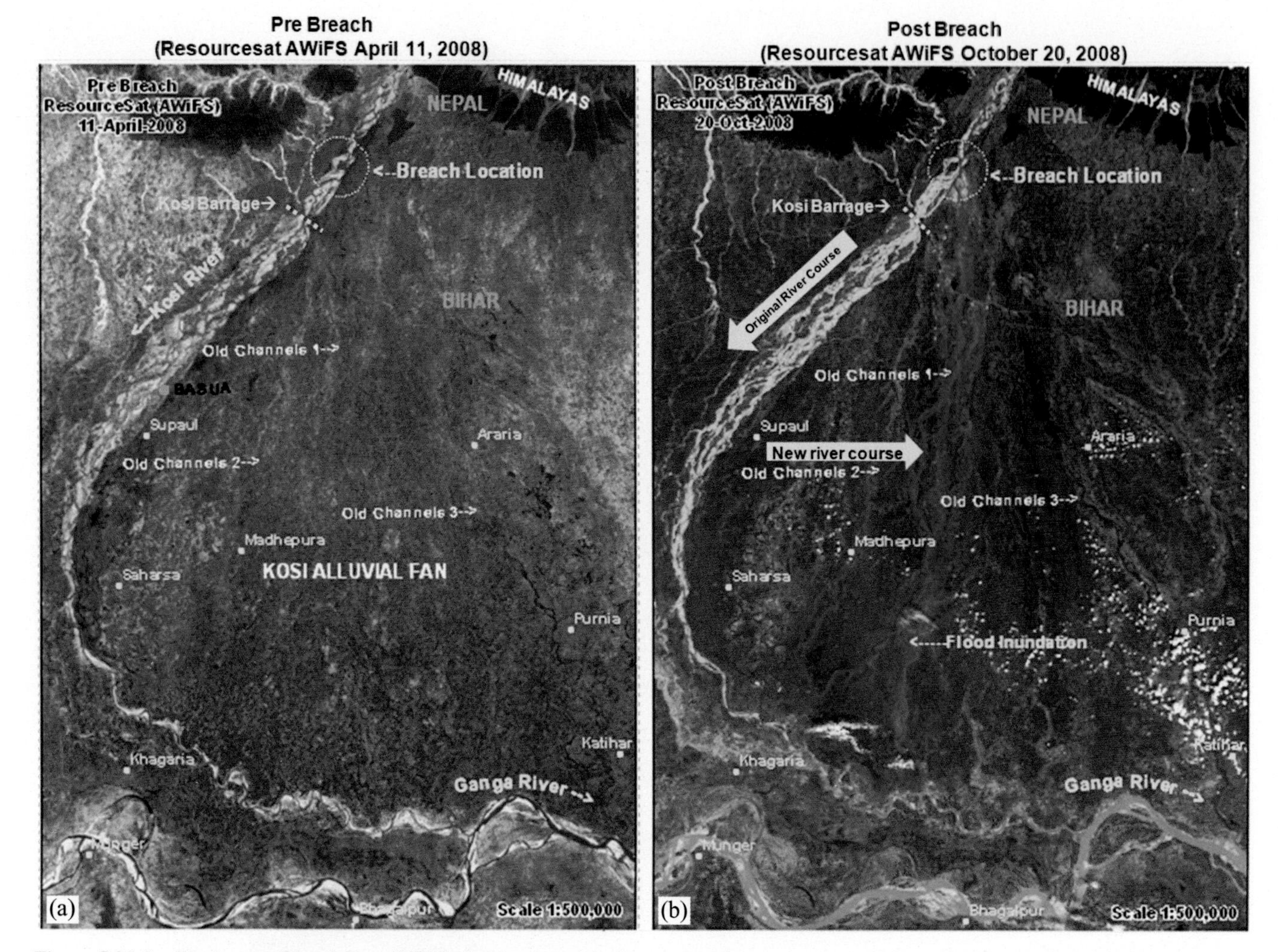

Figure 5.26 Satellite images (Resourcesat AWiFS) showing the original course (southwest trending in left panel) of the Kosi River, India, and that developed after the breach (southwards in right panel) due to flood in 2008 (Bhatt *et al.*, 2010).

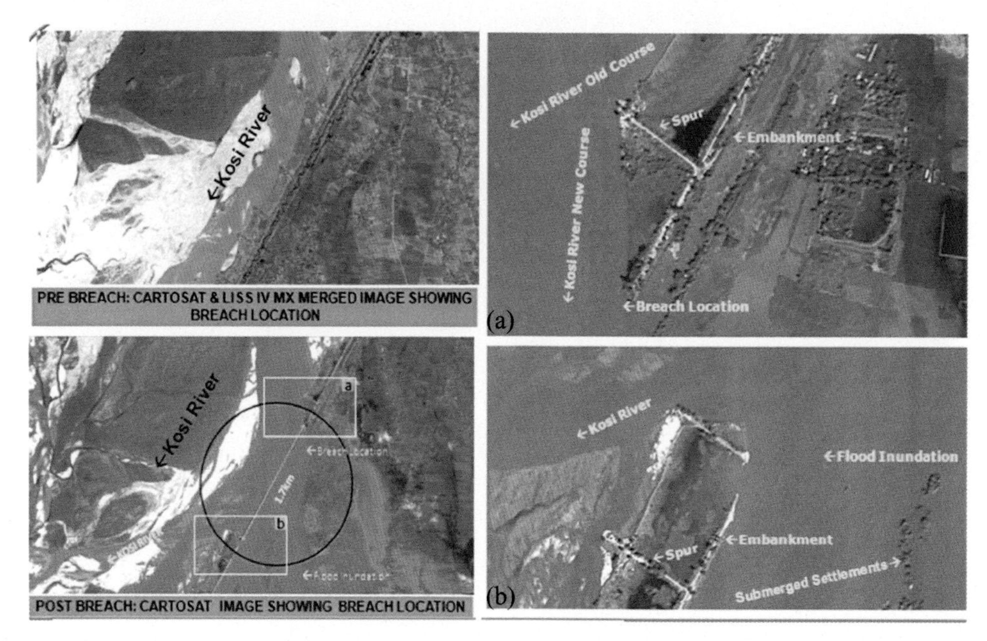

Figure 5.27 IRS CARTOSAT and LISS IV images showing before and after a flood event in the Kosi River, India (Bhatt *et al.*, 2010).

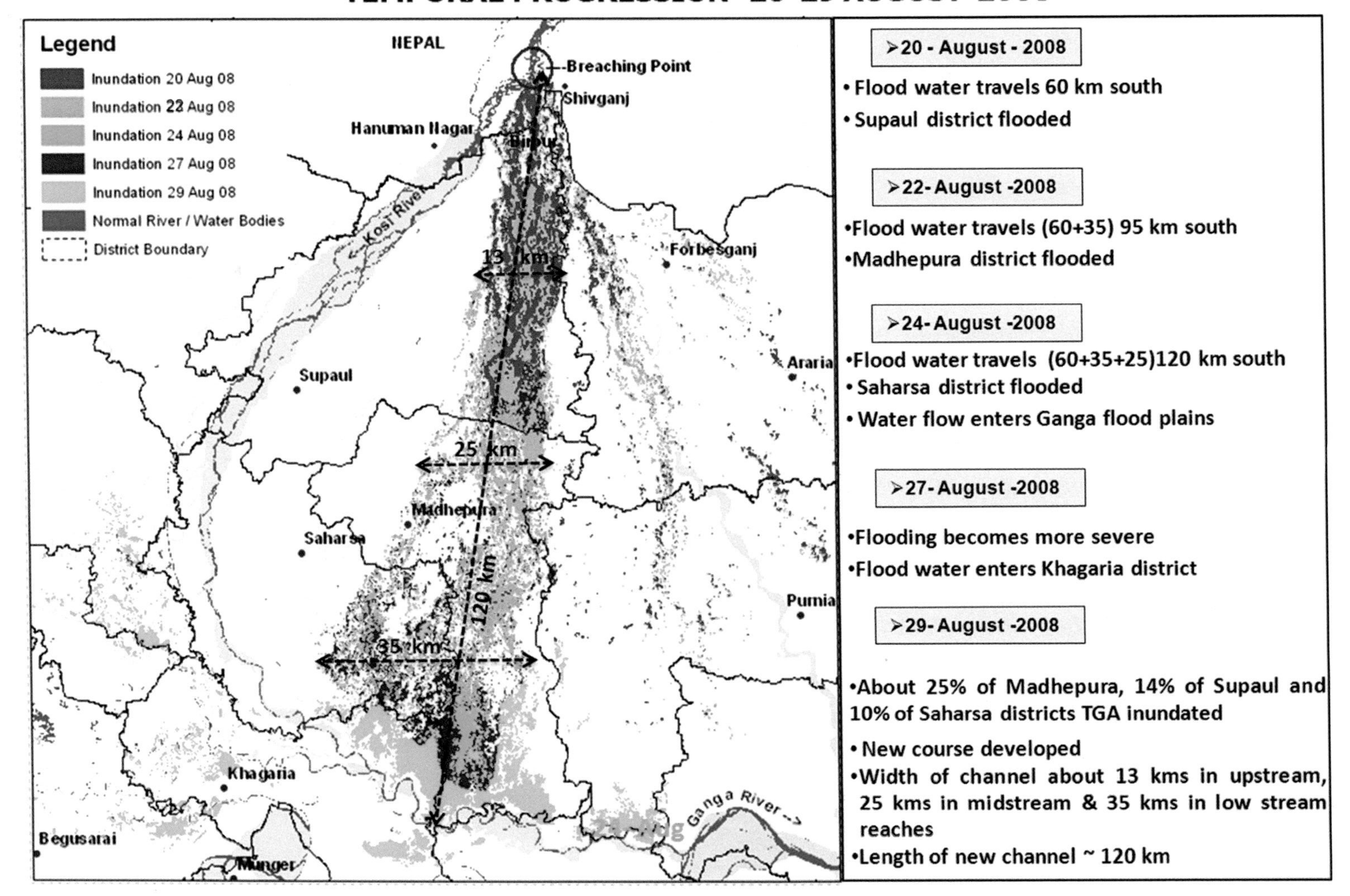

Figure 5.28 Temporal progression of Kosi River flood inundation in India during August 20–29, 2008 (Bhatt *et al.*, 2010).

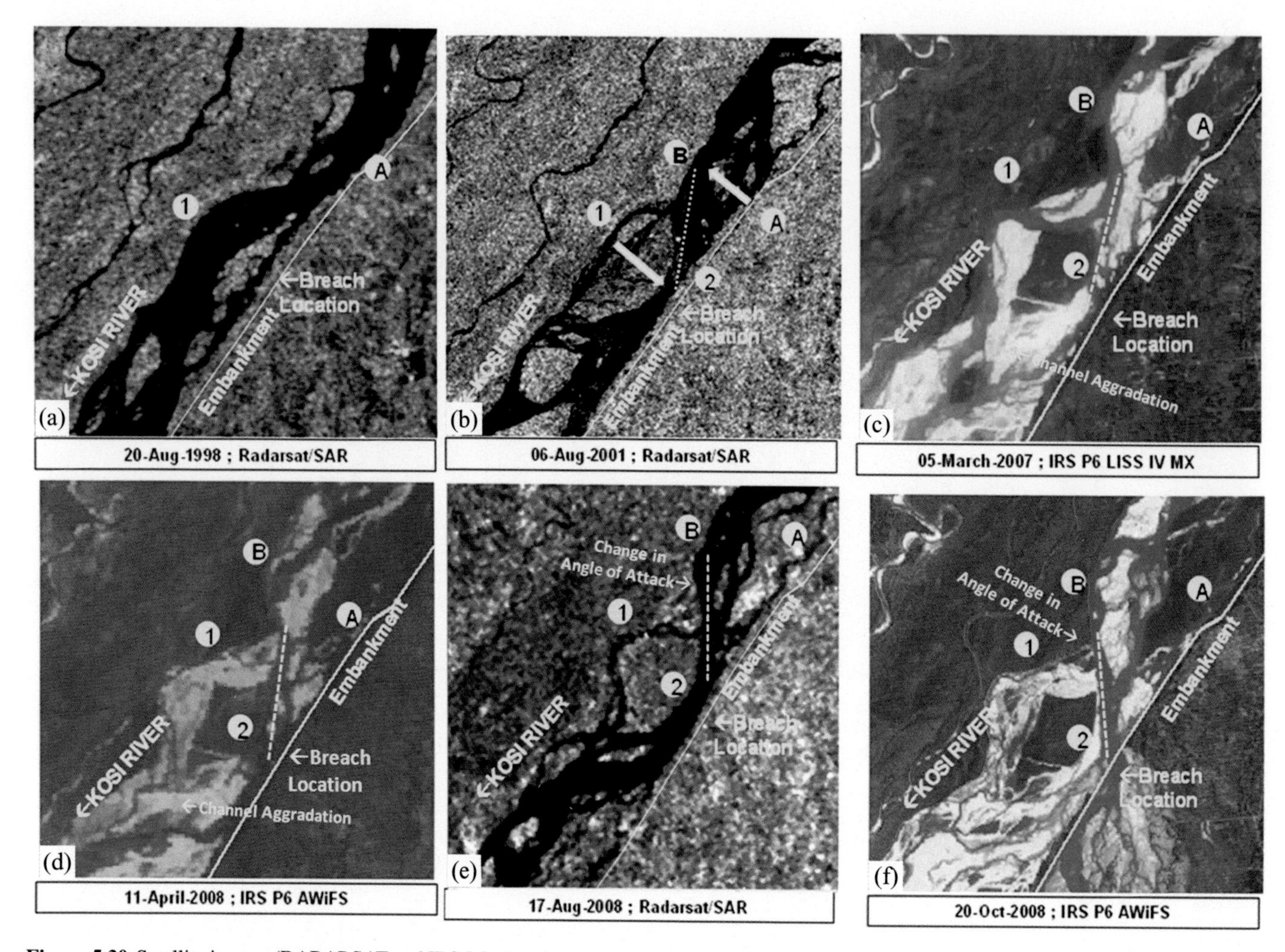

Figure 5.30 Satellite images (RADARSAT and IRS P6) showing the changes observed in the Kosi River course, India, near the breach location during the period 1998–2008. (Dashed lines in the images show changes in angle of attack, whereas numbers/letters encircled in yellow indicate the shift in river course) (Bhatt *et al.*, 2010).

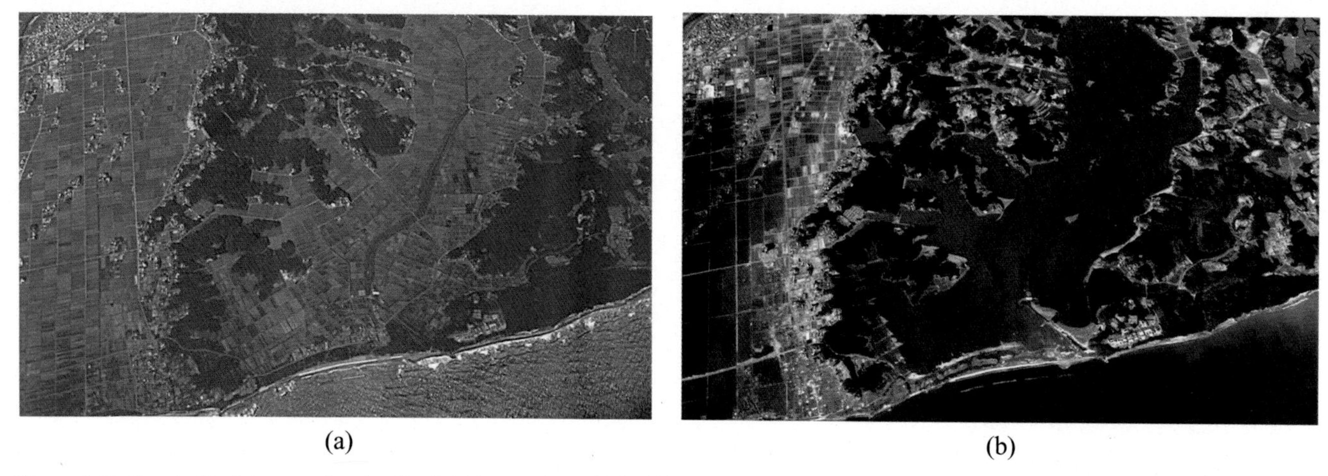

(a) (b)

Figure 5.31 Google Earth images showing Sendai Airport, Japan (a) in 2003 before the tsunami and (b) after the tsunami in 2011.

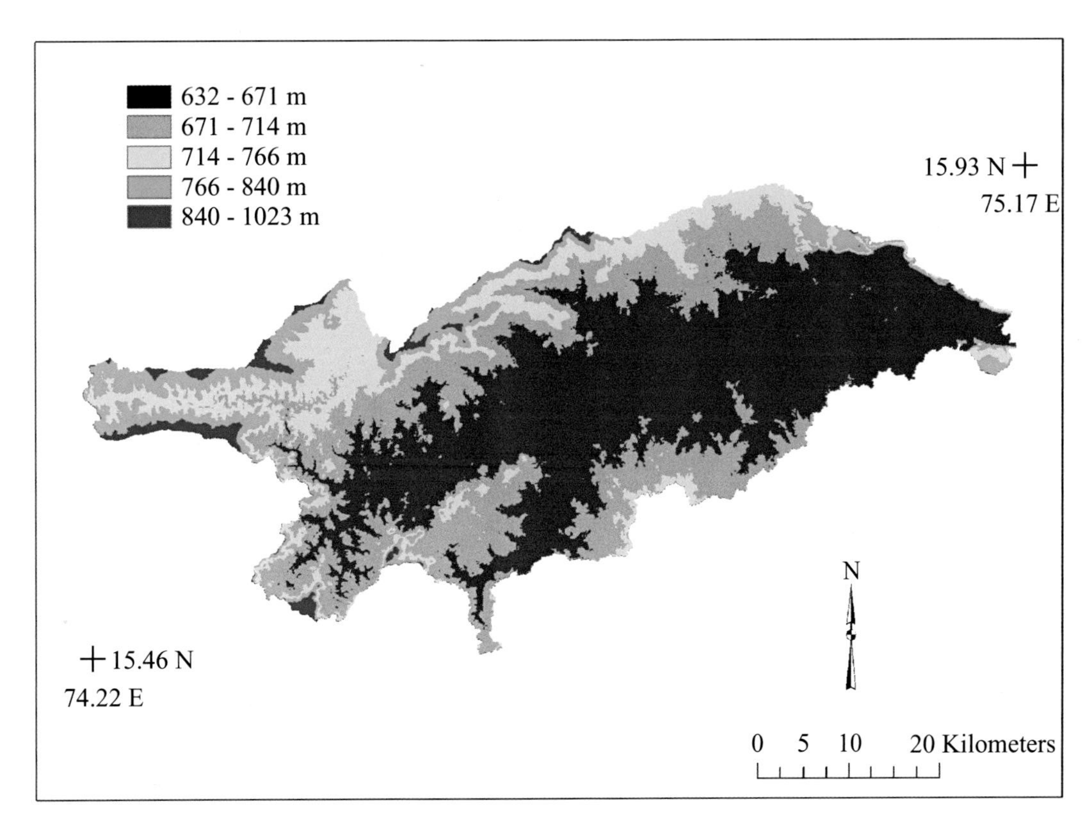

Figure 6.5 DEM of the catchment of the Malaprabha Reservoir.

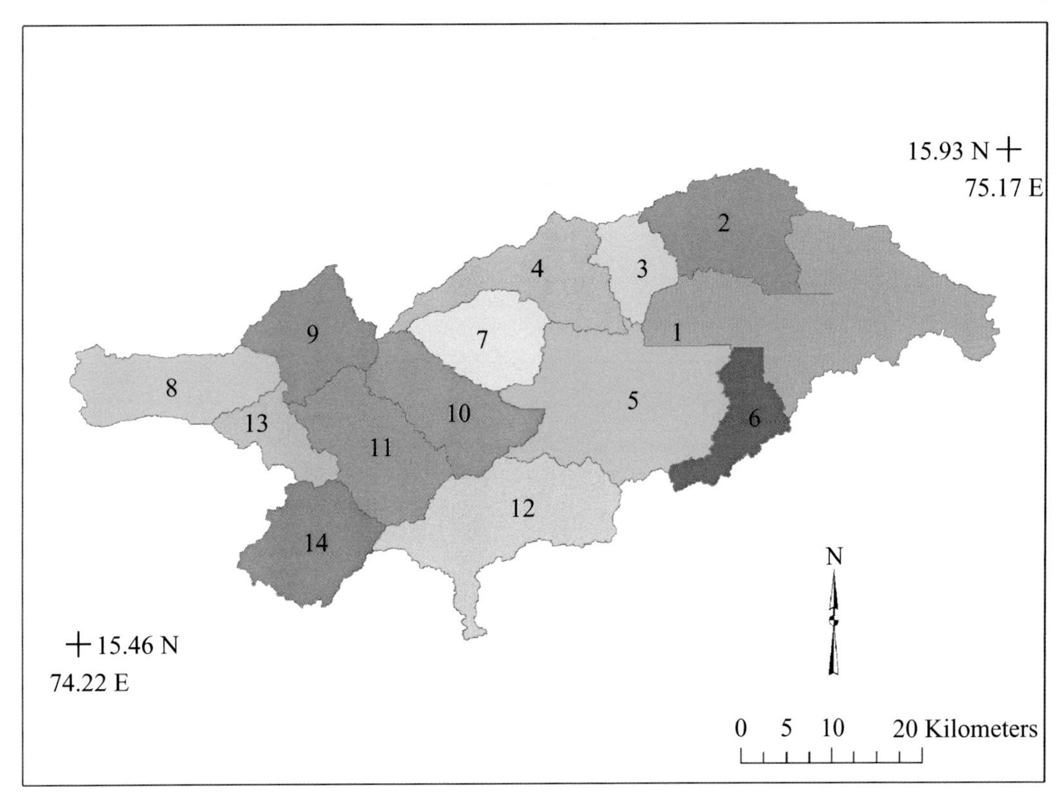

Figure 6.6 Sub-basins formed by AVSWAT for the outlets fixed in the catchment of the Malaprabha Reservoir.

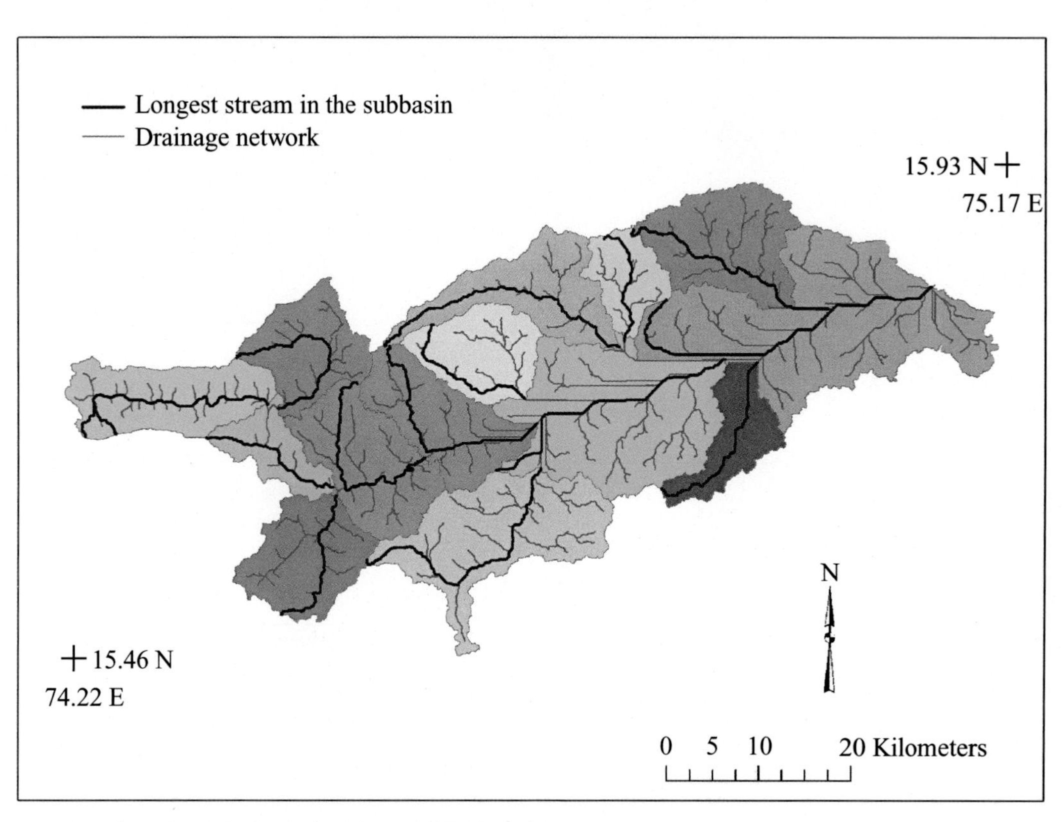

Figure 6.7 Longest stream in each sub-basin obtained from AVSWAT model.

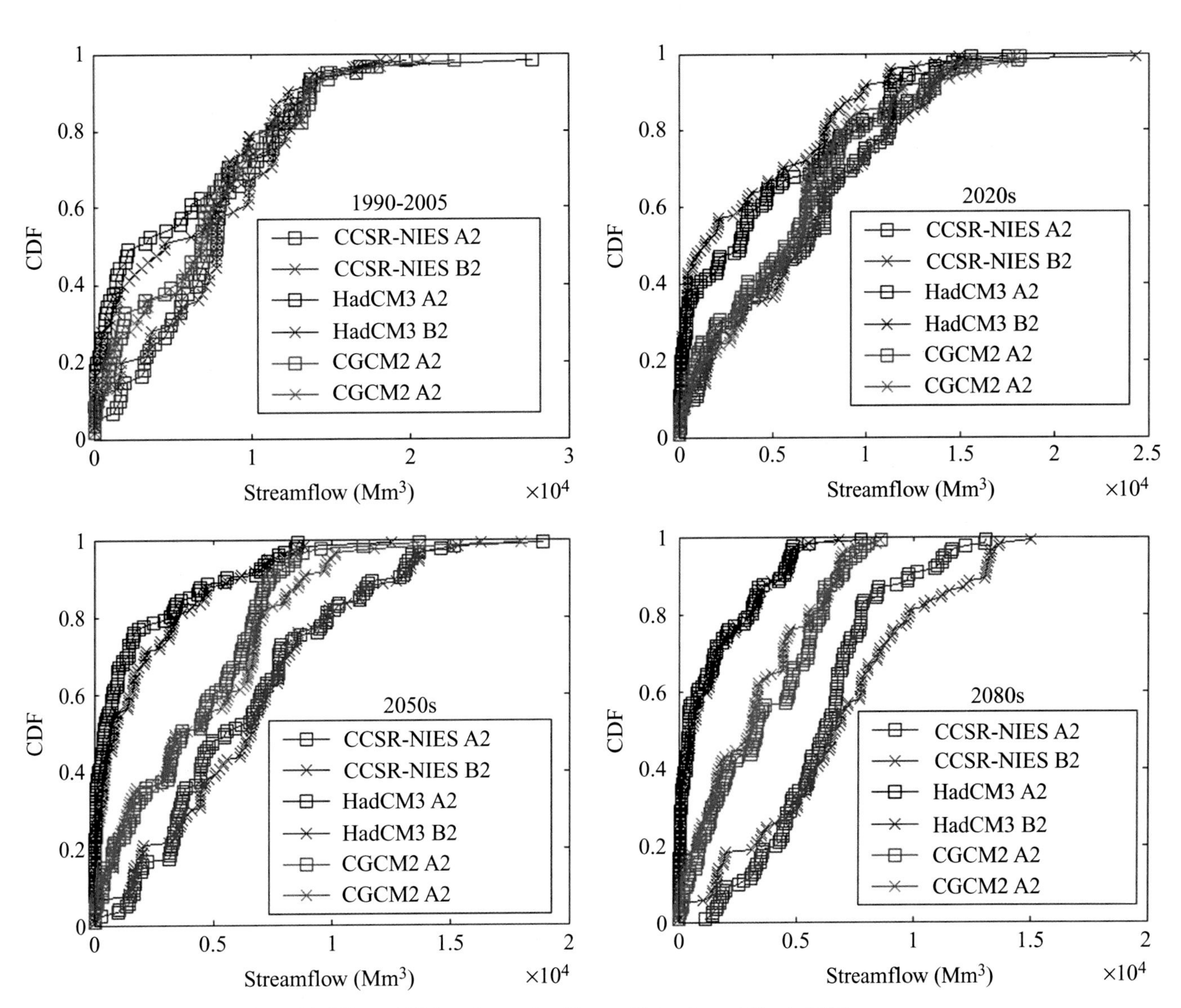

Figure 6.20 Predicted streamflow with different GCMs and scenarios (*Source*: Mujumdar and Ghosh, 2008).

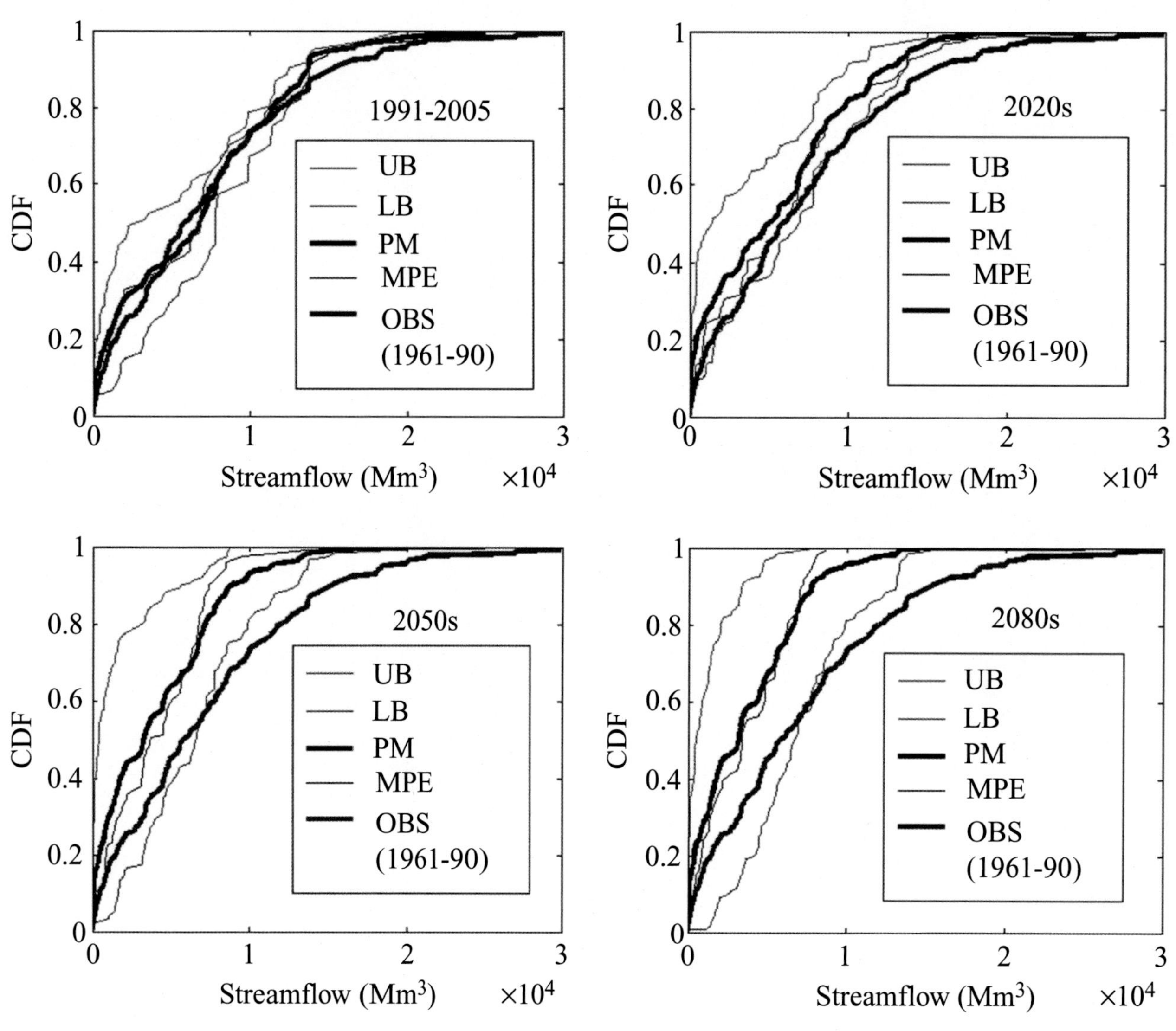

Figure 6.22 Upper bound (UB), lower bound (LB), and possibilistic mean CDF (*Source*: Mujumdar and Ghosh, 2008).

Table 2.10 *Inflow hydrograph*

Time (hr)	0	6	12	18	24	30	36	42	48	54
Inflow (m³/s)	10	20	50	60	55	45	35	27	2	15

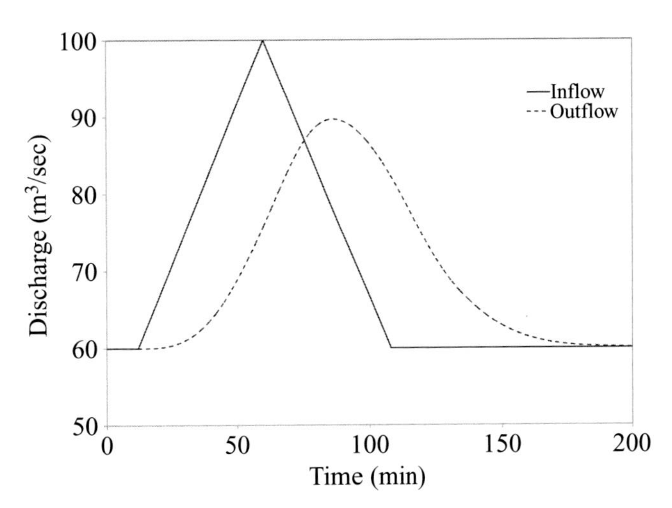

Figure 2.8 Inflow and outflow hydrographs for Example 2.4.

$$C_1 = \frac{7.03 \times 0.25 + 0.5 \times 6}{7.03 - 7.03 \times 0.25 + 0.5 \times 6} = 0.575098 = 0.575$$

$$C_2 = \frac{7.03 - 7.03 \times 0.25 - 0.5 \times 6}{7.03 - 7.03 \times 0.25 + 0.5 \times 6} = 0.2747053 = 0.275$$

For the first time interval, 0 to 6 hr,

$$I_1 = 10.0 \qquad C_1 I_1 = 5.75$$
$$I_2 = 20.0 \qquad C_0 I_2 = 3.00$$
$$Q_1 = 10.0 \qquad C_2 Q_1 = 2.75$$

From Equation 2.33: $Q_2 = C_0 I_2 + C_1 I_1 + C_2 Q_1$
$= 11.50\,\text{m}^3/\text{s}$

For the next time step, 6 to 12 hr, $Q_1 = 11.5$ m³/s. The procedure is repeated for the entire duration of the inflow hydrograph. The computations are carried out in a tabular form in Table 2.11. The resulting outflow hydrograph is shown in Figure 2.10, along with the inflow hydrograph.

A modified method, called the Muskingum–Cunge method, essentially uses the same structure of computations, but converts the routing method into a hydraulic routing by considering both the continuity and momentum equations, and is used as a distributed flow routing technique. Perumal *et al.* (2007, 2010) have developed the variable parameter Muskingum stage hydrograph (VPMS) routing method, using stage (depth of flow) as a routing variable. A main advantage of the enhanced VPMS

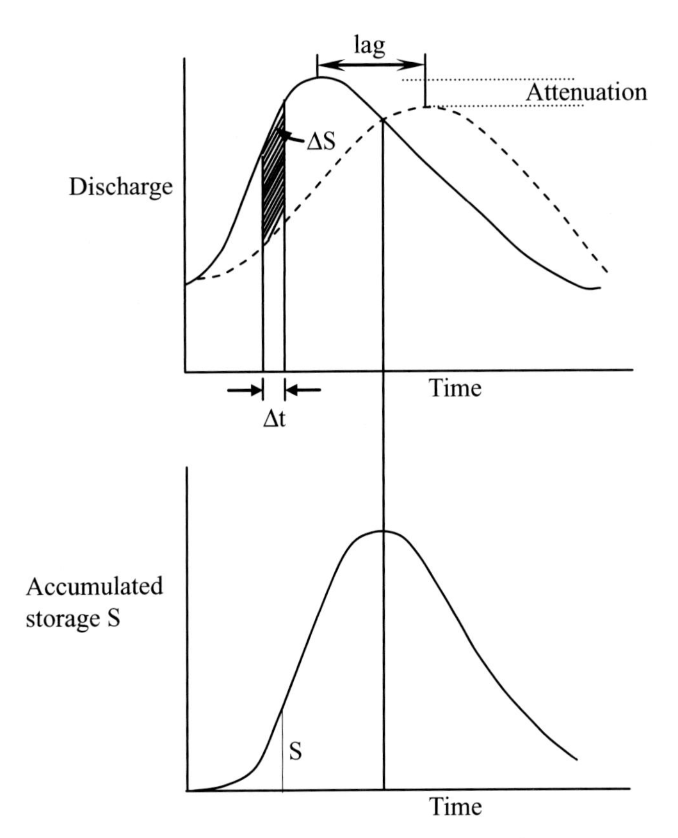

Figure 2.9 Channel routing.

(Perumal *et al.*, 2010) is that only two bounded natural channel cross sections of a reach are adequate for carrying out the hydraulic routing. This method is especially useful for carrying out flood routing in ungauged basins.

2.5 A BRIEF REVIEW OF COMMONLY USED HYDROLOGIC MODELS

A large number of hydrologic models exist today with varying degrees of data requirements that may be used for purposes such as estimation of flood runoff, routing of flood hydrographs, and assessment of flood inundation, which may be done with a GIS interface. An exhaustive review of hydrologic models is provided by Singh and Woolhiser (2002).

The hydrologic processes that occur in nature are *distributed* in the sense that the time and space derivatives of the processes are both important. The models can be classified as distributed and lumped depending on whether the models consider the space derivatives (distributed) or not (lumped). The semi-distributed models account for spatial variations in some processes while ignoring them in others. On the time scale, the models may be discrete or continuous time models.

Table 2.11 *Muskingum method of routing* $\Delta t = 6$ hr

Time (hr)	I (m³/s)	$0.15 * I_2$	$0.575 * I_1$	$0.275 * Q_1$	Q (m³/s)
1	2	3	4	5	6
0	10				10.00
		3.00	5.75	2.75	
6	20				11.50
		7.50	11.50	3.16	
12	50				22.16
		9.00	28.75	6.09	
18	60				43.84
		8.25	34.50	12.06	
24	55				54.81
		6.75	31.63	15.07	
30	45				53.45
		5.25	25.88	14.70	
36	35				45.82
		4.05	20.13	12.60	
42	27				36.78
		3.00	15.53	10.11	
48	20				28.64
		2.25	11.50	7.88	
54	15				21.63
		1.80	8.63	5.95	
60	12				16.37
		1.50	6.90	4.50	
66	10				12.90
		1.50	5.75	3.55	
72	10				10.80
		1.50	5.75	2.97	
78	10				10.22
		1.50	5.75	2.81	
84	10				10.06
		1.50	5.75	2.77	
90	10				10.02
		1.50	5.75	2.75	
96	10				10.00
		1.50	5.75	2.75	
102	10				10.00

Table 2.12, extracted from Singh and Woolhiser (2002), provides an overview of commonly used hydrologic models.

In this section, three models that are freely available on the Web, along with user-friendly manuals with tutorial examples, are discussed. These models are: HEC-HMS, HEC-RAS, both developed by the Hydrologic Engineering Center, US Army Corps of Engineers, and the Storm Water Management Model (SWMM) developed by the US Environmental Protection Agency (USEPA). Another freely available model with GIS interface is the soil and water assessment tool (SWAT), but that model is not reviewed here. An application of the SWAT model is discussed in Chapter 6. The Variable Infiltration Capacity (VIC) Model, a recently developed hydrologic model (Liang *et al.*, 1994; Liang and Xie, 2001; Gao *et al.*, 2010), is ideally suited for river-basin-scale climate change impact studies. It is a grid-based, macroscale hydrologic model, typically run at 1/8 degree to 2 degree grid size and has been used for climate change impact studies (e.g., Christensen *et al.*, 2004).

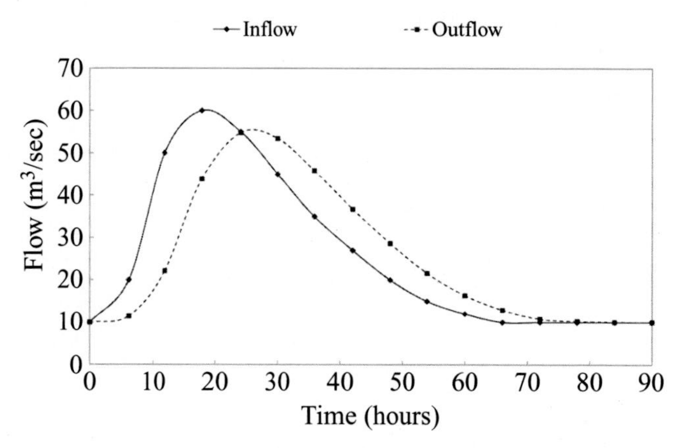

Figure 2.10 Inflow and outflow hydrographs for Example 2.5.

2.5.1 HEC-HMS

HEC-HMS is the Hydrologic Engineering Center's Hydrologic Modeling System, developed by the US Army Corps of Engineers (USACE). It is considered as the standard model in the private sector in the USA for the design of drainage systems, and for quantifying the effect of land use change on flooding (Singh and Woolhiser, 2002). It is designed to simulate the precipitation–runoff processes of dendritic watershed systems. The mass or energy flux in the hydrologic cycle in a watershed is represented with a mathematical model. In most cases, several model choices are available for representing each flux. Each mathematical model included in the program is typical of a particular environment. HEC-HMS is applicable in a wide range of geographic areas for solving a wide range of problems. The potential of the model includes applications for large river-basin water balance, flood hydrology, and for estimating small urban or natural watershed runoff (US Army Corps of Engineers, 2008).

2.5.1.1 ILLUSTRATION OF THE MODEL

Illustration of use of the HEC-HMS for hydrologic modeling is discussed here through a case study. The catchment area considered lies upstream of the Tunga-Bhadra Dam in Karnataka state, India, and covers an area of about 15,000 km². The upper catchments are characterized by undulating terrain and higher rainfall, while the middle portion of the basin has lower rainfall, drought conditions, and not so undulating a terrain. The average slope of the basin is 6%. The basin is dominated by clay loam soil with the soil composition of: 42% clay loam, 34% clayey soil,

Table 2.12 *A sample of commonly used hydrologic models*

Model name/acronym	Author(s) (year)	Remarks
Stanford Watershed Model (SWM)/Hydrologic Simulation Package–Fortran IV (HSPF)	Crawford and Linsley (1966) Bicknell *et al.* (1993)	Continuous, dynamic event or steady-state simulator of hydrologic and hydraulic and water quality processes
Catchment Model (CM)	Dawdy and O'Donnell (1965)	Lumped, event-based runoff model
Tennessee Valley Authority (TVA) Model	Tennessee Valley Authority (1972)	Lumped, event-based runoff model
US Department of Agriculture Hydrograph Laboratory (USDAHL) Model	Holtan and Lopez (1971), Holtan *et al.* (1974)	Event-based, process-oriented, lumped hydrograph model
US Geological Survey (USGS) Model	Dawdy *et al.* (1970, 1978)	Process-oriented, continuous/event-based runoff model
Utah State University (USU) Model	Andrews *et al.* (1978)	Process-oriented, event/continuous streamflow model
Purdue Model	Huggins and Monke (1970)	Process-oriented, physically based, event runoff model
Antecedent Precipitation Index (API) Model	Sittner *et al.* (1969)	Lumped, river flow forecast model
Hydrologic Engineering Center–Hydrologic Modeling System (HEC-HMS)	Feldman (1981), HEC (1981, 2000)	Physically based, semidistributed, event-based, runoff model
Streamflow Synthesis and Reservoir regulation (SSARR) Model	Rockwood (1982), US Army Corps of Engineers (1987), Speers (1995)	Lumped, continuous streamflow simulation model
National Weather Service River Forecast System (NWS-RFS)	Burnash *et al.* (1973), Burnash (1975)	Lumped, continuous river forecast system
University of British Columbia (UBC) Model	Quick and Pipes (1977), Quick (1995)	Process-oriented, lumped parameter, continuous simulation model
Tank Model	Sugawara *et al.* (1974), Sugawara (1995)	Process-oriented, semidistributed, or lumped continuous simulation model
Runoff Routing Model (RORB)	Laurenson (1964), Laurenson and Mein (1993, 1995)	Lumped, event-based runoff simulation model
Agricultural Runoff Model (ARM)	Donigian *et al.* (1977)	Process-oriented, lumped runoff simulation model
Storm Water Management Model (SWMM)	Metcalf and Eddy *et al.* (1971), Huber and Dickinson (1988), Huber (1995)	Process-oriented, semidistributed, continuous stormflow model
Xinanjiang Model	Zhao *et al.* (1980), Zhao and Liu (1995)	Process-oriented, lumped, continuous simulation model
Hydrological Simulation (HBV) Model	Bergstrom (1976, 1992, 1995)	Process-oriented, lumped, continuous streamflow simulation model
Great Lakes Environmental Research Laboratory (GLERL) Model	Croley (1982, 1983)	Physically based, semidistributed, continuous simulation model
Pennsylvania State University–Urban Runoff Model (PSU-URM)	Aron and Lakatos (1980)	Lumped, event-based urban runoff model
Chemicals, Runoff, and Erosion from Agricultural Management Systems (CREAMS)	USDA (1980)	Process-oriented, lumped parameter, agricultural runoff and water quality model
Areal Non-point Source Watershed Environment Response Simulation (ANSWERS)	Beasley *et al.* (1977), Bouraoui *et al.* (2002)	Event-based or continuous, lumped parameter runoff and sediment yield simulation model
Erosion Productivity Impact Calculator (EPIC) Model	Williams *et al.* (1984), Williams (1995a,b)	Process-oriented, lumped parameter, continuous water quantity and quality simulation model
Simulator for Water Resources in Rural Basins (SWRRB)	Williams *et al.* (1985), Williams (1995a,b)	Process-oriented, semidistributed, runoff and sediment yield simulation model

(*cont.*)

Table 2.12 *(cont.)*

Model name/acronym	Author(s) (year)	Remarks
Simulation of Production and Utilization of Rangelands (SPUR)	Wight and Skiles (1987), Carlson and Thurow (1992), Carlson *et al.* (1995)	Physically based, lumped parameter, ecosystem simulation model
National Hydrology Research Institute (NHRI) Model	Vandenberg (1989)	Physically based, lumped parameter, continuous hydrologic simulation model
Technical Report-20 (TR-20) Model	Soil Conservation Service (1965)	Lumped parameter, event based, runoff simulation model
Système Hydrologique Européen/Système Hydrologique Européen Sediment (SHE/SHESED)	Abbott *et al.* (1986a,b), Bathurst *et al.* (1995)	Physically based, distributed, continuous streamflow and sediment simulation
Institute of Hydrology Distributed Model (IHDM)	Beven *et al.* (1987), Calver and Wood (1995)	Physically based, distributed, continuous rainfall–runoff modeling system
Physically Based Runoff Production Model (TOPMODEL)	Beven and Kirkby (1976, 1979), Beven (1995)	Physically based, distributed, continuous hydrologic simulation model
Agricultural Non-Point Source Model (AGNPS)	Young *et al.* (1989, 1995)	Distributed parameter, event-based, water quantity and quality simulation model
Kinematic Runoff and Erosion Model (KINEROS)	Woolhiser *et al.* (1990), Smith *et al.* (1995)	Physically based, semidistributed, event-based, runoff and water quality simulation model
Groundwater Loading Effects of Agricultural Management Systems (GLEAMS)	Knisel *et al.* (1993), Knisel and Williams (1995)	Process-oriented, lumped parameter, event-based water quantity and quality simulation model
Generalized River Modeling Package–Système Hydrologique Européen (MIKE-SHE)	Refsgaard and Storm (1995)	Physically based, distributed, continuous hydrologic and hydraulic simulation model
Simple Lumped Reservoir Parametric (SLURP) Model	Kite (1995)	Process-oriented, distributed, continuous simulation model
Snowmelt Runoff Model (SRM)	Rango (1995)	Lumped, continuous snowmelt–runoff simulation model
THALES	Grayson *et al.* (1995)	Process-oriented, distributed-parameter, terrain analysis-based, event-based runoff simulation model
Constrained Linear Simulation (CLS)	Natale and Todini (1976a,b, 1977)	Lumped parameter, event-based or continuous runoff simulation model
ARNO (Arno River) Model	Todini (1988a,b, 1996)	Semidistributed, continuous rainfall–runoff simulation model
Waterloo Flood System (WATFLOOD)	Kouwen *et al.* (1993), Kouwen (2000)	Process-oriented, semidistributed, continuous flow simulation model
Topographic Kinematic Approximation and Integration (TOPIKAPI) Model	Todini (1995)	Distributed, physically based, continuous rainfall–runoff simulation model
Hydrological (CEQUEAU) Model	Morin *et al.* (1995, 1998)	Distributed, process-oriented, continuous runoff simulation model
Large Scale Catchment Model (LASCAM)	Sivapalan *et al.* (1996a,b,c)	Conceptual, semidistributed, large-scale, continuous, runoff and water quality simulation model
Mathematical Model of Rainfall–Runoff Transformation System (WISTOO)	Ozga-Zielinska and Brzezinski (1994)	Process-oriented, semidistributed, event-based or continuous simulation model
Rainfall–Runoff (R–R) Model	Kokkonen *et al.* (1999)	Semidistributed, process-oriented, continuous streamflow simulation model
Soil–Vegetation–Atmosphere Transfer (SVAT) Model	Ma *et al.* (1999), Ma and Cheng (1998)	Macroscale, lumped parameter, streamflow simulation system
Hydrologic Model System (HMS)	Yu (1996), Yu and Schwartz (1998), Yu *et al.* (1999)	Physically based, distributed-parameter, continuous hydrologic simulation system

Table 2.12 (*cont.*)

Model name/acronym	Author(s) (year)	Remarks
Hydrological Modeling System (ARC/EGMO)	Becker and Pfutzner (1987), Lahmer *et al.* (1999)	Process-oriented, distributed, continuous simulation system
Macroscale Hydrolgical Model–Land Surface Scheme (MODCOU-ISBA)	Ledoux *et al.* (1989), Noilhan and Mahfouf (1996)	Macroscale, physically based, distributed, continuous simulation model
Regional-Scale Hydroclimatic Model (RSHM)	Kavvas *et al.* (1998)	Process-oriented, regional scale, continuous hydrologic simulation model
Global Hydrology Model (GHM)	Anderson and Kavvas (2002)	Process-oriented, semidistributed, large-scale hydrologic simulation model
Distributed Hydrology Soil Vegetation Model (DHSVM)	Wigmosta *et al.* (1994)	Distributed, physically based, continuous hydrologic simulation model
Système Hydrologique Européen Transport (SHETRAN)	Ewen *et al.* (2000)	Physically based, distributed, water quantity and quality simulation model
Cascade Two-dimensional Model (CASC2D)	Julien and Saghafian (1991), Ogden (1998)	Physically based, distributed, event-based runoff simulation model
Dynamic Watershed Simulation Model (DWSM)	Borah and Bera (2000), Borah *et al.* (1999)	Process-oriented, event-based, runoff and water quality simulation model
Surface Runoff, Infiltration, River Discharge and Groundwater Flow (SIRG)	Yoo (2002)	Physically based, lumped parameter, event-based streamflow simulation model
Modular Kinematic Model for Runoff Simulation (Modular System)	Stephenson (1989), Stephenson and Randell (1999)	Physically based, lumped parameter, event-based runoff simulation model
Watershed Bounded Network Model (WBNM)	Boyd *et al.* (1979, 1996), Rigby *et al.* (1999)	Geomorphology-based, lumped parameter, event-based flood simulation model
Geomorphology-Based Hydrology Simulation Model (GBHM)	Yang *et al.* (1998)	Physically based, distributed, continuous hydrologic simulation model
Predicting Arable Resource Capture in Hostile Environments – The Harvesting of Incident Rainfall in Semi-arid Tropics (PARCHED-THIRST)	Young and Gowing (1996)	Process-oriented, lumped parameter, event-based agro-hydrologic model
Daily Conceptual Rainfall–Runoff Model (HYDROLOG)-Monash Model	Potter and McMahon (1976), Chiew and McMahon (1994)	Lumped, conceptual rainfall–runoff model
Simplified Hydrology Model (SIMHYD)	Chiew *et al.* (2002)	Conceptual, daily, lumped parameter rainfall–runoff model
Two Parameter Monthly Water Balance Model (TPMWBM)	Guo and Wang (1994)	Process-oriented, lumped parameter, monthly runoff simulation model
Water and Snow Balance Modeling System (WASMOD)	Xu (1999b)	Conceptual, lumped, continuous hydrologic model
Integrated Hydrometeorological Forecasting System (IHFS)	Georgakakos *et al.* (1999)	Process-oriented, distributed, rainfall and flow forecasting system
Stochastic Event Flood Model (SEFM)	Scaefer and Barker (1999)	Process-oriented, physically based event-based, flood simulation model
Distributed Hydrological Model (HYDROTEL)	Fortin *et al.* (2001a,b)	Physically based, distributed, continuous hydrologic simulation model
Agricultural Transport Model (ACTMO)	Frere *et al.* (1975)	Lumped, conceptual, event-based runoff and water quality simulation model
Soil Water Assessment Tool (SWAT)	Arnold *et al.* (1998)	Distributed, conceptual, continuous simulation model

Reproduced, with permission, from Singh and Woolhiser (2002) and 19% sandy clay soil. Farm land is the main land use/land cover as of 2004–5, accounting for more than 55% of the surface. Other cultivable areas, such as trees and groves, fallow land, and cultivable waste, add up to 13%. Forests and natural vegetation make up 16%, and around 5% is used as permanent pastures. About 11% of the territory is not available for cultivation or for natural vegetation (Source: STRIVER Task Summary Report No. 9.3, http://kvina.niva.no/striver/Disseminationofresults/STRIVERReports/tabid/80/Default.aspx). The basin receives an average of 1024 mm of rainfall in a year. The annual average temperature reaches 26.7 °C. In general, humidity is high during the monsoon period and comparatively low during the post-monsoon period. In summer the weather is dry and the humidity is low. The relative humidity in the basin ranges from 17 to 92%.

For use in the HEC-HMS model, the basin is divided into four sub-basins, depending on topographic and meteorologic features. The *deficit and constant loss* model option available in HEC-HMS is used to compute the losses from the watershed. This model uses a single soil layer to account for continuous changes in moisture content. It is a quasi-continuous model and is used in conjunction with a meteorological model that computes evapotranspiration. The parameters for this model include initial moisture deficit, maximum moisture deficit, constant loss rate, and impervious percentage. In order to compute direct runoff from excess precipitation, a transform method is used. In this exercise, the *SCS unit hydrograph* model was used to transform the flows. The Soil Conservation Service Unit Hydrograph (SCS UH) model is a parametric UH model, based on the averages of UH derived from gauged rainfall and runoff for a large number of small agricultural watersheds. The input parameter for this method is the basin lag, which is taken as 0.6 times the time of concentration of the flow. The *constant monthly baseflow* method is used to account for the baseflows. It represents base flow as a constant flow, which may vary monthly. The monthly baseflow values are obtained using a baseflow separation technique and the values are averaged over the monthly time span.

The spatio-temporal precipitation distribution is accomplished by the gauge weight method. This method uses separate parameter data for each gauge used to compute precipitation and also uses separate parameter data for each sub-basin in the meteorologic model. The meteorologic model uses the Priestley–Taylor evapotranspiration method as input for continuous hydrologic simulation. The data requirements in this method are the maximum and minimum temperature, solar radiation, crop coefficients, and dryness coefficient.

The control specifications model specifies the start and end of the computation period and the computation time interval. The computation time interval considered in this exercise is one day. The computation period is divided into calibration and validation periods. Time windows are created for calibration, validation, and any time period of required length.

2.5.1.2 MODEL CALIBRATION AND VALIDATION

Model calibration is a systematic process of adjusting the model parameter values until model results match acceptably with the observed data. Model validation is the process of testing the ability of the model to simulate observed data other than that used for the calibration, with acceptable accuracy. During this process, calibrated model parameters are not allowed to change. A split sample procedure is followed in the model testing. The available historical weather data (years 1972–2003) are divided into two sets: 20 years (1973–1992) for calibration and 10 years (1993–2002) for validation. The characteristics of the river basin, i.e., land use, properties of soil, etc., are held constant throughout the simulation period. The constant loss rate and maximum deficit parameters needed for the deficit and constant loss method, and the basin lag parameter in the SCS unit hydrograph transform method are provided as data in the simulation. The peak-weighted root-mean-square function is chosen as the objective function and the Nelder and Mead algorithm is used to minimize the objective function.

2.5.1.3 PERFORMANCE EVALUATION OF THE MODEL

The model is used to simulate, on a daily scale, the water balance components, streamflow, soil moisture, evapotranspiration, and infiltration. The statistics of the flows such as mean, standard deviation (SD), root-mean-squared error (RMSE), normalized root-mean-squared error (NRMSE), and covariance of root-mean-squared error (CVRMSE) are compared. The model performance efficiency criteria, such as coefficient of determination R^2, Nash–Sutcliffe model efficiency E (Nash and Sutcliffe, 1970), and percentage deviation D are used to evaluate the model simulations during the calibration and validation periods. The R^2 value indicates the correlation between the observed and simulated values and E measures how well the plot of the observed against the simulated flows fits the 1:1 line. The R^2 coefficient is calculated using the equation:

$$R^2 = \frac{\sum (Q_{obs} - Q_{obs\ avg}) \times (Q_{sim} - Q_{sim\ avg})}{\sqrt{\sum (Q_{obs} - Q_{obs\ avg})^2 \times (Q_{sim} - Q_{sim\ avg})^2}} \tag{2.36}$$

where Q_{sim} is the simulated value, Q_{obs} is the observed value, $Q_{sim\ avg}$ is the average simulated value, and $Q_{obs\ avg}$ is the average observed value. The range of values for R^2 is 1.0 (best) to 0.0 (unacceptable).

The E value is calculated using Equation 2.37. If the E value is less than or close to 0, the model simulation is unacceptable. The highest value of E is 1.0.

$$E = 1 - \sum \frac{(Q_{obs} - Q_{sim})^2}{\sum \left(Q_{obs} - Q_{obs\ avg}\right)^2} \tag{2.37}$$

The percentage deviation (D) of streamflows over a specified period with total days calculated from measured and simulated

Table 2.13 *Initial estimates of important parameters of the hydrologic model*

Parameter	Units	Estimate	Remarks
Initial moisture deficit	mm	0	Calibration begins after an initialization period. Initial deficit is assumed to be zero.
Maximum deficit	mm	150–235	Depends on soil type. Varies from sub-basin to sub-basin.
Constant loss rate	mm/hr	0.0016–2.1	Depends on the soil type of the sub-basin.
Impervious percentage	%	10.7–13.7	Based on the land use/land cover pattern of the sub-basin.
SCS basin lag	minutes	2000–3281	Estimated using empirical equation based on area and length of river.
Dryness coefficient	–	1.26	Recommended value for non-arid climates (Source: FAO).
Crop coefficient	–	0.9–1.5	Based on the crops grown in the sub-basin and their growing season.

values of the quantity in each model time step is determined using Equation 2.38.

$$D = 100 \times \frac{Q_{obs} - Q_{sim}}{Q_{obs}} \tag{2.38}$$

Table 2.14 *Streamflow statistics*

	Calibration period (1973–92)		Validation period (1993–2002)	
Statistic (m^3/s)	Observed	Simulated	Observed	Simulated
Maximum	7358.0	7402.0	5842.0	3238.0
Mean	233.1	233.0	225.8	221.3
Standard deviation	412.8	370.7	416.0	327.2

A value close to 0% is desirable for D. A negative value indicates overestimation and a positive value indicates underestimation.

Table 2.13 provides the initial estimates used in the model for this illustration. The model calibrates these parameters internally, starting with the initial values provided. Results of the model are relatively more sensitive to parameters such as the constant loss rate, impervious percentage, and the crop coefficients used across the time periods.

HEC-HMS is useful for simulating the overall water balance in a basin but, when used for this purpose, may not simulate the peak flows well. Figure 2.11 shows the simulated daily streamflow, indicating that the peak flows are not satisfactorily reproduced. However, the average monthly flows shown in Figure 2.12 indicate a better performance. Table 2.14 shows the results obtained for the calibration and validation periods. The performance statistics for daily and monthly streamflows for the calibration and validation periods are shown in Table 2.15.

The model output includes components of the water balance such as the actual evapotranspiration, deep percolation, and excess runoff, as shown in Figure 2.13.

The HEC-HMS model, being available freely for download on the Web, is ideally suited for climate change impact studies in developing countries (e.g., Meenu *et al.*, 2012).

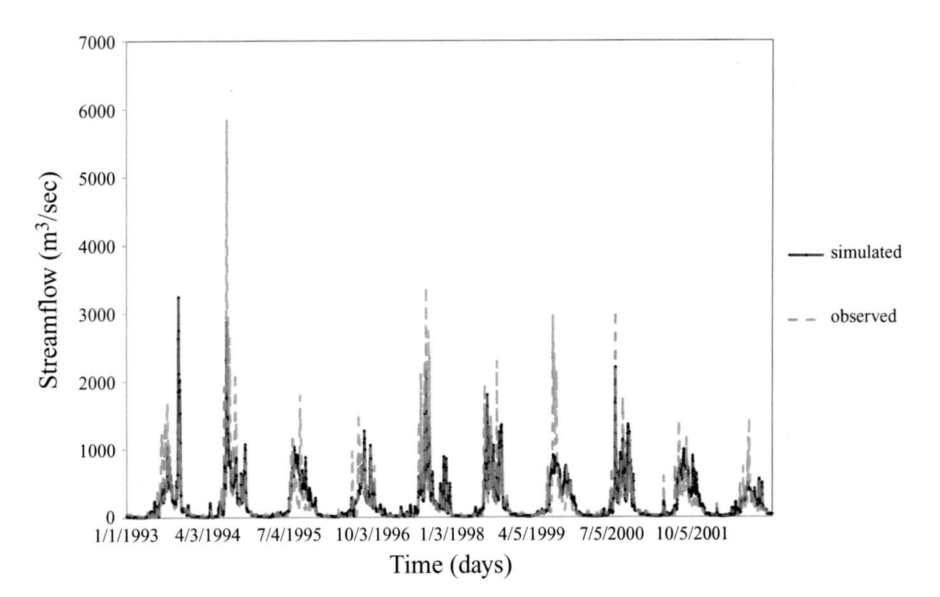

Figure 2.11 Simulated daily streamflows in the validation period (*Source*: Meenu *et al.*, 2012).

Table 2.15 *Model performance*

	Daily streamflow		Monthly streamflow	
Performance statistic	Calibration period (1973–92)	Validation period (1993–2002)	Calibration period (1973–92)	Validation period (1993–2002)
R^2	0.72	0.77	0.87	0.88
E	0.48	0.59	0.75	0.78
D (%)	0.07	1.96	0.07	1.96
Index fit (%)	99.93	98.04	99.93	98.03
RMSE (m^3/s)	297.62	267.75	4729	4596
NRMSE	0.04	0.05	0.09	0.07
CVRMSE	1.28	1.19	0.67	0.67

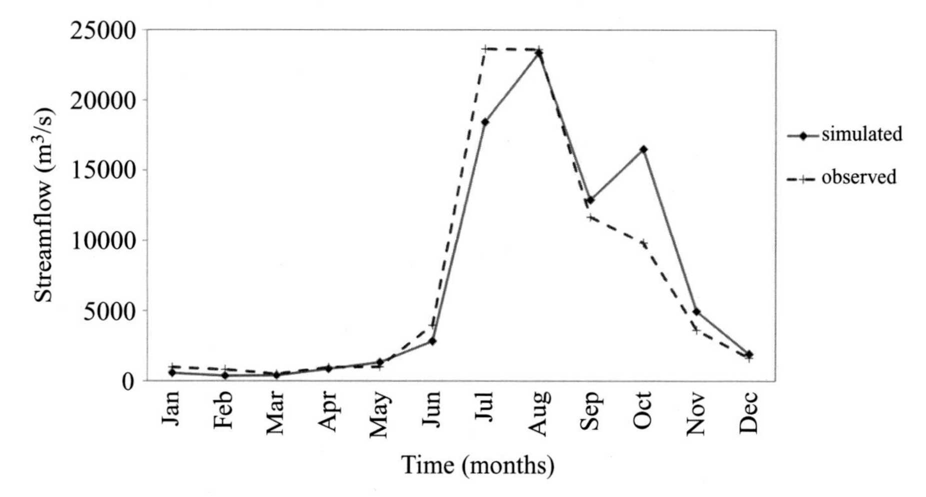

Figure 2.12 Simulated average monthly streamflow in the validation period (*Source*: Meenu *et al.*, 2012).

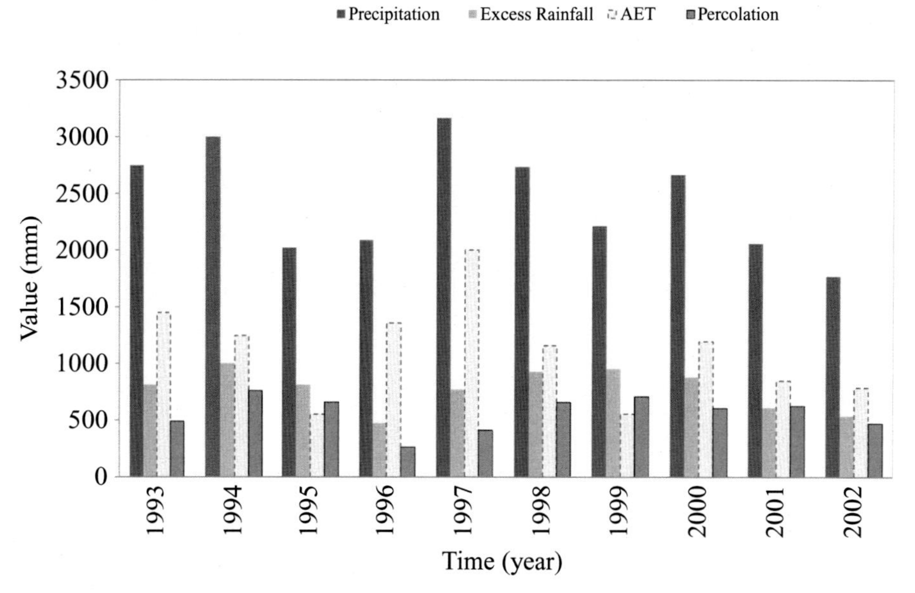

Figure 2.13 Simulated water balance components in the validation period.

2.5.2 HEC-RAS

HEC-RAS (HEC-River Analysis System) is an integrated system of software, designed for interactive use in a multi-tasking environment. It is developed by the Hydrologic Engineering Services of the USACE. HEC-RAS is useful in flood hydrology as a tool to carry out the flood routing, computations of backwater profiles, and gated operation of reservoirs to discharge flood waters.

The HEC-RAS system contains three one-dimensional hydraulic analysis components for:

1. steady flow water surface profile computations;
2. unsteady flow simulation;
3. movable boundary sediment transport computations.

All three components use a common geometric data representation and common geometric and hydraulic computation routines. In addition to the three hydraulic analysis components, the system contains several hydraulic design features that can be invoked once the basic water surface profiles are computed.

The geometric data of the study area used in the model consist of connectivity information for the stream system, cross-section data, and data on hydraulic structures (bridges, culverts, weirs, etc.). The geometric data are drawn first in the river system schematic. Cross sections are ordered within a reach from the highest river station upstream to the lowest river station downstream. The flow data to be used in the model depend on the type of analysis to be performed: backwater computations, flood routing, reservoir operation, etc.

The options for the results include graphic plots for cross sections, water surface profile, rating curves, X–Y–Z perspective, hydrographs, and tabular outputs at user specified locations.

2.5.2.1 ILLUSTRATION

A stretch of 20 km length for a river is considered in this section to illustrate the use of HEC-RAS for flood routing. The cross section of the river is given in Table 2.16; the bed elevation with respect to chainage is given in the table. The cross-section data are entered from the cross-section tool on the geometric data window (Figure 2.14).

Three types of calculations can be performed in HEC-RAS: steady flow analysis, unsteady flow analysis and hydraulic design functions. In the steady flow analysis option, the flow data are given from upstream to downstream for each reach (at least one flow value must be entered for each reach in the river system) as input. Once a flow value is given at the upstream end of a reach, it is assumed that the flow remains constant (steady-state conditions) until another flow value is encountered within the reach.

In the unsteady flow analysis, a variable flow within the reach can be considered. Initially, the boundary conditions at all of the external boundaries of the system, as well as any desired internal

Table 2.16 *River cross section used in the illustration of HEC-RAS*

Chainage (m)	Elevation (m)
0	536
990	531.33
1800	528.575
2400	530.945
3210	526.135
3450	520.035
3600	511.535
3690	526.07
4080	529.865
4800	529.24
5600	529.905
6420	530
7858	535
8058	536

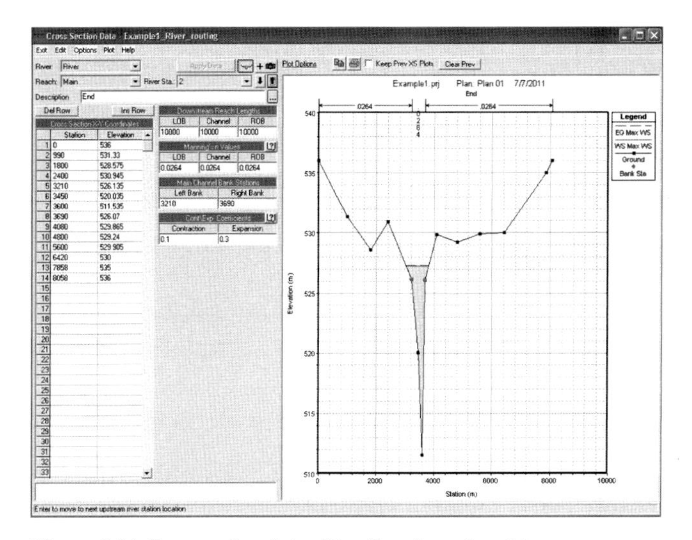

Figure 2.14 Cross section data editor. See also color plates.

locations are given. Then the initial flow, storage area conditions at the beginning of the simulation are assigned.

The bridge scour computations are performed by the hydraulic design functions option. The program automatically goes to the output file and gets the computed output for the approach section, the section just upstream of the bridge, and the sections inside of the bridge for scour computations.

The unsteady flow analysis option is used in this illustration. A symmetric triangular hydrograph with a peak discharge of 5000 m^3/s occurring at $t = 10$ hr is provided as an input hydrograph as shown in Figure 2.15. The input hydrograph at the upstream boundary is given as the boundary condition and the initial flow of 500 m^3/s is given as an initial condition of the river.

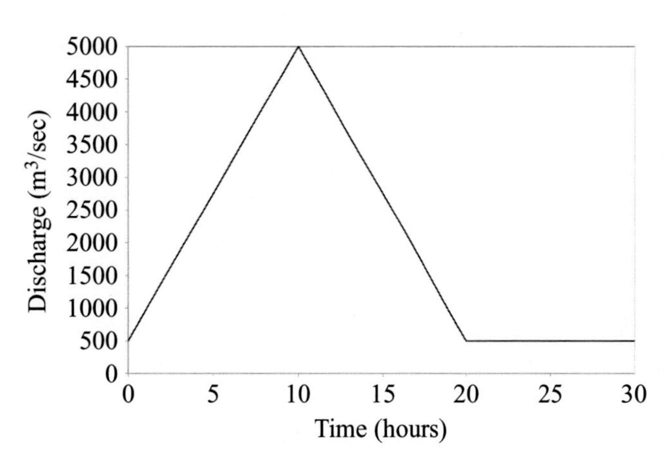

Figure 2.15 Input hydrograph.

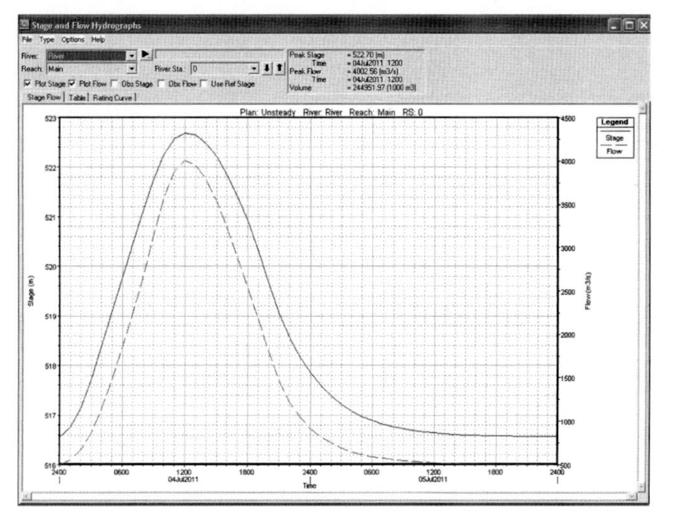

Figure 2.16 Outflow hydrographs at downstream boundary. See also color plates.

The outflow hydrograph at the downstream boundary condition is viewed from the view menu. The outflow hydrograph is shown in Figure 2.16.

HEC-RAS is particularly useful for examining implications of control structures, such as dams, weirs, barrages, etc., on the water surface profiles, for routing flood hydrographs with such control structures specified through boundary conditions, and for obtaining reservoir operating rules for flood control.

2.5.3 Storm Water Management Model

The US Environmental Protection Agency Storm Water Management Model (EPA SWMM) is a dynamic rainfall–runoff simulation model (Huber and Dickinson, 1988) used for simulation of runoff quantity and quality from predominantly urban areas. Although most commonly used in urban flood studies, SWMM may also be employed for riverine flood studies, especially for runoff hydrograph generation and for flood routing (e.g., Camorani *et al.*, 2005). The runoff component of SWMM operates on a collection of sub-catchment areas that receive precipitation and generate runoff and pollutant loads. The routing module of SWMM transports this runoff through a system of channels, storage/treatment devices, pumps, and regulators. SWMM tracks the quantity and quality of runoff generated within each sub-catchment, and the flow rate, flow depth, and quality of water in each pipe and channel during a simulation period comprising multiple time steps.

SWMM was first developed in 1971 and has undergone several major upgrades since then. The version 5.0.022 (which may be downloaded from www.epa.gov/nrmrl/wswrd/wq/models/swmm/index.htm, accessed on December 26, 2011) has modules for simulating continuous runoff hydrographs and chemographs, hydrodynamic streamflow routing, pollutant transport, storage and treatment, subsurface flow routing, snowmelt and evapotranspiration simulations. It is widely used for planning, analysis, and design related to storm water runoff, combined sewers, sanitary sewers, and other drainage systems primarily in urban areas.

The options available in the model for accounting for infiltration losses are the Horton model, the Green Ampt model, or use of the SCS curve number. Either the kinematic wave or the dynamic wave routing method may be chosen for flow routing. Details of rain gauges, sub-catchments, aquifers, and hydraulics (nodes, links, etc.) may be specified in the model.

The status report on a successful run of the model contains total precipitation quantity, losses, surface runoff, node depths, maximum depth of water and depth of water at different time steps in the simulation, nodes and conduits flooded in case of flooding, and time of flooding.

While extremely useful in urban flooding studies, SWMM may also be used for small watersheds where hydrologic homogeneity may be assumed.

2.5.3.1 ILLUSTRATION

The river cross section modeled in HEC-RAS is considered in this illustration. The objective is to route the input flood hydrograph through the river stretch using the SWMM software.

The junctions and conduits are created according to the plan of the river. The image of the river can be loaded as the backdrop image and junctions/conduits can be added by clicking the tool on the object toolbar or from the category list of the data browser. The properties of junctions (elevation, maximum depth, etc.) and conduits (cross section, invert elevation, etc.) are assigned to complete the geometric model.

Then the input hydrograph is given as input to the upstream junction as "inflows." The analysis options (simulation date and time, number of routing time steps, etc.) are assigned and the model is run for analysis.

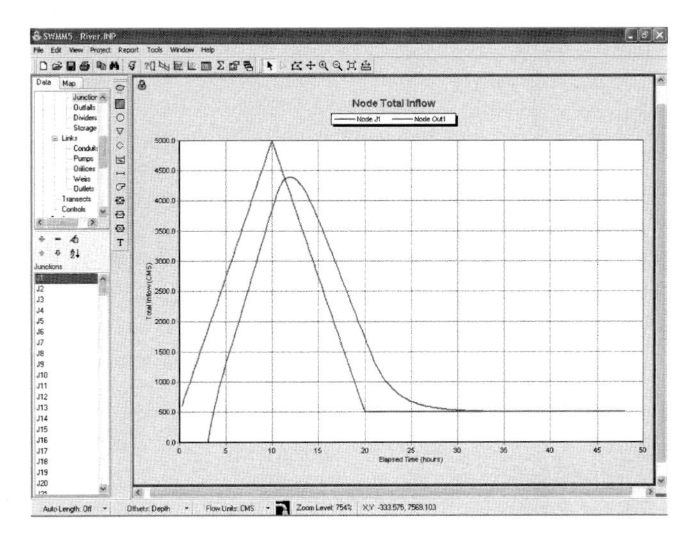

Figure 2.17 Inflow and outflow hydrographs in SWMM. See also color plates.

Any error or warning messages will appear in a Status Report window. The results can be viewed from the Status Report window.

Results of analysis can be viewed using different types of graphs. A time series plot graphs the value of a particular variable at up to six locations against time. When only a single location is plotted, and that location has calibration data registered for the plotted variable, then the calibration data are plotted along with the simulated results.

The inflow and outflow hydrographs at the upstream junction and the downstream junction are shown in Figure 2.17.

2.5.4 Requirements of data for hydrologic models

The extensive data requirements of physically based models such as HEC-HMS, HEC-RAS, and SWAT for calibration and validation are a major limitation in their use, especially in the developing countries where most data are not readily available. A comprehensive data set for a fully distributed hydrologic model would consist of (Singh and Woolhiser, 2002): (i) hydrometerorologic data such as rainfall, temperature, snowfall, radiation, humidity, vapor pressure, sunshine hours, wind velocity, and pan evaporation; (ii) agricultural data including land use and vegetation cover; (iii) soil data including the soil type, texture, soil condition, particle size distribution, porosity, moisture content, hydraulic conductivity, antecedent moisture content, etc.; and (iv) geological data, to be provided at various spatial grids of the models.

Global data sets may be made use of in many situations as proxies to the actual observed data. Table 2.17 lists some of the global data sets that may be useful in hydrologic models. In addition, DEMs and remotely sensed land use/land cover maps will be useful for hydrologic models. These are separately discussed in Chapter 4.

2.6 EMPIRICAL MODELS

The hydrologic models discussed in the previous section are physically based conceptual models. In flood hydrology, empirical models such as regression models, artificial neural networks (ANNs), and fuzzy logic based models are also gaining popularity. In this section, ANNs and fuzzy logic based models are discussed.

2.6.1 Artificial neural network models

Artificial neural networks are a relatively new technique used in hydrologic and water resources systems modeling. An ANN is structured to resemble a biological neural network in two aspects: knowledge acquisition through a learning process, and storage of knowledge through connections, known as synaptic weights. ANNs are particularly useful as pattern recognition tools for generalization of input–output relationships. Applications of ANNs in hydrology include those for rainfall–runoff relationships and streamflow/flood forecasting.

In this section an introduction to ANNs is provided and an application to streamflow forecasting is discussed briefly. For a more rigorous understanding of ANNs the reader may refer to Bishop (1995) and Haykin (1994). A comprehensive review of ANN applications in hydrology is available in ASCE (2000a,b).

2.6.1.1 BASIC PRINCIPLES

An ANN consists of interconnected neurons receiving one or more inputs and resulting in an output. Figure 2.18 shows one such neuron.

Each input link i ($i = 1, 2, 3$) has an associated external input signal or stimulus x_i and a corresponding weight w_i (Figure 2.18). A schematic black box equivalent of the neuron is illustrated in Figure 2.19. The neuron behaves as an activation or mapping function $f(.)$ producing an output $y = f(z)$, where z is the cumulative input stimulus to the neuron and f is typically a non-linear function of z. The neuron calculates the weighted sum of its inputs as

$$z = w_1 x_1 + w_2 x_2 + w_3 x_3 \tag{2.39}$$

The function $f(z)$ is typically a monotonic non-decreasing function of z. Some examples of commonly used activation functions $f(z)$ are shown in Figure 2.20.

ANNs are classified, based on the direction of information flow, as feed-forward and recurrent networks. A feed-forward ANN generally consists of several layers of nodes (see Figure 2.21), starting from the input layer and ending with the output layer. The intermediate layers are called the hidden layers and a feed-forward network may consist of several such hidden layers. The nodes in one layer are connected to all the nodes in the next. The information passes from the input layer through the hidden layers to the output layer, in a forward direction. In a recurrent ANN, the information passes in both directions.

Table 2.17 *A sample of global data sets and related resources useful in hydrologic modeling*

Set no.	Description	Website	Remarks
1	Precipitation	http://lwf.ncdc.noaa.gov/oa/wmo/wdcamet-ncdc.html (Accessed on December 26, 2011)	This data set is prepared by the Global Precipitation Climatology Project (GPCP). Daily, global 1 × 1-deg gridded fields of precipitation totals for the period October 1996 to the present date are available.
2	Streamflow	http://www.gewex.org/grdc.html (Accessed on December 26, 2011)	The Global Runoff Data Centre (GRDC), located in Germany, has compiled a global database of streamflow data. The database is updated continually and contains daily and monthly discharge data information for over 2900 hydrologic stations in river basins located in 143 countries.
3	Soil cover	Harmonized World Soil Database (HWSD) http://www.iiasa.ac.at/Research/LUC/External-World-soil-database/HTML/HWSD_Data.html?sb=4 (Accessed on December 26, 2011)	Contains worldwide soil cover data.
4	Digital elevation data	http://srtm.csi.cgiar.org/ (Accessed on December 26, 2011)	SRTM 90 m DEM data.
5	Continental watersheds and river networks	http://www.ngdc.noaa.gov/ecosys/cdroms/graham/graham/graham.htm (Accessed on December 26, 2011)	A data set of watersheds and river networks, which is derived primarily from the TerrainBase 5′ Global DTM with additional information from the World Data Bank. Using this data set, the runoff produced in any grid cell, when coupled with a routing algorithm, can be transported to the appropriate water body and distributed across that water body as desired. The data set includes watershed and flow direction information, as well as supporting hydrologic data at 5′, 1/2°, and 1° resolutions globally
6	Terrain data	http://eros.usgs.gov/#/Find_Data/Products_and_Data_Available/gtopo30/hydro (Accessed on December 26, 2011)	Topographically derived data sets derived from the USGS 30 arc-second DEM of the world (GTOPO30) containing aspect, drainage data, elevation, flow accumulation, flow direction, slope, compound topographic index, streams data.
7	Sources of global data sets	http://iridl.ldeo.columbia.edu/docfind/databrief/ (Accessed on December 26, 2011)	Provides categorywise list of data sets. The categories of Atmosphere, Hydrology and Climate Indices are useful for hydrologic modeling. Additionally the category Historical Model Simulations, which includes links to NCEP/NCAR reanalysis data, is useful in climate change impact studies.
8	Estimates of parameter values required in HEC-HMS hydrologic model	http://www.scwa2.com/Documents/Hydrology%20Manual%20for%20Web%20Page/Chapter%203.4%20HEC-q%20&%20HEC-HMS.pdf (Accessed on December 26, 2011)	Constant loss rates for each of the NRCS soil groups, impervious percentage for common land uses and estimates of initial loss rates.
9	Roughness coefficients, runoff curve numbers, and soil texture class	Flood runoff analysis, 7–19, Table http://cleveland2.ce.ttu.edu/documents/public/A-AUTHORS/USACOE_1110-2-1417/Flood-Runoff%20Analysis.pdf (Accessed on December 26, 2011)	Roughness coefficients for overland flow suggested by Hjelmfelt (1986), curve numbers for selected agricultural, suburban, and urban land uses and values of various soil properties are listed, such as conductivity, porosity, etc., estimated by Rawls and Brakensiek (1983) and Rawls *et al.* (1983). Useful in peak flow estimates.

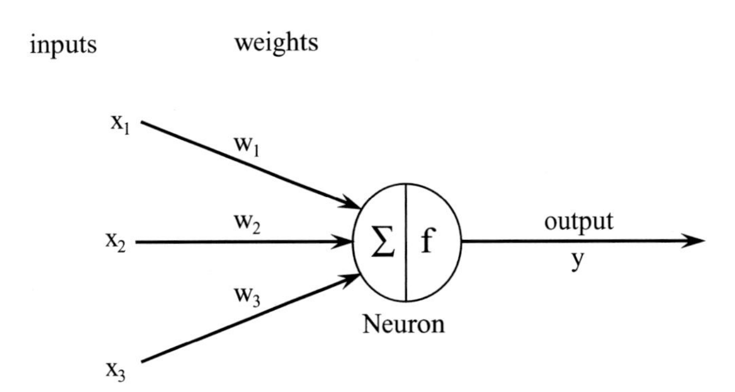

Figure 2.18 A simplified neuron model.

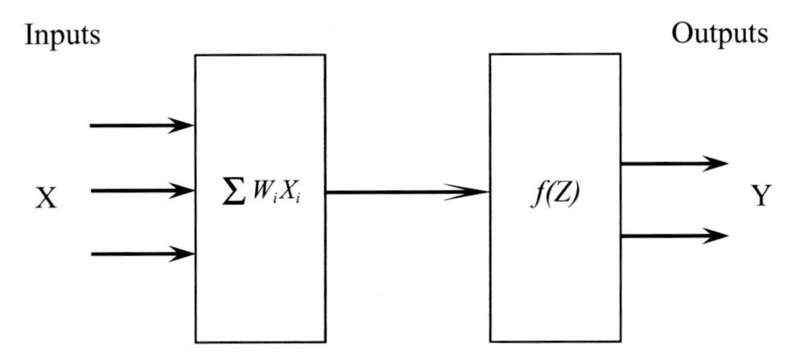

Figure 2.19 Schematic diagram of a single neuron.

In an application to develop rainfall–runoff relationships, the output layer may consist of a single node representing the runoff during a period, while the input layer may consist of several nodes representing rainfall in the catchment area at a number of locations and/or in a number of preceding time periods. In the case of real-time flood forecasting, the input layer may simply consist of nodes representing observed streamflows during the previous few time periods, and the output layer may be a single node representing streamflow during the current period. An ANN is trained to learn the relationship between the input and the output. To achieve the training, a number of sets of inputs and outputs are presented to the ANN. A training algorithm achieves the training by determining the optimal weights. The back propagation (BP) and the radial basis function (RBF) are the most commonly used network training algorithms in hydrologic applications, and are discussed here.

2.6.1.2 BACK PROPAGATION

Back propagation is perhaps the most popular algorithm for training the feed-forward networks. In the BP algorithm, the input layer receives the input data patterns, the output layer produces the result, and the hidden layers sequentially transform the input into the output. A maximum of three hidden layers has been found to be adequate for most problems in water resources applications. The number of neurons (nodes) in the hidden layers is usually fixed by trial and error.

The transfer function used in BP networks is usually a sigmoid function (Figure 2.21). During training, BP networks process the patterns in a two-step procedure. In the first or forward phase of BP learning, an input pattern is applied to the network, with initially assumed weights to provide the output at the output layer. The error is estimated from the corresponding output value provided in the training set. In the second, or backward, phase, the error from the output is propagated back through the network to adjust the interconnection weights between layers. This process is repeated until the network's output is acceptable. Back-propagation learning is this process of adapting the connection weights. When presented with an unknown input pattern the trained network produces a corresponding output pattern. Details of the BP method are described by Freeman and Skapura (1991) and ASCE (2000a).

2.6.1.3 RADIAL BASIS FUNCTION

Radial basis function (RBF) neural networks (Figure 2.22) are feed-forward neural networks with only one hidden layer and radial basis functions as activation functions. The hidden layer

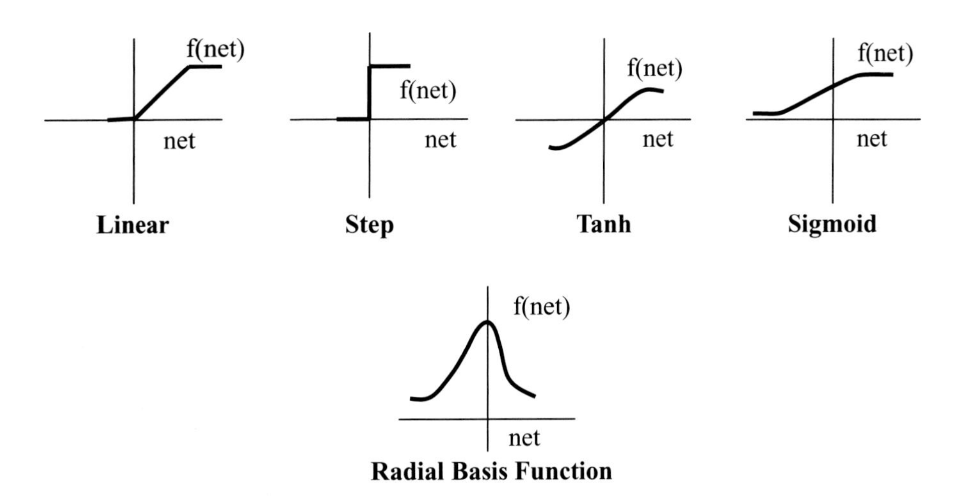

Figure 2.20 Typical neural network activation functions.

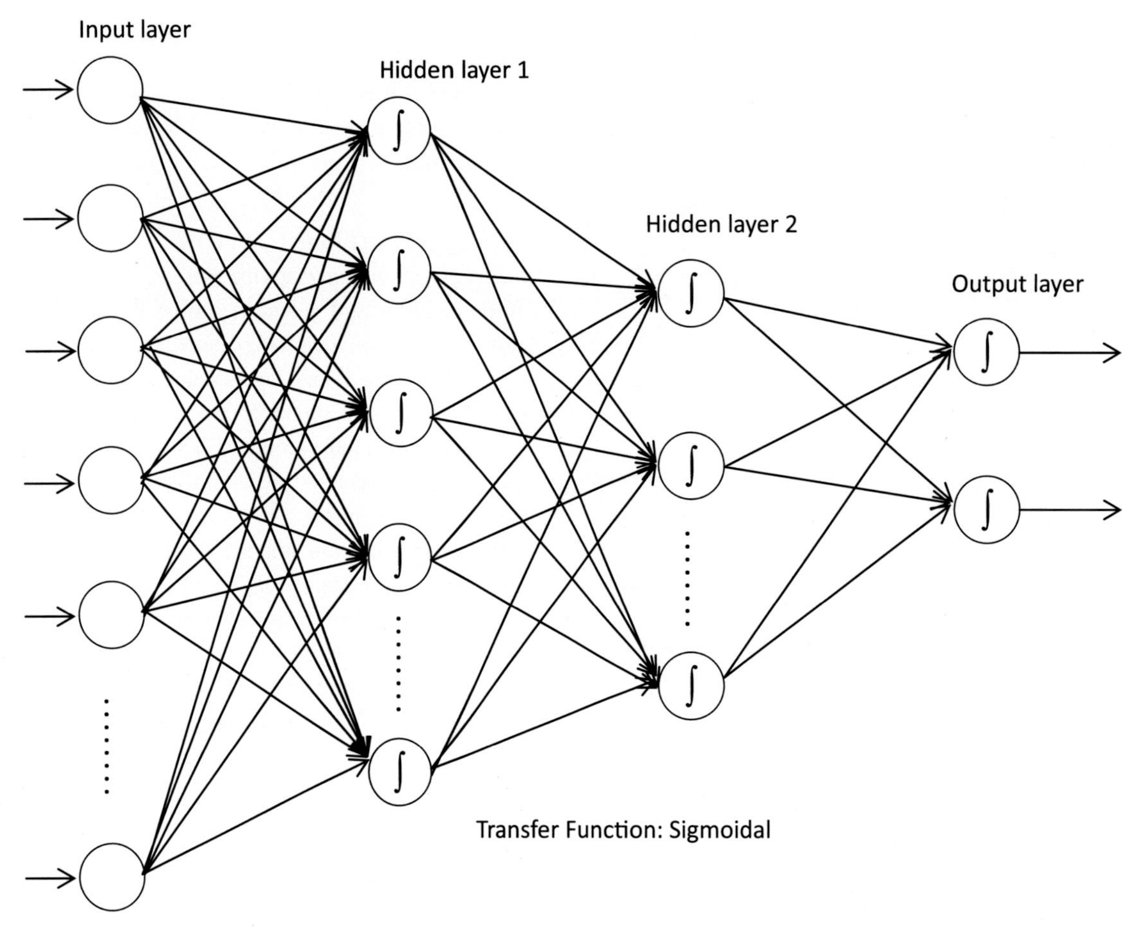

Figure 2.21 Structure of a feed-forward network.

performs a fixed non-linear transformation with no adjustable parameters. The RBF networks are trained with a modified form of gradient descent training. The training algorithm of RBF networks may be found in Leonard *et al.* (1992) and Haykin (1994). The primary difference between the RBF network and the BP training of a feed-forward network lies in the nature of the non-linearities associated with the hidden node. The non-linearity in BP is implemented by a fixed function such as a sigmoid. The RBF method, on the other hand, bases its non-linearities on the data in the training set (ASCE, 2000a). For most problems, the BP algorithm requires more training time than the RBF networks. With this advantage, the RBF networks are ideally suited for applications involving real-time forecasting.

2.6.1.4 ADVANTAGES AND LIMITATIONS OF ANNs

Some advantages and limitations of ANNs discussed in Nagesh Kumar (2003) are listed here.

1. *Non-linearity:* Non-linearity is a useful property of the neural networks, particularly if the underlying physical mechanism responsible for generation of the output is inherently non-linear (e.g., runoff from a watershed).
2. *Input–output mapping:* A popular paradigm of learning called *supervised learning* involves modification of the synaptic weights of a neural network by applying a set of labeled *training samples* or *task examples*. Thus, the network learns from the examples by constructing an *input–output mapping* for the problem at hand.
3. *Adaptivity:* A neural network trained to operate in a specific environment can be easily *retrained* to deal with minor changes in the operating environmental conditions. Moreover, when it is operating in a *non-stationary* environment (i.e., where statistics change with time), a neural network can be designed to change its synaptic weights in real time (e.g., modeling non-stationary hydrologic time series).
4. *Evidential response:* In the context of pattern classification, a neural network can be designed to provide information not only about which particular pattern to *select*, but also about the *confidence* in the decision made. This latter information may be used to reject ambiguous patterns, and thereby improve the classification performance of the network.

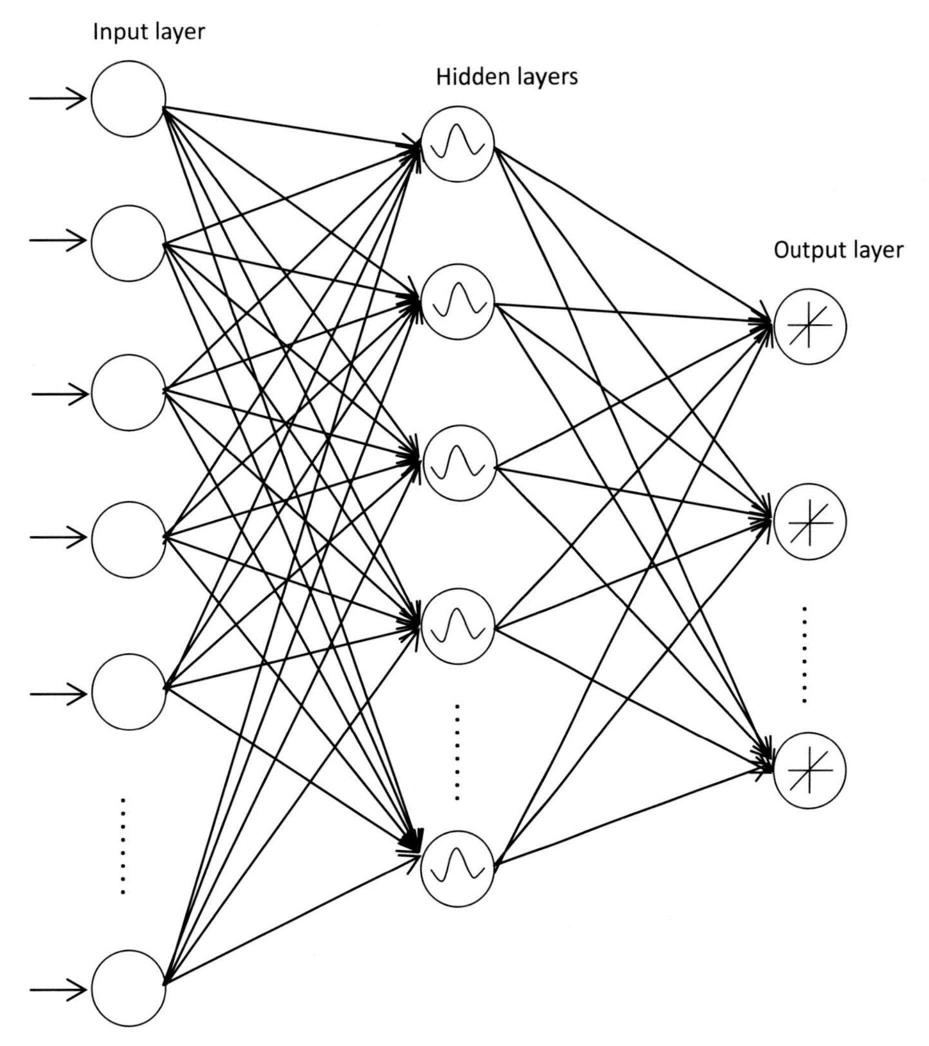

Figure 2.22 Structure of the radial basis function network.

A major limitation of ANNs arises due to their inability to learn the underlying physical processes. Another difficulty in using ANNs is that there is no standardized way of selecting the network architecture. The choice of network architecture, training algorithm, and definition of error are usually determined by the user's experience and preference, rather than the physical features of the problem. Very often, ANN training is performed with limited length of hydrologic data. Under these circumstances, it is difficult to ascertain when the generalization will fail, to decide the range of applicability of the ANN. In addition, the following limitations of ANNs must be kept in view.

- *Long training process*: For complex problems, the direction of error reduction is not very clear and may lead to some confusion regarding the continuation of the training process. The training process can be very long and sometimes futile for a bad initial weight space.
- *Moving target*: The problem of a moving target arises as the weights have to continuously adjust their values from one output to another for successive patterns. The change in a weight during one training pass may be nullified in the next pass because of a different training pattern.

2.6.1.5 AN ILLUSTRATION OF USE OF ANNs FOR FLOW FORECASTING

In the following example, ANNs are used to predict streamflow in the River Narmada at Station Barman, for time leads of 1 day, 2 days, and 3 days, based on the last 10 days' measured streamflow values. The input layer thus consists of ten neurons representing the streamflow during the previous 10 days. Daily streamflow data from January 1, 1972 to December 31, 1974, plotted in Figure 2.23, are used in the ANN real-time runoff forecasting model.

A two-layer feed-forward network with 30 hidden neurons and linear output neurons is used in the ANN model for each of the three different time lead cases. Of the data, 70% are used for training the network, while 15% are used for validation, and 15% for testing. The network is trained with the Levenberg–Marquardt BP

Table 2.18 *Performance of ANN model for streamflow forecasting at different lead times*

	Training		Validation		Testing	
ANN model	MSE	*R*	MSE	*R*	MSE	*R*
1-day lead	21650.3	0.95	190719.3	0.88	711302.4	0.89
2-days lead	767711.5	0.85	1243893.4	0.88	995805.9	0.81
3-days lead	553512.9	0.85	1307481.6	0.59	2067783.6	0.63

MSE: mean-square error; *R*: correlation coefficient

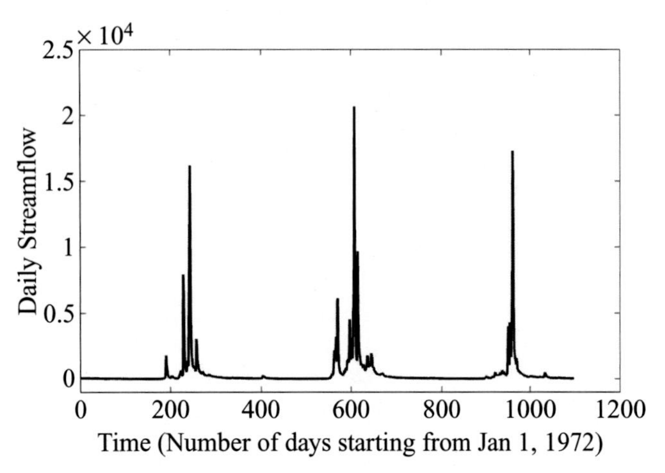

Figure 2.23 Daily streamflow (m^3/s) of the Narmada at Barman from January 1, 1972 to December 31, 1974.

algorithm. The creation, training, and evaluation of the network are performed using the neural network toolbox in MATLAB® 7.8.0 (R2009a). The ANN model performance is assessed based on two parameters: the mean-squared error and the regression values. These are presented for all the cases in Table 2.18. The accuracy of the ANN model decreases as the lead time of the forecast increases, as may be expected. Nonetheless, the reasonably acceptable values of the performance measures reported establish the viability of ANN models for real-time streamflow forecasting.

Figures 2.24, 2.25, and 2.26 show the observed and ANN predicted daily streamflows at the different lead times.

The ANN models, with their capability to capture the underlying non-linear patterns in the data series presented, are very useful as "black boxes" (in the sense that there is no physical meaning associated with the non-linear relationships guiding the neural networks) in flood forecasting. The ANNs also serve as useful tools in climate change impact studies to address the scale issues, as discussed in the next chapter.

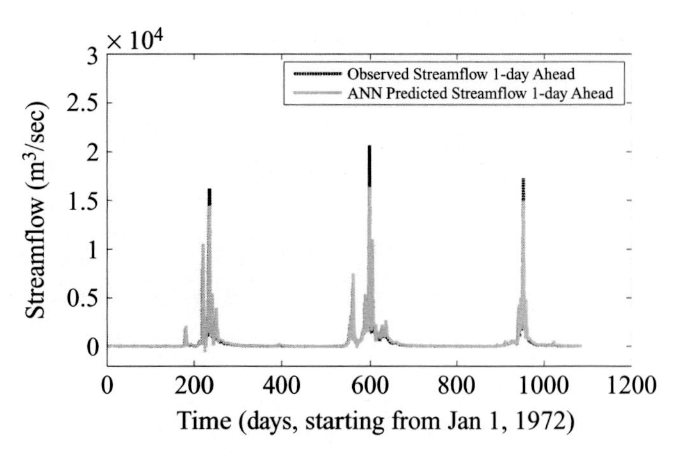

Figure 2.24 Observed and ANN predicted 1-day ahead streamflow.

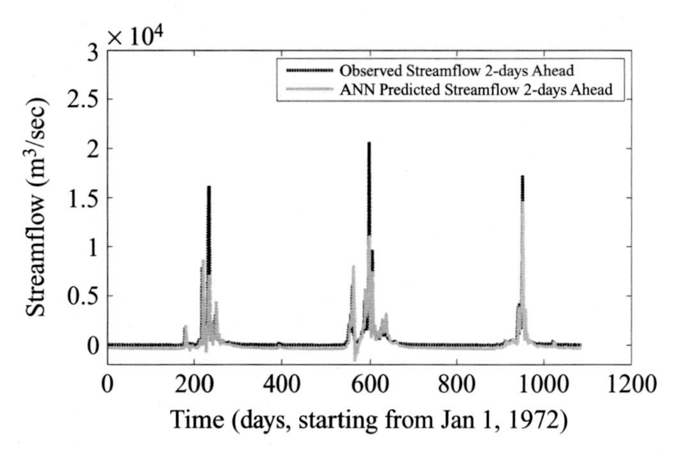

Figure 2.25 Observed and ANN predicted 2-days ahead streamflow.

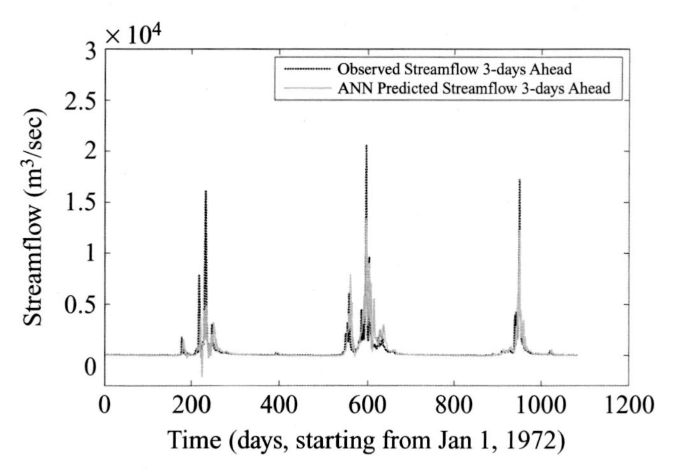

Figure 2.26 Observed and ANN predicted 3-days ahead streamflow.

2.6.2 Fuzzy logic based models

Fuzzy logic based models are suitable when information about the physical process is vague and the data available are scarce. In flood hydrology, uncertainty due to parameters and non-availability of data may be addressed, to some extent, with fuzzy logic. Use of fuzzy logic for flood forecasting and flood routing is of a relatively recent origin. Some very useful applications of fuzzy logic based models (including the neuro-fuzzy models, which use both neural networks and fuzzy systems in an integrated framework) for flood forecasting/routing are available in Chen *et al.* (2006), Wu and Chau (2006), Shu and Ouarda (2008), Chu (2009), and Khatibi *et al.* (2011).

In this section a brief introduction to fuzzy logic is provided, followed by an illustration of application of fuzzy logic to a flood routing problem.

2.6.2.1 FUZZY SETS AND FUZZY LOGIC

The concept of fuzzy logic was first introduced by Zadeh (1965). Details of fuzzy sets, fuzzy numbers, membership functions, and fuzzy rules can be found in Zimmermann (1996) and Ross, (1997). Physical information about a system is used in development of fuzzy rules, which are then tuned using the available data. The fuzzy rule based models are suitable for non-linear input–output mapping. In fuzzy rule based modeling, given input data undergo the processes of fuzzification, application of fuzzy operators, implication, aggregation, and defuzzification, before finally forming into output from the model (MATLAB, 1995; Panigrahi and Mujumdar, 2000).

2.6.2.2 MEMBERSHIP FUNCTION

A *membership function* (MF) is a function – normally represented by a geometric shape – that defines how each point in the input space is mapped to a membership value between 0 and 1. If X is the input space (e.g., inflow to a reservoir) and its elements are denoted by x, then a fuzzy set A in X is defined as a set of ordered pairs:

$$A = \{x, \mu_A(x) \mid x \in X\} \tag{2.40}$$

where $\mu_A(x)$ is called the membership function of x in A. Thus, the membership function maps each element of X to a membership value between 0 and 1. A membership function can be of any valid geometric shape. Some commonly used membership functions are triangular, trapezoidal, and bell shaped.

An important step in applying methods of fuzzy logic is the assessment of the membership function of a variable. For the purpose of hydrologic modeling of floods, for example, the membership functions required are typically those of stage or discharge at several locations along the channel/stream. When the standard deviation of a variable is not large, it is appropriate to use a simple membership function consisting of only straight lines, such as a triangular or a trapezoidal membership function. The fuzzy logic based models attain their best performance when there is an overlap of the adjacent membership functions. A good rule of thumb is that the adjacent fuzzy values should overlap approximately 25% (Kosko, 1996).

2.6.2.3 FUZZY RULES

A fuzzy rule system is defined as the set of rules that consists of sets of input variables or premises A, in the form of fuzzy sets with membership functions μ_A, and a set of consequences B also in the form of a fuzzy set. Typically a fuzzy if–then rule assumes the form:

if x is A then y is B,

where A and B are linguistic values defined by fuzzy sets of the variables X and Y, respectively. The "if" part of the rule "x is A" is called the *antecedent* or *premise*, and the "then" part of the rule "y is B" is called the *consequence*. In the case of binary or two-valued logic, if the premise is true then the consequence is also true. In a fuzzy rule, if the premise is true to some degree of membership, then the consequence is also true to that same degree.

The premise and consequence of a rule also can have several parts, such as:

if x is A and y is B and z is C, then m is N and o is P, etc.

For example, we may have rules such as "IF discharge at location A is low and discharge at location B is very high, THEN stage at location C is low–medium".

2.6.2.4 FUZZY DYNAMIC FLOOD ROUTING MODEL FOR NATURAL CHANNELS

Gopakumar and Mujumdar (2009) presented a fuzzy inference model for dynamic wave routing. Major advantages of the procedure are that it does not require extensive data for flood routing in natural channels and that it is very simple for application. The methodology is presented in this section, followed by a numerical example.

Flood routing computations involve determination of stage and discharge hydrographs at various locations along a river from the known flood hydrograph at the upstream end based on specified initial and boundary conditions.

The dynamic wave routing involves solution of the following governing equations (Saint-Venant equations, Chow *et al.*, 1988):

Continuity equation:

$$\frac{\partial A}{\partial t} + \frac{\partial Q}{\partial x} = 0 \tag{2.41}$$

Momentum equation:

$$\frac{\partial Q}{\partial t} + \frac{\partial}{\partial x}\left[\frac{Q^2}{A}\right] + gA\frac{\partial h}{\partial x} - gAS_0 + gAS_f = 0 \tag{2.42}$$

where A is flow cross-sectional area, h is flow depth, Q is discharge, S_0 is bed slope, S_f is friction slope, g is acceleration due to gravity, t is time variable, and x is space variable. Here it is assumed that lateral inflow is negligible. Equations 2.41 and 2.42 are generally solved using numerical methods, as shown in Section 2.4.2, to obtain variation of flow depth (h) and discharge (Q), with respect to time, at various locations along the channel. However, estimation of the friction slope (S_f) in the momentum equation (2.42) causes a major difficulty in applications. The friction slope is usually computed with Manning's equation with the roughness coefficient (n) estimated based on several channel characteristics such as surface roughness, size, shape, vegetation, irregularity, alignment, stage, discharge, etc. (Chow, 1959). The roughness coefficient is one of the main variables used in calibrating the numerical methods (US Army Corps of Engineers, 2002). A slight error in the estimated value of n can result in large errors in the computed stage and discharge values. Accurate estimation of n for natural channels is difficult due to its dependence on the several factors stated above. Gopakumar and Mujumdar (2009) have demonstrated that it is possible to overcome these difficulties by replacing the momentum equation (2.42) by a fuzzy rule based model, while retaining the continuity equation (2.41) in its original form.

The fuzzy rule based model is constructed based on physical characteristics of the channel, such as its geometric characteristics (represented by stage–area relationships) and flow characteristics (represented by stage–discharge relationships). Even though hydrologic routing methods, such as the Muskingum method (Chow *et al.*, 1988), that also do not require explicit specification of the roughness coefficient are available, the advantage of the fuzzy logic based model is that it gives spatial distribution of the flow parameters also in addition to their temporal variation.

Major advantages of the fuzzy logic based model are that it requires fewer parameters and is not very sensitive to parameter changes (Bardossy and Disse, 1993). However, the fuzzy logic based routing model is not a replacement for the complete dynamic wave routing model, and may be used as a good approximation of the flood wave movement, in the absence of data required to estimate the roughness coefficient.

2.6.2.5 MODEL DEVELOPMENT

Development of a fuzzy dynamic flood routing model (FDFRM) to simulate flood movement in a natural channel is demonstrated here, assuming that the flood wave can be approximated to a monoclinal wave (Chow, 1959; Henderson, 1966). A uniform channel reach is considered for development of the model. The continuity equation (2.41) is discretized at node (i, j), in the x–t computational domain, using the Lax diffusive scheme (Chaudhry, 1993), which is an explicit finite difference scheme, as follows:

$$\frac{1}{\Delta t}\left[A_i^{j+1} - 0.5 \times \left(A_{i-1}^{j} + A_{i+1}^{j}\right)\right] + \frac{1}{2\Delta x}\left[Q_{i+1}^{j} - Q_{i-1}^{j}\right] = 0 \tag{2.43}$$

where i is node number in the x-direction and j is node number in the t-direction.

As mentioned earlier, the momentum equation (2.42) is replaced by a fuzzy rule based model. This is done based on the principle that during unsteady flow the disturbances in the form of discontinuities in the gradient of the flow parameters, which appear in the governing equations (2.41 and 2.42), will propagate along the channel with a velocity equal to that of the velocity of the shallow water wave (V_w) (Cunge *et al.*, 1980). The discontinuities in the gradient of the flow parameters A and Q are obtained as the differences between their final steady flow gradient and present gradient. At any node ($i, j+1$) in the computational domain (Figure 2.27), these discontinuities are computed by adopting the following discretization:

Discontinuity in the gradient of area (A):

$$\frac{\Delta A_i^{j+1}}{\Delta x} = \frac{\left[A_i^{F.S} - A_{i-1}^{F.S}\right] - \left[A_i^{j+1} - A_{i-1}^{j+1}\right]}{\Delta x} \tag{2.44}$$

Discontinuity in the gradient of discharge (Q):

$$\frac{\Delta Q_i^{j+1}}{\Delta x} = \frac{\left[Q_i^{F.S} - Q_{i-1}^{F.S}\right] - \left[Q_i^{j+1} - Q_{i-1}^{j+1}\right]}{\Delta x} = \frac{-\left[Q_i^{j+1} - Q_{i-1}^{j+1}\right]}{\Delta x} \tag{2.45}$$

Note that $Q_i^{F.S} = Q_{i-1}^{F.S}$ during the final steady flow condition.

In Equations 2.44 and 2.45, the notations with superscript "*F.S*" indicate the values of the flow parameters corresponding to the final steady condition. The ratio of discontinuity in the gradient of Q and discontinuity in the gradient of A, at the node ($i, j+1$), gives the wave velocity (V_w) at this node (Gopakumar and Mujumdar, 2009), i.e.,

$$\text{wave velocity } (V_w) \text{ at the node } (i,\ j+1) = \frac{\Delta Q_i^{j+1}}{\Delta A_i^{j+1}} \tag{2.46}$$

As the flood wave is approximated to a monoclinal wave, the wave velocity (V_w) can also be obtained from the discharge– area relationship, based on the Kleitz–Seddon principle (Chow, 1959). From Equation 2.46, it may be seen that ΔA corresponding to the same ΔQ will be different at different flow depths and therefore crisp relations do not exist between ΔA and ΔQ. All values of ΔA corresponding to each ΔQ are determined as the training sets to develop the fuzzy rule based model as explained below.

The fuzzy rules are developed relating the antecedent (ΔA) and the consequent (ΔQ). Development of the fuzzy rule based model involves determination of the membership function parameters and construction of the fuzzy rule base. The membership functions are assumed to be triangular.

Let (ΔQ_1, ΔQ_2, ΔQ_3) be a typical fuzzy subset developed, where (ΔQ_1, ΔQ_3) is the support (range) and ΔQ_2, which is the mean of ΔQ_1 and ΔQ_3, is the prototype point. ΔA_1 is the lower bound ordinate corresponding to ΔQ_1, ΔA_2 is the mean ordinate corresponding to ΔQ_2 and ΔA_3 is the upper bound ordinate corresponding to ΔQ_3. Similarly, all corresponding fuzzy subsets of the antecedent and consequent are determined. Then they are assigned linguistic names such as *low*, *medium*, *high*, etc., and are related through the fuzzy rules. The Mamdani implication method of inference (Mamdani and Assilian, 1975) and the centroid method of defuzzification (Ross, 1997) are used in the fuzzy rule based model.

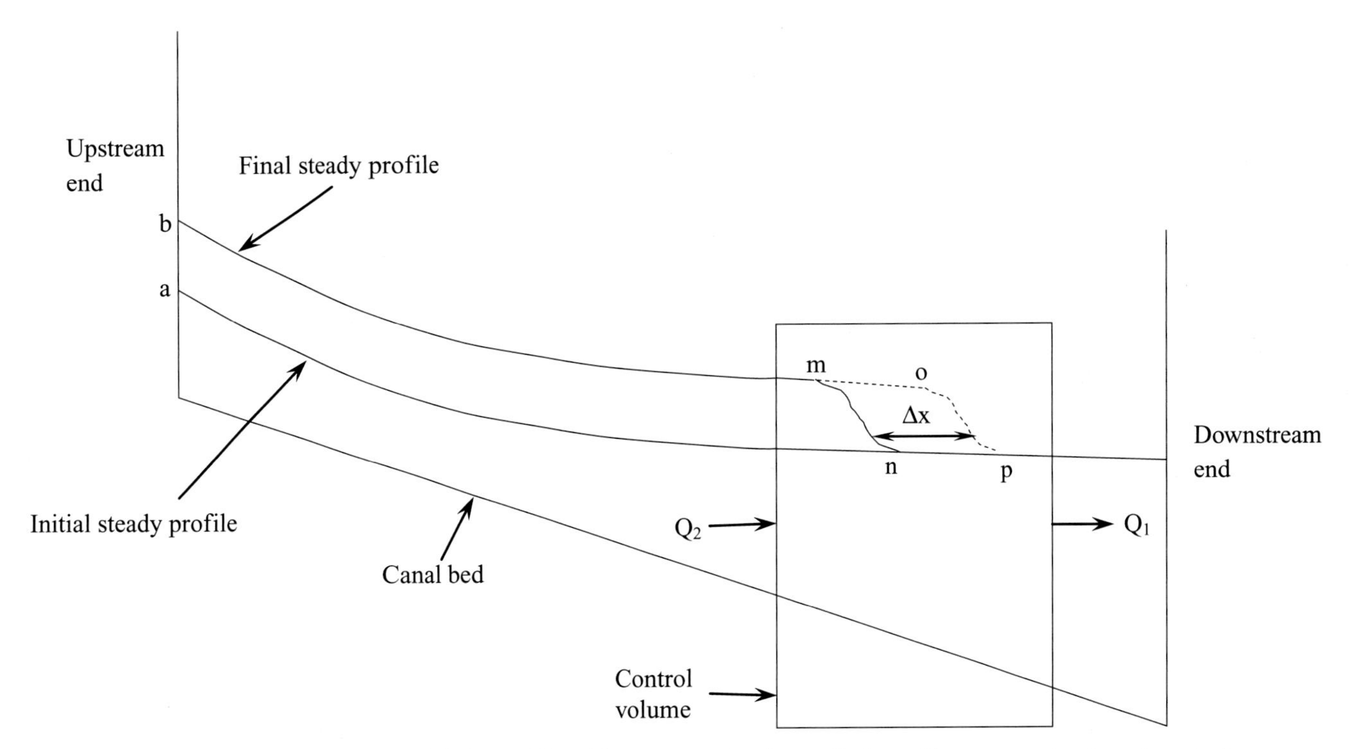

Figure 2.27 Discontinuities in the gradient of discharge (Q) and flow cross-sectional area (A) of a positive wave moving downstream in the presence of backwater (*Source*: Gopakumar and Mujumdar, 2009).

If the channel reach is not exactly uniform then it is divided into a number of uniform sub-reaches and the fuzzy rule based model for each sub-reach is developed separately. As natural channels are under consideration, the sub-reaches will be only approximately uniform. Therefore the Q–A relationship for the mean cross section of each sub-reach is used for development of the fuzzy rule based model for that sub-reach. If the heterogeneity of the sub-reach with respect to its cross sections is very high, then an optimized fuzzy rule based model, based on Q–A relationships of different cross sections, is obtained by means of a neuro-fuzzy algorithm, explained subsequently through an application.

2.6.2.6 METHOD OF COMPUTATION

Values of flow depth (h) and discharge (Q) at the beginning of the time step are specified at all the computational nodes, along the channel, as initial conditions. The three boundary conditions required by the FDFRM are the inflow discharge hydrograph at the upstream boundary, the stage hydrograph at the upstream boundary, and the stage hydrograph at the downstream boundary. The data in terms of h are converted into terms of area (A), while applying the FDFRM. This means that the h–A relationships should be available at all the computational nodes. Also, for every sub-reach, the stage–discharge relationships should be available at the boundary nodes and at the mean section. Stage–discharge relationships at the other computational nodes are obtained by interpolation.

In the case of a channel with a single uniform reach, the following are the steps involved in the flood routing computations:

Step 1: From the known initial conditions and using the discretized continuity equation, Equation 2.43, the values of A at the end of the time step (Δt) are obtained at all the nodes. As Equation 2.43 is based on an explicit finite difference scheme, the time step should be chosen satisfying the C.F.L. (Courant–Frederich–Levy) condition (Chaudhry, 1987).

Step 2: For the considered inflow flood discharge into the reach during the present time step, the corresponding final steady water surface profile along the reach is obtained from the stage–discharge relationships at the computational nodes.

Step3: Using the A values obtained in Step 1 along with corresponding values of the final steady profile, and applying the upstream and downstream boundary stage hydrograph ordinates, the ΔA values at the nodes are calculated using Equation 2.44, proceeding from upstream to downstream.

Step 4: Corresponding ΔQ values at the nodes are obtained by running the fuzzy rule based model.

Step 5: The values of Q at each node at the end of the time step are obtained from the ΔQ values obtained in Step 4, by using Equation 2.45 and the ordinates of the upstream boundary discharge hydrograph.

Step 6: Thus, values of Q and A at the end of the time step are obtained at all the nodes. These are used as the new initial conditions and Steps 1 to 5 are repeated for the next time step. This procedure is continued until the entire time period is covered.

For the case of a channel with two uniform sub-reaches, the steps involved in the computations are as follows:

Step 1: From the known initial conditions and using the discretized continuity equation, Equation 2.43, the values of A at the end of the time step (Δt) are obtained at all the internal nodes of both the sub-reaches. The flow depth/stage at the junction node of the sub-reaches is obtained by interpolation between the flow depth/stage values of the last internal node of the upstream sub-reach and the first internal node of the downstream sub-reach.

Step 2: For the considered inflow flood discharge during the present time step, the corresponding final steady water surface profile along the upstream sub-reach is obtained from the stage–discharge relationships at the computational nodes of this sub-reach. The values of ΔA and ΔQ, as well as Q, at all nodes of this sub-reach, corresponding to the end of the time step (Δt), are obtained by adopting the same computational procedure as that described previously for the single-reach channel.

Step 3: The value of Q at the junction of the sub-reaches is taken as the inflow flood discharge to the downstream sub-reach. The corresponding final steady profile for this sub-reach is obtained and ΔA and ΔQ, as well as Q, at all nodes of this sub-reach, corresponding to the end of the time step (Δt), are computed in the same way as that for the upstream sub-reach.

Step 4: Thus values of Q and A at the end of the time step are obtained at all the nodes. These are used as the new initial conditions and Steps 1 to 3 are repeated for the next time step. This procedure is continued until the entire time period is covered.

The above procedure of computation, for the channel with two sub-reaches, can be extended to a channel with any number of sub-reaches.

2.6.2.7 ILLUSTRATION: FLOOD ROUTING IN A NATURAL CHANNEL

The procedure is illustrated with the problem of flood routing in the channel "Critical Creek," described in HEC-RAS applications guide (US Army Corps of Engineers, 2002). The channel consists of only one heterogeneous reach named Upper Reach. The length of the channel is around 2000 m with bed slope 0.01. The Manning's roughness coefficient (n) for the main channel is 0.04 and for the flood plain it is 0.1. Thus, the hydraulics involved in the flood movement within the main channel are different from those over the flood plain and so they are taken as two separate channels. Therefore, theoretically speaking, the FDFRM should adopt separate fuzzy rule based models for the main channel and flood plain. Such a case, however, is not considered for the FDFRM.

The purpose of this illustration is to demonstrate that the FDFRM is capable of producing results comparable to those of HEC-RAS, for this natural channel, by taking heterogeneity in the shape of its cross sections into account. The entire cross section is considered as a single channel with a uniform roughness coefficient of 0.07, being the average of the roughness coefficients of the main channel and flood plain. Both HEC-RAS and FDFRM models are run under this condition. The expansion/contraction losses are considered negligible. Twelve cross sections are considered along the channel, which are numbered serially from 1 to 12, cross section 12 being the most upstream one. These cross sections are located wherever there is significant change in the shape of the cross section. Hence the sub-reach between any two adjacent cross sections can be assumed to be approximately uniform. A mean cross section is interpolated in every sub-reach, between adjacent cross sections, and they are numbered in the same way. For example, the mean cross section interpolated between the cross sections 12 and 11 is numbered as 11.5, the one interpolated between cross sections 11 and 10 is numbered as 10.5, etc. The stage–area as well as discharge–area relationships for all these cross sections are obtained from steady flow analysis using HEC-RAS. This is done for a series of discharge values ranging from 20 m^3/s to 310 m^3/s. Flow in the channel is under the sub-critical regime and so the normal depth condition is applied at the downstream boundary. The stage–area and discharge–area relationships for all the cross sections are developed from the HEC-RAS outputs corresponding to the steady flow analysis. As mentioned previously, the fuzzy rule based model for each sub-reach is derived from the Q–A relationship for the mean cross section. The ranges of values of ΔA and ΔQ of the fuzzy rule based model for all the sub-reaches are given in Table 2.19. A total of 31 membership functions are derived for both ΔA and ΔQ and the two variables are related through 31 rules in each fuzzy rule based model.

For flood routing, the following discharge hydrograph, which is based on a log-Pearson type III distribution (Murty *et al.*, 2003), is applied at the upstream boundary:

$$Q(t) = Q_b + (Q_p - Q_b)e^{-(t-t_p)/(t_g-t_p)}\left(\frac{t}{t_p}\right)^{t_p/(t_g-t_p)} \tag{2.47}$$

where Q_b is the initial base flow existing in the channel, taken equal to 200 m^3/s, Q_p is the peak flow rate, taken as 300 m^3/s, t_p is the time to peak, taken equal to 21,600 s (6 hr), and t_g is the time to centroid of the hydrograph, which is obtained from the relation $(t_p/t_g) = 0.95$. The time interval used in the simulation is 10 s and the total simulation period is 46,800 s (13 hr). The flood routing is carried out using both the HEC-RAS and the FDFRM. Flood routing using the HEC-RAS is based on its unsteady flow analysis capability. The same normal-depth downstream boundary condition as that used for the steady flow analysis is applied in the unsteady flow analysis also. The HEC-RAS stage hydrograph outputs, at the upstream and downstream ends of the study reach, are shown in Figure 2.28. They are used as the

Table 2.19 *Ranges of input variable (ΔA) and output variable (ΔQ) of the fuzzy rule based model for all sub-reaches of the Critical Creek*

	Range of input variable (ΔA)		Range of output variable (ΔQ)	
Sub-reach	Minimum (m^2)	Maximum (m^2)	Minimum (m^3/s)	Maximum (m^3/s)
12–11	0	211.99	0	310
11–10	0	152.39	0	310
10–9	0	228.66	0	310
9–8	0	229.22	0	310
8–7	0	218.74	0	310
7–6	0	190.32	0	310
6–5	0	129.98	0	310
5–4	0	209.23	0	310
4–3	0	213.04	0	310
3–2	0	167.09	0	310
2–1	0	215.01	0	310

boundary conditions for flood routing using the FDFRM, in addition to the upstream discharge hydrograph, Equation 2.47. The FDFRM outputs (discharge and stage hydrographs) for three typical cross sections are shown in Figure 2.29, along with the corresponding HEC-RAS outputs. It is observed that the FDFRM

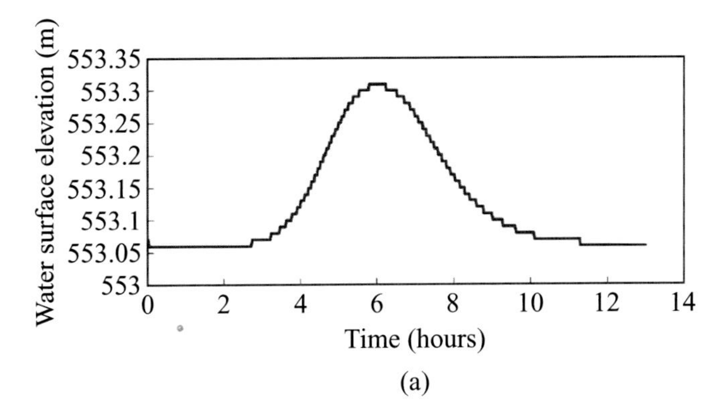

(a)

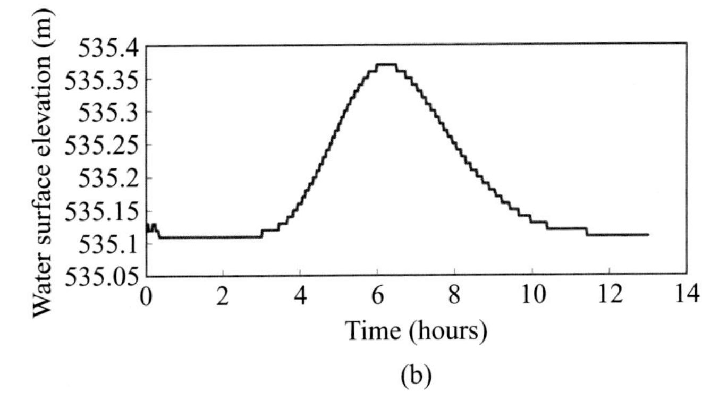

(b)

Figure 2.28 Boundary conditions for flood routing in Critical Creek. (a) Stage hydrograph at upstream boundary; (b) stage hydrograph at downstream boundary (*Source*: Gopakumar and Mujumdar, 2009).

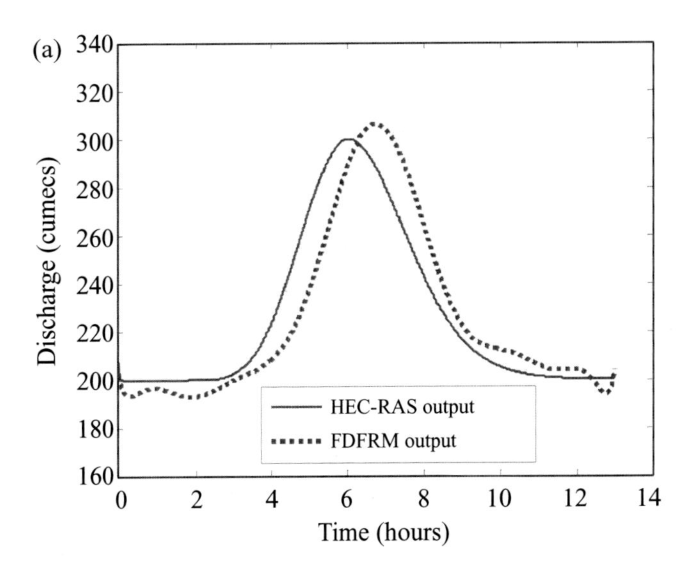

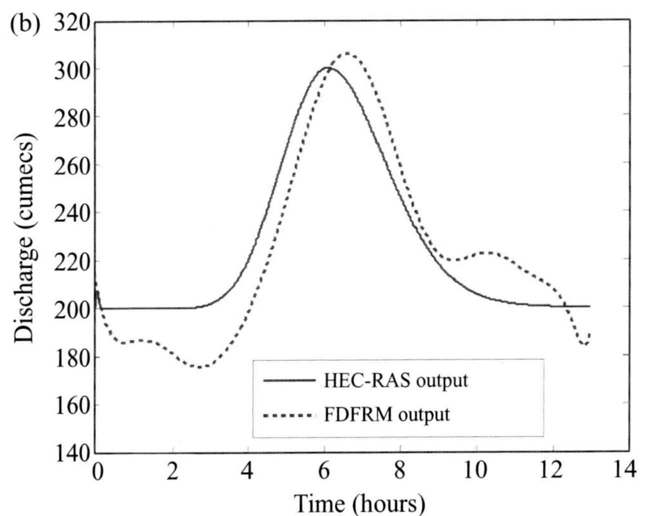

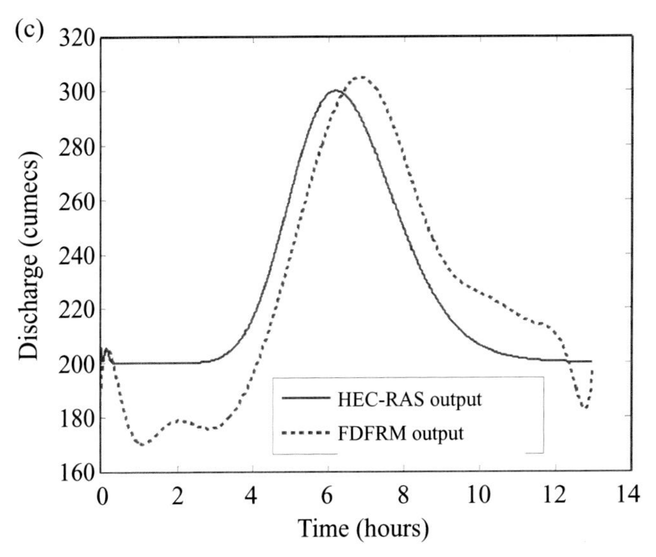

Figure 2.29 Simulation results of flood routing in Critical Creek. (a) Discharge hydrograph at cross section 10; (b) discharge hydrograph at cross section 7; (c) discharge hydrograph at cross section 2 (*Source*: Gopakumar and Mujumdar, 2009).

outputs are comparable to those of the HEC-RAS, except for cross sections 10, 7, and 2. This is due to the significant heterogeneity in cross sections of the sub-reaches 11–10, 8–7, and 3–2, and therefore the fuzzy rule based models, which were originally derived on the assumption that the sub-reaches are approximately uniform, are not able to provide satisfactory results. Thus, there is a need for deriving optimum fuzzy rule based models for these sub-reaches, which will be capable of taking the heterogeneity into account. This may be achieved by using a neuro-fuzzy algorithm, as described in Gopakumar and Mujumdar (2009).

2.7 SUMMARY

In this chapter, the basic hydrologic processes are introduced and various methods for flood runoff generation and flood routing are discussed. A brief review of commonly used hydrologic models is provided with examples. Use of three freely available hydrologic models, namely, HEC-HMS, HEC-RAS, and SWMM, is demonstrated with illustrations. Information on availability of global data sets for use in hydrologic models is provided. To overcome the limitations due to the extensive data requirements of the models, empirical models are now gaining popularity. Two classes of empirical models, the artificial neural networks and the fuzzy logic based models are introduced with examples of flow forecasting and dynamic flood wave routing.

EXERCISES

2.1 The cross sections along a river stretch of 60 km length are given in the table below.

The highlighted chainages correspond to main channel bank stations as required in HEC-RAS. C, chainage of the point in the river cross section (running length across the river) in m; E, bed elevation of the river in m.

Route the river stretch using HEC-RAS for flood peak discharges of 20,000 m^3/s, 22,000 m^3/s, and 25,000 m^3/s for 100 years, 500 years, and 1000 years return periods, respectively. The initial flow in the river is 9000 m^3/s. Consider a symmetric triangular hydrograph, with time to peak of 10 hr. Assume friction slope $= 1.43 \times 10^{-5}$ and Manning's $n =$ 0.0245.

2.2 A small tributary of a river is divided into four reaches as shown in the figure. The tributary joins the river downstream of the last reach.

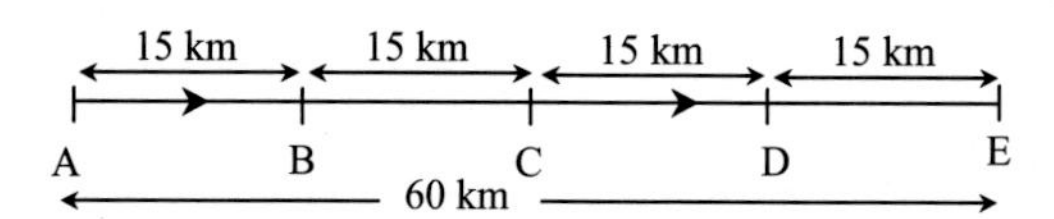

Cross section A		Cross section B		Cross section C		Cross section D		Cross section E	
C	E	C	E	C	E	C	E	C	E
0	545.27	0	536	0	532	0	530.6	**0**	**528.87**
900	539.39	1100	534	1225	526	420	524.03	180	518.12
1770	540	1962	530	1463	528	**810**	**520.37**	360	504.22
1845	**538.23**	2500	531	3038	527	900	519.57	550	500.7
1880	531.39	3617	529.53	4413	526	990	511.39	650	498.12
1950	524.58	4217	524.4	4913	522	1200	507.76	740	492.31
2010	527.68	**4337**	**527.6**	5215	527.1	1440	514	906.5	489.26
2020	**538.4**	4457	520.02	**5440**	**526.17**	1620	518.91	1094.5	489.58
2220	538.89	4607	517.23	5490	520.82	**2190**	**521.01**	1217	492.37
2310	539.52	4712	512.33	5535	513.01	2490	520.95	1419.6	509.84
2700	542.71	**4817**	**528.45**	5795	513.56	2820	521.79	**1564.8**	**527.11**
3060	541.38	5297	531.82	5830	517.67	3210	525.24		
4230	545.03	5777	530.07	**5880**	**524.83**	3600	528.65		
4320	546.32	6467	529.8	6300	535.26	4020	530.53		
		6887	532.1	7370	543.61	4200	528.32		
		7787	532.38						
		8537	533.85						
		9017	535.87						
		9677	540.12						

Details of cross sections and sub-catchments are given in the following tables

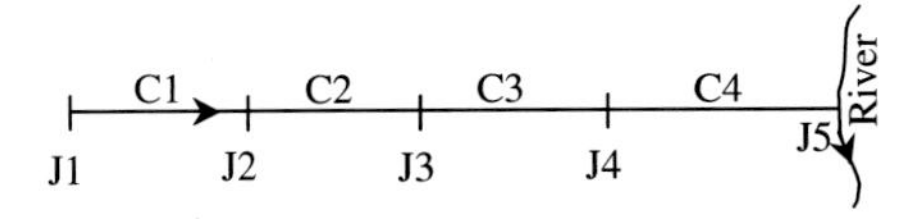

Junction	Reach	Chainage[a] (m)	Bed level (m)	Width[b] (m)	Depth[b] (m)
J1		0	911.34		
	C1			9.5	2.25
J2		2200	895.46		
	C2			17.5	2.1
J3		4500	885.62		
	C3			22	2.48
J4		7075	879.1		
	C4			15	1.83
Outfall (J5)		10275	872.68		

[a] Cross section approximated as rectangular
[b] Running length along the tributary

Sub-catchment ID	Area of sub-catchment (hectares)	Slope of sub-catchment	Weighted imperviousness of sub-catchment (%)	Width of sub-catchment (m)	Sub-catchment outlet
A1a	95.38	0.0143	59.0	705.7	J1
A1b	49.22	0.0133	55.6	356.1	J2
A2a	59.93	0.0214	75.0	337.8	J2
A2b	99.64	0.0105	75.0	632.4	J3
A3a	87.97	0.0117	75.0	454.2	J3
A3b	58.60	0.0133	72.8	261.0	J4
A4a	95.44	0.0053	72.6	656.1	J4
A4b	59.48	0.0067	42.3	345.3	J5

Using SWMM, obtain the flood hydrographs at the downstream junction J5 for the following three cases. Assume a free outfall at the junction J5.

Case 1: Rainfall of 181.5 mm occurring over duration of 24 hours, uniformly in all sub-catchments.

Case 2: Rainfall of 90 mm occurring over duration of 1 hour, uniformly in all sub-catchments.

Case 3 : Rainfall intensity of 100 mm/hr over a duration of 20 minutes in the sub-catchments A1a, A1b, and A2a, and a rainfall intensity of 120 mm/hr over a duration of 30 minutes in the other sub-catchments.

2.3 The following table gives discharge values, Q (m^3/s), for 122 days (N) during a flood season in a catchment. Develop feed-forward neural network models for forecasting the flow with lead times of 1 day, 2 days, and 3 days, and compare the performance of the models with input layers consisting of nodes representing flow during the preceding (a) 2 days, (b) 5 days, and (c) 10 days.

N	Q	N	Q	N	Q	N	Q	N	Q	N	Q
1	2527.7	22	976.4	43	3726.4	63	1560	83	940	103	280
2	3557.1	23	1133.3	44	2696.7	64	1318.4	84	696	104	273.5
3	4000	24	930.7	45	2360	65	2181.2	85	663.8	105	320
4	4162.5	25	1655.9	46	1898.1	66	1493.9	86	491.6	106	244.9
5	5703.2	26	1980	47	1517.2	67	2537.4	87	1130	107	304.5
6	5533.1	27	2093.7	48	2030.9	68	2469.5	88	772.1	108	316.8
7	4596	28	3245.2	49	1509.1	69	2670	89	682.7	109	324
8	4395.4	29	14068.7	50	1517.2	70	3530	90	416.5	110	306.9
9	4432.7	30	21893.4	51	1329.8	71	3230	91	500	111	287.4
10	3240	31	21424.5	52	2125.1	72	3880	92	279.4	112	350
11	2177.1	32	17193.3	53	4367	73	1631.8	93	335	113	298.2
12	2278.8	33	6600	54	2439.8	74	1279.4	94	420	114	244.5
13	3908.6	34	3276.2	55	1606.1	75	1023.3	95	340.7	115	387.9
14	4229.2	35	3045.3	56	1119.4	76	780	96	307	116	970
15	3360	36	2939.8	57	1195.4	77	2550	97	374.5	117	586.3
16	2746.8	37	4919.6	58	934.8	78	5853.6	98	375	118	678.4
17	2400	38	7680	59	680	79	2436.8	99	295.5	119	580
18	2240	39	3827.9	60	814.3	80	1825	100	291	120	475.7
19	1000	40	2391.2	61	760.7	81	1540	101	279.7	121	475
20	1320	41	2289.1	62	1362	82	1140	102	247.7	122	447.2
21	1133	42	4856.2								

2.4 Use the data provided in Exercise 2.1 to develop a fuzzy logic based model for flood routing. Compare the hydrographs obtained at the downstream end of the river reach from the HEC-RAS and the fuzzy logic based models.

2.5 Choose a watershed in a region of your interest. Using the global data sets (e.g., from the web links provided in Table 2.17), calibrate and validate the HEC-HMS model for this watershed to simulate the flow (runoff) at a location in the watershed, at daily time steps. With this calibrated model, simulate the runoff for the next 50 years for the following climate change scenarios: (a) rises in air temperature of 0 °C, 1 °C, 1.5 °C and 2 °C; (b) increase/decrease in precipitation of 0%, 10%, and 20%. Assume that all other variables remain unchanged in the calibrated model. From the results, draw inferences on the likely impact of climate change, modeled with the scenarios considered, on the peak flow (maximum of the daily flow among all days in a year).

3 Climate change impact assessment[1]

3.1 INTRODUCTION

In this chapter, some methodologies for assessing hydrologic impacts due to climate change are described. The main focus of the chapter is on explaining general procedures for projection of local hydrometeorological variables such as temperature and rainfall under climate change scenarios, using statistical downscaling techniques. The projections may then be used as inputs to hydrologic models to obtain streamflow, evapotranspiration, soil moisture, and other variables of interest. This chapter also provides a description of methodologies used for flood modeling under climate change scenarios. Quantification of uncertainty in hydrologic projections under climate change is discussed. On understanding this chapter, the reader should be in a position to use hydrologic models for projecting changes in river basin hydrologic variables, and flood frequencies and magnitudes under climate change. The reader should also be able to understand and quantify the uncertainties associated with the projections.

The chapter is organized as follows. An introduction to emissions scenarios used in climate change studies is provided first with a discussion on limitations in the direct use of GCM-simulated climate variables for hydrologic impact assessment. An overview of dynamical downscaling methods is provided next, followed by a detailed description of various statistical downscaling techniques used for projecting hydrologic impacts. Disaggregation techniques necessary for projections at shorter time scales for hydrologic studies, such as flood risk assessment are then discussed. Use of macroscale hydrologic models (MHMs) and hypothetical scenarios for hydrologic impact assessment is discussed. Specific methodologies for modeling of floods under climate change are described. Techniques for uncertainty quantification, along with discussions on propagation of uncertainties through time, are provided. It is expected that, by training through this chapter, the user will be introduced to the important concepts and methodologies of assessment of climate change impacts on floods and uncertainties associated with the climate change projections.

[1] with a significant contribution from Dr. Deepashree Raje, Research Associate (2010), IISc, Bangalore

3.1.1 Climate change: emissions scenarios

Recent studies now concur that climate change impacts on regional hydrologic regimes are likely to be significant. Changes in precipitation patterns and frequencies of extreme precipitation events, along with changes in soil moisture and evapotranspiration will affect runoff and river discharges at various time scales from sub-daily peak flows to annual variations. At sub-daily and daily time scales, flooding events are likely to cause enormous socio-economic and environmental damage, which necessitates the use of robust and accurate techniques for projections of flood frequencies and magnitudes under climate change, and development of flood protection measures to adapt to the likely changes. Modifications in return periods of floods have a significant implications for hydrologic designs of structural measures for flood control in which gradual adaptation over a long period may not be possible. On the other hand, large changes in monthly or annual flows have implications for water resources planning and storage structures, as well as agriculture.

The most widely used tools for projecting future climate changes are GCMs. These are three-dimensional mathematical models based on the physical principles of fluid dynamics, thermodynamics, and radiative heat transfer. The climate system is represented in a simplified form in GCMs, using combinations of models for different components of the climate system. Usually these include an atmospheric model for wind speeds, temperature, clouds, and other atmospheric properties; an ocean model for predicting temperature, salt content, and circulation of water; models for ice cover on land and sea; and a model of heat and moisture transfer from soil and vegetation to the atmosphere. Some models also include treatments of chemical and biological processes (IPCC, 2007).

In GCMs, physical laws and processes are represented on a grid and integrated forward in time. Processes with scales too small to be resolved on the spatial scale of the model grid, or those incompletely understood or difficult to formulate, are represented through modules based on observations and physical theory, called parameterizations. For example, land surface processes are described by the land surface (parameterizations)

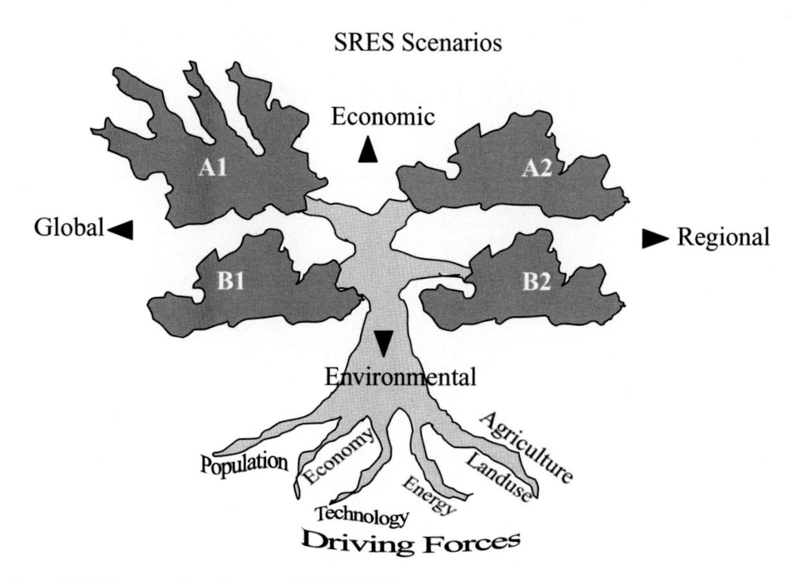

Figure 3.1 Schematic illustration of SRES scenarios (*Source*: IPCC, 2007).

schemes (LSSs or LSPs) or land surface models (LSMs) embedded within the GCM. Thus, some physical processes, such as cloud–aerosol interactions or land–atmosphere exchanges, are largely represented through parameterizations. New generations of GCMs show significant skill in representing important climate features, such as large-scale distributions of atmospheric temperature, radiation and wind, and oceanic temperatures. GCMs can also simulate many of the important patterns of climate variability for time scales ranging from seasonal to annual to decadal, such as the monsoons, seasonal shifts of temperatures, El Niño–Southern Oscillation (ENSO), Pacific Decadal Oscillation (PDO), and northern and southern annular modes (IPCC, 2007). This gives substantial confidence in the use of GCMs as tools for impact assessment.

Future greenhouse gas (GHG) emissions are determined by complex systems and their interactions, such as population growth, energy use, and socio-economic and technological development. Scenarios used in climate change studies are alternative mappings or paths for future changes, and are widely used to analyze future emissions outcomes. Such scenarios project future socio-economic drivers of development, population growth, and energy consumption patterns, and thus involve various assumptions of future GHG emissions, land use changes, and other climate forcings. The IPCC in 2000 developed a new set of emissions scenarios in the Special Report on Emissions Scenarios (SRES), which formed an input to the IPCC Third Assessment Report (TAR) in 2001 and Fourth Assessment Report (AR4) in 2007. There are 40 SRES scenarios developed by the IPCC, each making different assumptions for future GHG pollution, land use, and other driving forces. The 40 SRES scenarios fall into different groups – the three scenario families A2, B1, and B2, and three groups within the A1 scenario family.

Scenario storylines describe developments in many different social, economic, technological, environmental, and policy dimensions. Different assumptions about future technological development as well as the future economic development are made for each scenario. Most scenarios include an increase in the consumption of fossil fuels. The emissions scenarios are organized into families, which contain scenarios that are similar to each other in some respects. Scenario families contain individual scenarios with common themes. The six families of scenarios discussed in AR4 (IPCC, 2007) are A1FI, A1B, A1T, A2, B1, and B2. Each scenario family is based on a common specification of the main driving forces. The four scenario families are illustrated as branches of a two-dimensional tree, as shown in Figure 3.1. The two dimensions indicate global and regional scenario orientation and development and environmental orientation, respectively.

- The A1 storyline and scenario family projects very rapid economic growth, low population growth, and quick use of more efficient technologies. Subsets to the A1 family are based on their technological emphasis: A1FI – an emphasis on fossil fuels, A1B – a balanced emphasis on all energy sources, and A1T – an emphasis on non-fossil energy sources.
- The A2 storyline and scenario family projects a heterogeneous pattern of development across the world and high population growth due to slow convergence of fertility patterns. Economic development is primarily regionally oriented, and per capita economic growth and technological change are more fragmented and slower than in other storylines.
- The B1 storyline and scenario family projects a convergent world with low population growth, as in the A1 storyline, but with rapid changes toward a service and information economy, reductions in material intensity, and the introduction of clean and resource-efficient technologies.

- The B2 storyline and scenario family projects a world in which the emphasis is on local solutions to economic, social, and environmental sustainability. It projects moderate population growth, intermediate levels of economic development, and less rapid and more diverse technological change than the B1 and A1 storylines.

The quantitative inputs for each scenario are measures such as regionalized measures of population, economic development, and energy efficiency, the availability of various forms of energy, agricultural productivity, and local pollution controls (IPCC, 2007).

The following sections explain how GCM simulations of future IPCC scenarios are used for projecting hydrologic impacts.

3.2 PROJECTION OF HYDROLOGIC IMPACTS

GCMs were initially developed to simulate average, large-scale (of the order of 10^4–10^6 km^2) atmospheric circulation patterns for specified external forcing conditions. The size of the climate system and the time range of climate experiments (several decades to thousands of years) is a constraint for the design of GCMs. GCMs deal most proficiently with fluid dynamics at the continental scale and operate on horizontal grid resolutions ranging from 200 to 600 km. Hence, it is not possible to explicitly model local geographic factors that impact basin hydrology, such as topography, land/water distribution, or vegetation type. Also, although GCMs use short time steps, commonly 10–30 min, most verifications of the models are based on long-term mean simulations, hence only seasonal to monthly temporal scales are considered reliable. The ability of GCMs to predict spatial and temporal distributions of climatic variables declines from global to regional to local catchment scales, and from annual to monthly to daily amounts.

GCMs are more skillful in simulating the free troposphere climate than the surface climate. Variables such as wind, temperature, and air pressure field can be predicted quite well, whereas precipitation and cloudiness are less well predicted variables. Other variables of key importance in the hydrologic cycle, such as runoff, soil moisture, and evapotranspiration, are not well simulated by GCMs. Runoff predictions in GCMs are over-simplified and there is no lateral transfer of water within the land phase between grid cells (Xu, 1999a; Fowler *et al.*, 2007). GCM simulation of rainfall has been found to be especially poor.

Hydrologic processes are affected by spatial inhomogeneities in land-surface processes due to topography, soil properties, and vegetation cover, as well as in precipitation, at scales of 1 km or less. Hydrologic models are mainly concerned with small, sub-catchment scale processes, occurring on spatial scales much smaller than those resolved in GCMs. For example, the Hadley Centre's HadCM3 model has a spatial resolution of 2.5° latitude by 3.75° longitude, whereas a spatial resolution of about 0.125° latitude and longitude is required for hydrologic simulations of monthly flow in mountainous catchments. Moreover, hydrologic responses to climate change, such as changes in annual or monthly runoff, are important, but may give very little information on the changes in the flow regime, especially extremes. Continuous rainfall–runoff simulation at daily, hourly, or even sub-hourly time steps is necessary to model the flood regime of a catchment correctly. Hydrologic impact models typically use a time step of one day, routing rainfall through two to three soil layers to produce output for hydrologic variables.

The scale issues discussed above result in a spatio-temporal scale mismatch between the output provided by GCMs and the needs of hydrologic models. Figure 3.2 shows the major approaches currently used for projecting hydrologic impacts of climate change. Addressing the gap between climate model output and local-scale climate change necessitates the use of additional methods, such as downscaling, for hydrologic impact assessments. Downscaling techniques use high-resolution dynamical models or statistical techniques to simulate weather and climate at finer spatial resolutions than is possible with GCMs. The two main types of downscaling approaches currently in use for deriving regional-scale information from GCMs are: the dynamical approach, where a higher resolution climate model is embedded within a GCM; and the statistical approach to establish empirical relationships between GCM-simulated climate variables and regional-scale hydrologic variables. Apart from downscaling, approaches such as the use of MHMs, which can operate on large spatial scales, and hypothetical climate scenarios have also been used in the literature. These approaches are described in the sections below.

3.3 DYNAMICAL DOWNSCALING APPROACHES

Dynamical downscaling is the use of high-resolution regional climate models (RCMs) to provide simulated climate outputs at finer resolutions. They operate currently at scales of 20–50 km and typically make use of observed, reanalysis, or lower-resolution GCM data to provide boundary conditions. Higher-resolution simulations may be performed within coarser-scale RCM output (e.g., Dankers *et al.*, 2007), using double-nesting of RCMs, in order to achieve very high resolution simulations (i.e., <10–20 km). The main advantage of dynamical downscaling is its potential for capturing meso-scale non-linear effects and providing information for many climate variables, while ensuring internal consistency with respect to the physical constraints. It is found that RCMs are able to realistically simulate regional climate features such as

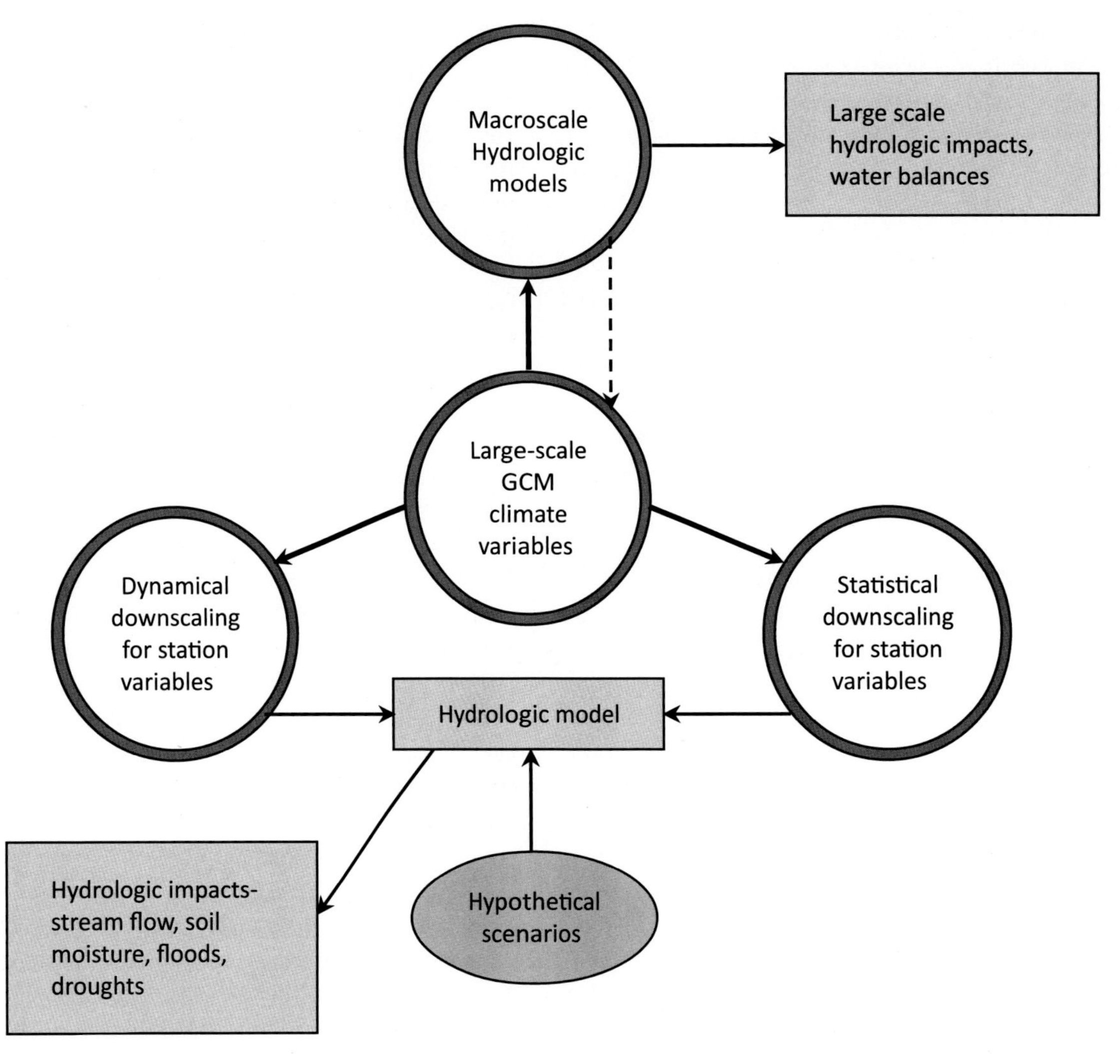

Figure 3.2 Projecting hydrologic impacts of climate change.

orographic precipitation, extreme events, and regional-scale climate anomalies, such as those associated with the ENSO. This increases confidence in their ability to realistically downscale future climates.

The main drawbacks of dynamical downscaling models are their computational cost and the lack of provision of information at the point or station scale. RCM model skill is dependent on biases present in the driving GCM and regional forcings such as orography, land–sea contrast, and vegetation cover. Since boundary conditions are derived from a specific GCM, use of different GCMs will result in different projections. Due to the computational cost of RCM simulations, these are usually restricted to time slices. Producing scenarios for other periods is achieved using "pattern scaling," where changes are scaled according to the temperature signal modeled for the intervening period, assuming a linear pattern of change (e.g., Prudhomme *et al.*, 2002). Using an RCM provides additional uncertainty to that inherent in GCM output. For temperature projections, the uncertainty introduced by the RCM is less than that from the emissions scenario, but for precipitation projections, scenario uncertainty is larger than RCM uncertainty. However, the largest source of uncertainty derives from the structure and physics of the formulation of the driving GCM.

Study applications to geographically diverse regions and model intercomparison studies have revealed some strengths and weaknesses of dynamical downscaling. Wang *et al.* (2004) have reported that studies within the western USA, Europe, and New Zealand, where topographic effects on temperature and precipitation are important, show more skillful dynamical downscaling than in regions such as the continental USA and China, where regional forcings are weaker. Dynamical downscaling can also provide improved simulation of meso-scale precipitation processes and higher moment climate statistics (Schmidli *et al.*, 2006), and more reliable climate change scenarios for extreme events and climate variability at the regional scale. For this purpose, longer duration, higher spatial resolution and ensemble RCM simulations are becoming more common.

3.4 STATISTICAL DOWNSCALING APPROACHES

Statistical downscaling derives a statistical or empirical relationship between the variables simulated by the GCMs, called predictors, and station-scale hydrologic variables, called predictands. It is seen that accurate estimates of the local variables are strongly dependent on the quality and the length of the data series used for the calibration and on the performance of the models in capturing the variability of the observed data. Statistical downscaling methods have been preferred in hydrologic impact assessment because they provide good fits to observed data while being computationally inexpensive, provide quick results, and their domain of application can be easily transferred from one region to another. Statistical downscaling is based on the following assumptions (Wilby and Wigley, 2000): (i) suitable relationships can be developed between the large-scale predictors and local predictands; (ii) the relationships are valid under future climate conditions; and (iii) the predictor variables and their changes are well simulated by GCMs. A diverse range of empirical/ statistical downscaling techniques have been developed over the past decade. Each method generally lies in one of three major categories, namely, regression or transfer function methods, stochastic weather generators, and weather typing schemes. A review of downscaling techniques may be found in Prudhomme *et al.* (2002) and Fowler *et al.* (2007). Identification and selection of appropriate downscaling methods for impact assessment and adaptation planning depend on many factors, such as ease of accessibility, resource requirements, and type of output, in addition to the performance of the method itself.

The following sections discuss statistical downscaling approaches in detail. First, data needs for the methods and issues of data scarcity are discussed in Section 3.4.1. Then, methodologies employed for choice of predictor variables are discussed in Section 3.4.2, and preprocessing of data is explained in Section 3.4.3. Sections 3.4.4–3.4.6 discuss the three categories of statistical downscaling methods in detail, with worked-out examples.

3.4.1 Data needs and sources

Statistical downscaling methods require observed values of atmospheric predictors and observed station variables, for a sufficiently long duration, for the model to be well trained and validated. When adequate observed climatological data are not available, reanalysis data are used in climate change studies. These data are outputs from a high-resolution climate model, run using assimilated data from sources such as surface observation stations and satellites, and can hence be considered as outputs from an ideal GCM. Four currently available reanalyses are from the National Centers for Environmental Prediction (NCEP), the European Centre for Medium Range Forecasting (ERA40), the North American Regional Reanalysis (NARR) project, and the Japanese 25-year Reanalysis (JRA25) project. The NARR data set has a much higher spatial and temporal resolution and improved precipitation and land-surface data assimilation compared to other global data sets. For climate projections, GCM output data from the IPCC AR4 data set or the multi-model data set of the World Climate Research Programme's Coupled Model Intercomparison Project (WRCP CMIP5) can be used for prediction, as well as data from individual GCMs from the respective climate modeling centers. Large-scale daily atmospheric variables simulated for the next one to three centuries for the SRES scenarios at necessary time resolution and for the specific geographical study region can be extracted for the purpose of downscaling.

In the case of downscaling for hydrologic extremes, it is important to choose statistical methods that can provide outputs (or events) beyond the current observed distribution, which is a drawback observed with some stochastic downscaling methods. In projection of extremes such as floods, there are fewer data points for rarer events. Hence, it is difficult to identify long-term changes in these events, without a sufficiently long time series. Trend analyses of extremes traditionally use standard and robust statistics for moderately extreme events that occur a few times a year. The time scale of extreme events to be projected determines the required temporal resolution, such as heavy hourly or daily precipitation versus a multi-year drought. Longer time resolution data, such as monthly, seasonal, and annual values for temperature and precipitation, are available for most parts of the world for the past century, facilitating analysis of drought and wet periods defined on those time scales. However, data scarcity is an issue for analysis of hourly or sub-hourly resolution extreme events. Flood modeling can make use of information transfers and pooled frequency analysis to tackle this issue. Pooled frequency analysis can make use of data from physically similar catchments in similar climatic zones in order to model the study catchment.

3.4.2 Choice of predictor variables

The choice of predictor variables is a key step in downscaling. However, there is no one method among downscaling studies to determine the right atmospheric predictor variables. The choice of predictors could vary from region to region depending on the characteristics of the large-scale atmospheric circulation and the predictand to be downscaled. Any variable or index can be used as predictor, where it is reasonable to expect that there exists a relationship between the predictor and the predictand. Generally, predictors are chosen as variables that are reliably simulated by GCMs and easily available from archives of GCM output and reanalysis data sets, while being strongly correlated with the predictand, and used in previous studies.

Table 3.1 *Choice of predictor variables in downscaling*

Set no.	Author(s)	Predictor(s)	Predictand(s)	Technique(s)
1	Bardossy and Plate (1992)	500 hPa geopotential height	Daily precipitation	Weather classification
2	Conway *et al.* (1996)	Vorticity	Daily precipitation	Regression and resampling
3	Goodess and Palutikof (1998)	Mean sea level pressure, airflow indices	Daily precipitation	Weather classification
4	Benestad (2001)	Surface temperature at 2 m, sea level pressure	Monthly mean temperature	Empirical orthogonal function (EOF)
5	Dibike and Coulibaly (2005)	Surface V-wind, geopotential height at 500 hPa, surface specific humidity, specific humidity at 850 hPa, surface air temperature at 2 m	Maximum temperature	Statistical downscaling model (SDSM) (multiple linear regression)
6	Tatli *et al.* (2005)	Geopotential height at 500 hPa, geopotential height thickness between 1000 hPa and 500 hPa	Monthly mean, maximum, and minimum temperature	Singular spectrum analysis (SSA), canonical correlation analysis (CCA), principal component analysis (PCA)
7	Ghosh and Mujumdar (2006)	Mean sea level pressure	Monthly mean precipitation	Linear regression with fuzzy clustering
8	Raje and Mujumdar (2009)	Mean sea level pressure, specific humidity at 500 hPa, precipitation flux, mean, maximum and minimum surface air temperatures at 2 m, surface U- and V-winds	Daily precipitation	Conditional random field (CRF)

Some studies have examined the role of predictor selection techniques in downscaling (e.g., Wilby and Wigley, 2000; Anandhi *et al.*, 2009). Variables of importance for hydrologic impact studies include those that may be used to predict rainfall and runoff in downscaling studies, such as precipitation rate and precipitable water, and those required for evaporation estimation, such as temperature, net radiation, wind speed, pressure, and specific humidity. Circulation-related predictors, such as sea level pressure, are also suitable, as GCM skill in simulating these variables is high. However, recent studies have found that circulation predictors fail to capture key precipitation mechanisms, based on thermodynamics and vapor content. Hence, humidity and moisture variables have increasingly been used to downscale precipitation, particularly as they may be important predictors under a changed climate. The selection of predictor variables should be referenced to the specific regional process for occurrence. For example, the summer monsoon rainfall over Orissa, India, occurs mostly due to low-pressure systems (LPS) developing over the Bay of Bengal. Hence, sea level pressure would be expected to be an important predictor of monsoon rainfall in the region. Preparation of correlation field patterns between the predictand (e.g., precipitation) and potential predictor variable (e.g., specific humidity) as maps over the region of interest has also been suggested (Wilby and Wigley, 2000) for choice of predictor variable as well as a spatial domain.

Table 3.1 shows the range of downscaling predictors used in some studies of daily precipitation and temperature. It is observed from this table that earlier studies tend to use circulation data as predictors, although some recent work has incorporated atmospheric moisture and other variables in downscaling schemes.

The explanatory power of any given predictor varies spatially and temporally. Johnson and Sharma (2009) defined a skill score, termed the variable convergence score (VCS), to rank variables based on the coefficient of variation of the ensemble of available GCMs, assumed to indicate the likely state of that variable in the future. A relative ranking of the variables averaged across Australia showed that surface variables with the highest VCS scores for the region are pressure, temperature, and humidity.

Example 3.1 Predictor selection

An example for predictor selection is provided from Anandhi *et al.* (2009). The probable predictor variables for downscaling maximum and minimum temperature are to be identified. The study region is the Malaprabha River catchment, upstream of the Malaprabha Reservoir in Karnataka state of India, as shown in

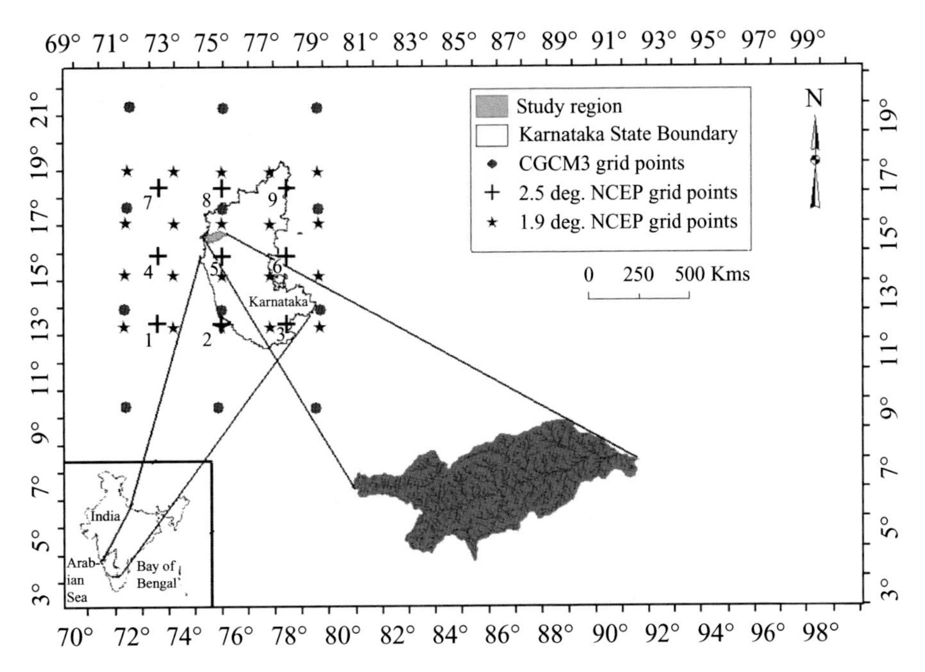

Figure 3.3 Location map of the study region in Karnataka state, India (*Source*: Anandhi *et al.*, 2008).

Figure 3.3. It has an area of 2564 km^2 situated between 15°30′ N and 15°56′ N latitudes and 74°12′ E and 75°15′ E longitudes. It receives an average annual precipitation of 1051 mm. It has a tropical monsoon climate where most of the precipitation is confined to a few months of the monsoon season.

Solution

Step 1: Identify probable predictor variables based on physical reasoning or expert judgment

Probable predictor variables are extracted at nine grid points over the study region from the NCEP reanalysis data set (Kalnay *et al.*, 1996) for the period 1978–2000. These data are publicly available at the website http://www.esrl.noaa.gov/psd/data/gridded/data.ncep.reanalysis.html of the Earth System Research Laboratory. The data are available from year 1948 to the current time, at several time resolutions, from four times daily to monthly, at several pressure levels, and with a horizontal grid spatially covering the entire globe. Specific time as well as latitude and longitude ranges can be entered and data for the specified ranges can be downloaded from the website. The GCM data used in the study are simulations obtained from CGCM3 of the Canadian Center for Climate Modeling and Analysis (CCCma), through the website http://www.cccma.bc.ec.gc.ca/. The data consist of present-day (20C3M) and future simulations forced by four emissions scenarios, namely, A1B, A2, B1, and COMMIT for the period 1978–2100. Probable predictor variables are identified as the atmospheric variables of air temperature (denoted Ta 925), zonal (denoted Ua 925), and meridional (denoted Va925) wind velocities at 925 mb, and surface flux variables such as latent heat (LH), sensible heat (SH), shortwave radiation (SWR), and longwave radiation fluxes (LWR), which control temperature of the Earth's surface.

Step 2: Classify the variables into groups

The predictor variables are classified into three groups. Large-scale atmospheric variables Ta 925, Ua 925, andVa925, which are often used for downscaling temperature, are considered as predictors in Group A. Surface flux variables such as LH, SH, SWR, and LWR, which control temperature of the Earth's surface, are tried as plausible predictors in Group B. Group C comprises all the predictor variables in both the Groups A and B.

Step 3: Identify inter-dependencies between predictor variables and predictor–predictand dependencies

Scatter plots and cross-correlations are examined to investigate the linear/non-linear dependence structure, (i) between the predictor variables in NCEP and GCM data sets (Figure 3.5) and (ii) between the predictor variables in the NCEP data set and each of the predictands (Figure 3.6). The cross-correlations between each of the predictor variables in NCEP and GCM data sets shown in Figure 3.4 can indicate if the predictor variables are realistically simulated by the GCM. The results from the correlation analysis indicate that the Group A predictors are better simulated by the GCM than Group B predictors, since they show higher correlations. The zonal wind velocity at 925 mb (Ua 925) is the most realistically simulated variable with a CC greater than 0.9, while LH flux is the least correlated variable between NCEP and GCM data sets (CC = 0.56). Generally, the correlations are not very high due to the differences in the simulations of GCM (e.g., for different runs) and possible errors in reanalysis data.

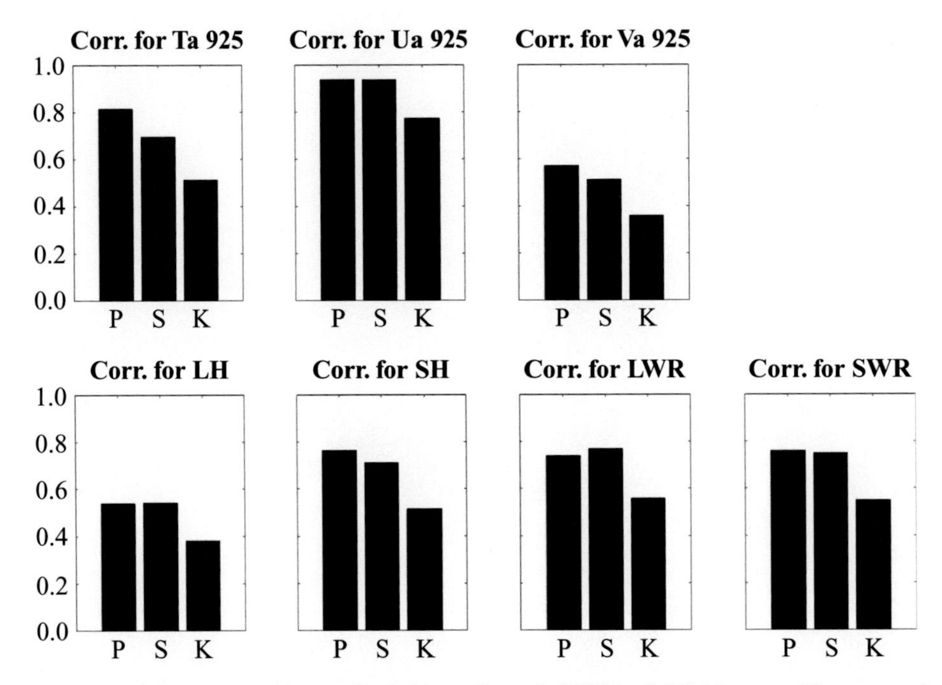

Figure 3.4 Bar plots for cross-correlation computed between probable predictors in NCEP and GCM data sets. The top row shows plots based on Group A and the bottom row shows plots for Group B predictors. P, S, and K represent product moment correlation, Spearman's rank correlation, and Kendall's tau, respectively (*Source*: Anandhi *et al.*, 2008).

To investigate the relationship between the probable predictors and predictands, cross-correlation bar plots between the probable predictor variables in NCEP data and each of the predictands (Tmax and Tmin) are presented in Figure 3.5. It can be observed that Ta 925 and SWR flux have high correlation with both the predictands, while Ua 925, Va 925, LH, and LWR fluxes have lower correlation. Ta 925, Ua 925, SH, and LWR have a positive correlation with both Tmax and Tmin. LH, Va 925, and SWR have a negative correlation with both the predictands. Between the two predictands, the Tmax is more correlated with the predictors.

Step 4: Choose predictors based on measures of dependence
The predictors can be ranked based on the relative magnitude of cross-correlations estimated by each measure of dependence. Results show similar rank for any chosen predictor by all the three dependence measures considered, indicating that the results are reliable. The results of this analysis indicate that Ta 925 is a better predictor in Group A, while SWR and SH are better predictors in Group B, while all these three (Ta 925, SWR, and SH) are better predictors in Group C, since Group C is a combination of predictors in Groups A and B. These results give an overall picture of relationships between predictors and predictands over the study area. The predictors for this case study can thus be chosen to be Ta925, SWR, and SH, on the basis of their cross-correlations with the predictands.

The next section explains the preprocessing of predictor variables, prior to their use as inputs to downscaling models.

3.4.3 Data preprocessing

Some preprocessing of the large-scale atmospheric data is necessary before its use for training or projection in the downscaling model. In general, preprocessing may involve some or all of the following steps: interpolation, bias removal, and dimensionality reduction. These steps are described in detail in the sections below.

3.4.3.1 INTERPOLATION

Since the locations of grid points of reanalysis data used as predictors and GCM output variable grid points usually do not match, an interpolation procedure is necessary before using GCM outputs for prediction. For example, Figure 3.6 shows the grids of NCEP reanalysis data variables and the MIROC3.2 GCM from the Center for Climate System Research (CCSR, Japan) output variables. The data from one grid hence have to be interpolated over the other grid before their use in downscaling. Methods for transferring data between different geo-spatial locations (or supports) belong to a class of problems called change of support problems (COSP; Gelfand *et al.*, 2000). In the context of downscaling, many methods have been studied and used for COSP. Undesirable effects of any proposed transformation are the change of scale and change of shape of the support, which need to be considered in a suitable choice of method.

Linear, spherical, planar projection, and other interpolation methods have been used in downscaling studies. Spherical methods for spatial interpolation, in particular, have been widely used. For interpolation on grids over the Earth's surface, interpolating

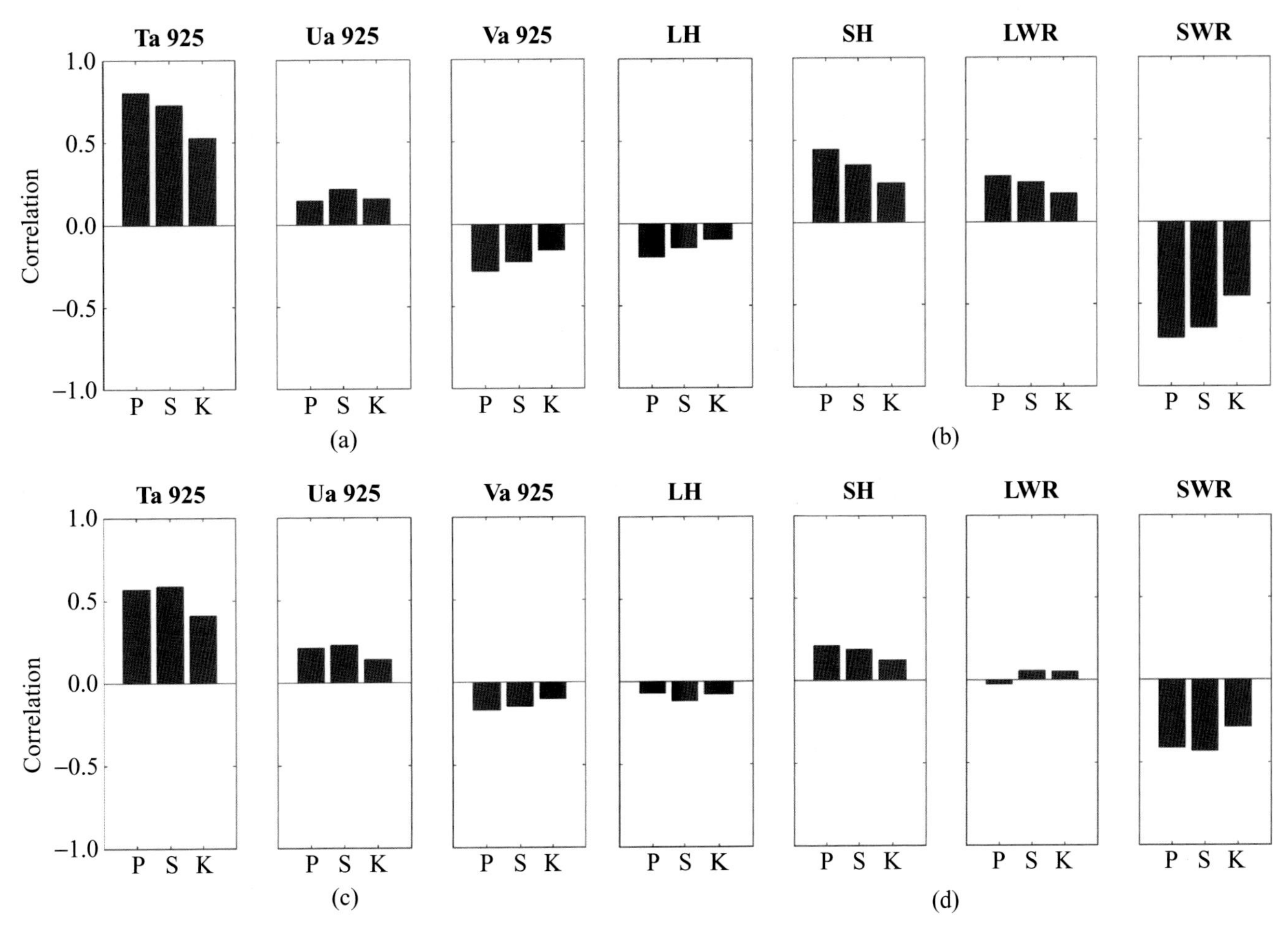

Figure 3.5 Bar plots for cross-correlation computed between probable predictors in NCEP data and observed Tmax and Tmin. (a) and (b) denote plots based on Group A and Group B predictors, respectively, for the predictand Tmax, while (c) and (d) denote plots based on Group A and Group B predictors, respectively, for the predictand Tmin. P, S and K represent product moment correlation, Spearman's rank correlation, and Kendall's tau, respectively (*Source*: Anandhi *et al.*, 2008).

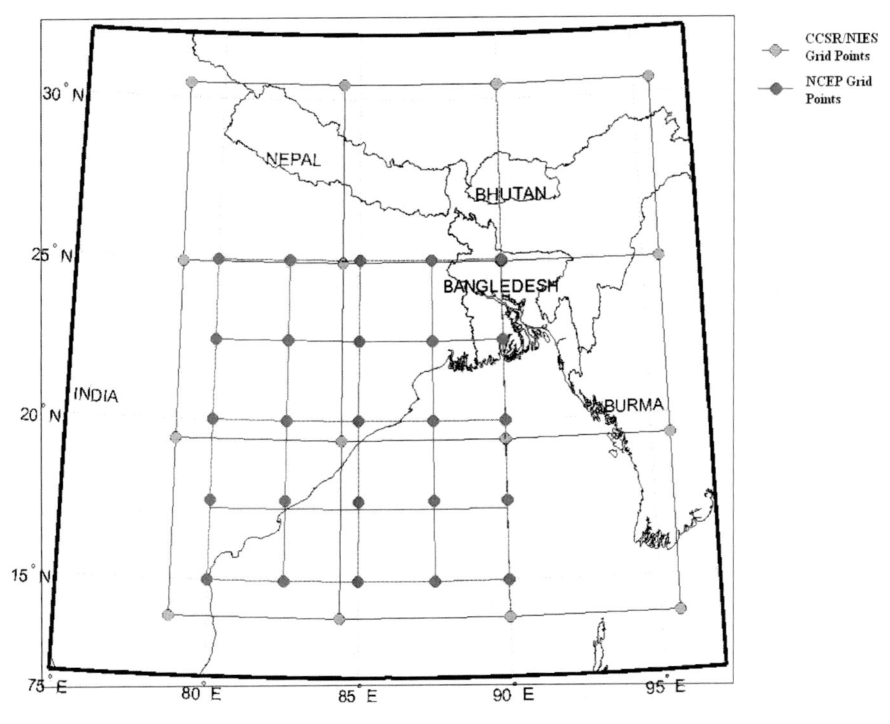

Figure 3.6 Mismatch between locations of reanalysis data and GCM grid points (*Source*: Ghosh and Mujumdar, 2007).

on the surface of a sphere, ellipsoid, or geoid is a more geometrically consistent and accurate approach. For large areas of the Earth, planar interpolation methods, i.e., interpolation within a cartographic projection, can produce large interpolation errors. In a comparison of spherical and planar interpolation, Willmott *et al.* (1985) found that planar interpolation methods can produce interpolation errors as large as 10 degrees Celsius in air temperature fields. A spatial interpolation toolkit developed by the National Center for Geographic Information and Analysis (NCGIA; University of California, Santa Barbara, USA), called Spherekit, is currently freely available for users. The user can select from several interpolation algorithms that have been adapted to the sphere including inverse distance weighting, thin plate splines, multiquadrics, triangulation, and kriging in this toolkit.

3.4.3.2 BIAS REMOVAL

Many processes are parameterized in GCMs due to incomplete knowledge about the underlying physics. Parameterization leads to differences between observed and GCM-simulated climate variables for the twentieth century. The difference between observed and GCM-simulated variables for the current climate is known as bias. In order to project future climatic scenarios correctly, the bias has to be removed. Standardization is used prior to statistical downscaling to reduce systematic biases in the mean and variance of GCM predictors relative to those of the observations or reanalysis (e.g., NCEP) data. In this step, subtraction of mean and division by standard deviation of a predictor variable for a predefined baseline period for both NCEP and GCM outputs is performed. The period 1960–90 is generally used as a baseline, and is the standard World Meteorological Organization baseline period. Data from the twentieth century experiment of the specific GCM used, from the baseline period (1960–90), are used to calculate the mean μ and standard deviation σ of each variable at each grid point for this period. Similarly, the mean and standard deviation of NCEP variables for the same period is computed. The bias-corrected, or standardized, value for the kth predictor variable at time t, $v_{stan,t}(k)$ is then computed as

$$v_{star,t}(k) = \frac{v_t(k) - \mu_{v,1960-1990}(k)}{\sigma_{v,1960-1990}(k)} \tag{3.1}$$

where $v_t(k)$ is the original value of the kth predictor variable at time t, $\mu_{v,1960-1990}(k)$ is the mean value of the kth predictor variable for the period 1960 to 1990 and $\sigma_{v,1960-1990}(k)$ is its standard deviation for this period.

3.4.3.3 DIMENSIONALITY REDUCTION

Training a downscaling model with a large number of predictors is not desirable in most cases due to the effects of overfitting, computer time and memory constraints, and existence of significant correlations between predictors. Principal component analysis (PCA) is generally used to reduce dimensionality while retaining the variability of the original data. PCA is a vector space transform of a multivariate set of data. The principal components (PCs) of a multivariate set of data are computed from the eigenvalues and eigenvectors of either the sample correlation or sample covariance matrix. If the variables of the multivariate data are measured in widely differing units (large variations in magnitude), it is usually best to use the sample correlation matrix in computing the PCs. Another alternative is to standardize the variables of the multivariate data prior to computing PCs. Standardizing the variables essentially makes them all equally important by creating new variables that each have a mean of zero and a variance of one. Proceeding in this way allows the PCs to be computed from the sample covariance matrix. The original data are projected along orthogonal unit vectors, which are eigenvectors of the covariance /correlation matrix of the original variables, ordered in decreasing magnitude of associated eigenvalues. The PCs are hence a set of ordered orthogonal vectors, such that the first PC is the direction along which the data have the most variance, the second PC is orthogonal to the first with the next highest variance, and so on. To summarize the data, the first k PCs are chosen and the data are projected onto these first k PCs. The first few PCs of each variable accounting for a sufficiently large percentage of the variance, or beyond which the variance accounted for rises very slowly with number of PCs, are retained. These are applied to the reanalysis (e.g., NCEP) and GCM data to get the projections in the principal directions.

A simple example illustrating the application of PCA for a sample data set is described below.

Table 3.2 *Sample data for PCA*

Data	X	Y	Z
Sample 1	2.0	1.0	3.0
Sample 2	4.0	2.0	3.0
Sample 3	4.0	1.0	0.0
Sample 4	2.0	3.0	3.0
Sample 5	5.0	1.0	9.0

Example 3.2 Principal component analysis

Sample data on which PCA is to be applied are shown in Table 3.2. The data have three variables (or three dimensions), measured using five samples. The samples may be data measured at different times, in which case they would correspond to observations for time $t = 1$ to 5. Principal components of the data are to be identified, such that the maximum variance lies along the first PC, the next highest along the second PC, and so on.

Table 3.3. *Sample statistics*

	X	Y	Z
Mean	3.4	1.6	3.6
Variance	1.8	0.8	10.8

Table 3.4 *Standardized data with mean zero and variance one*

X	*Y*	*Z*
– 1.043	– 0.670	– 0.182
0.447	0.447	– 0.182
0.447	– 0.670	– 1.095
– 1.043	1.565	– 0.182
1.192	– 0.670	1.643

Solution

Step 1: Standardize the data to zero mean and unit variance

Mean (X):

$$EX = \bar{X} = \frac{1}{n}\sum_{i=1}^{n} X_i \tag{3.2}$$

Variance (X):

$$E(X - \bar{X})^2 = \frac{1}{n-1}\sum_{i=1}^{n}(X_i - \bar{X})^2 \tag{3.3}$$

Standardized data:

$$X_s = \frac{X_i - \bar{X}}{std(X)} \tag{3.4}$$

The mean and variance of the sample data are shown in Table 3.3 and the standardized data are shown in Table 3.4.

Step 2: Compute the covariance matrix of standardized data

Note: If variables had not been standardized (with equal variances of 1), the correlation matrix would have to be used here.

$$\begin{aligned} Cov(X, Y) &= E[(X - EX)(Y - EY)] \\ &= \frac{1}{n-1}\sum_{i=1}^{n}(X_i - \bar{X})(Y_i - \bar{Y}) \end{aligned} \tag{3.5}$$

Hence, the covariance matrix $\Psi = \frac{1}{n-1}\boldsymbol{X}\boldsymbol{X}^T$ where $\boldsymbol{X}$ is the $n \times p$ (here 5×3) zero mean data matrix with p (= 3) variables and n (= 5) observations.

$$\boldsymbol{X} = \begin{bmatrix} -1.043 & -0.670 & -0.182 \\ 0.447 & 0.447 & -0.182 \\ 0.447 & -0.670 & -1.095 \\ -1.043 & 1.565 & -0.182 \\ 1.192 & -0.670 & 1.643 \end{bmatrix}$$

Here,

$$\Psi = \begin{bmatrix} 1.000 & -0.458 & 0.442 \\ -0.458 & 1.000 & -0.153 \\ 0.442 & -0.153 & 1.000 \end{bmatrix}$$

Step 3: Compute the eigenvalues and eigenvectors of the covariance matrix

Solving for $\Psi \boldsymbol{v} = \lambda \boldsymbol{v} \Rightarrow \det(\Psi - \lambda \boldsymbol{I}) = 0$. Expanding the determinant gives the characteristic polynomial, whose roots are the eigenvalues. Substituting the obtained eigenvalues into the matrix equation $(\Psi - \lambda_i \boldsymbol{I})\boldsymbol{v}_i = 0$ gives p simultaneous homogeneous equations, which give the set of eigenvectors $\boldsymbol{v}_i$.

In this case,

$$\lambda_1 = 0.435 \text{ corresponding eigenvector is } v_1 = \begin{bmatrix} 0.748 \\ 0.483 \\ -0.454 \end{bmatrix}$$

$$\lambda_2 = 0.847 \text{ corresponding eigenvector is } v_2 = \begin{bmatrix} -0.009 \\ 0.692 \\ 0.721 \end{bmatrix}$$

$$\lambda_3 = 1.718 \text{ corresponding eigenvector is } v_3 = \begin{bmatrix} 0.663 \\ -0.535 \\ 0.522 \end{bmatrix}$$

Verify that the trace of the covariance matrix ψ (which is the sum of its main diagonal elements) is equal to the sum of the eigenvalues. The largest eigenvalue is 3, accounting for 1.718/3 = 57.26% of the total variance. The second largest eigenvalue is 2, accounting for 0.847/3 = 28.23% of the total variance, and eigenvalue 1 accounts for 0.435/3 = 14.5% of the variance. Note that the eigenvectors are orthogonal and linearly independent of each other. Hence, the first PC will account for 57.26% of the variance of the original data and so on.

Step 4: Get the weights or loadings for the data

These are the eigenvectors, arranged in order from highest to lowest corresponding eigenvalues. Hence the components are

$$\boldsymbol{W} = \begin{bmatrix} 0.663 & -0.009 & 0.748 \\ -0.535 & 0.692 & 0.483 \\ 0.522 & 0.721 & -0.454 \end{bmatrix}$$

Note: The weights for the first PC are: 0.663, –0.535 and 0.522. These are the weights associated with the X, Y, and Z variables, respectively, to arrive at the projection in the direction of the first PC as below.

Step 5: Get the transformed data using projection

This is equivalent to a dot product obtained as

$$\boldsymbol{H} = \boldsymbol{X}\boldsymbol{W} \tag{3.6}$$

Here,

$$\boldsymbol{H} = \begin{bmatrix} -0.428 & -0.586 & -1.022 \\ -0.037 & 0.173 & 0.633 \\ 0.082 & -1.259 & 0.507 \\ -1.625 & 0.962 & 0.059 \\ 2.009 & 0.708 & -0.179 \end{bmatrix}$$

Each component of $\boldsymbol{H}$ is a linear combination of components of $\boldsymbol{X}$. Hence, for the first PC $H1 = 0.663{*}X - 0.535{*}Y + 0.522{*}Z$ and so on.

$$H1 = (0.663)\begin{bmatrix} 2.0 \\ 4.0 \\ 4.0 \\ 2.0 \\ 5.0 \end{bmatrix} + (-0.535)\begin{bmatrix} 1.0 \\ 2.0 \\ 1.0 \\ 3.0 \\ 1.0 \end{bmatrix} + (0.522)\begin{bmatrix} 3.0 \\ 3.0 \\ 0.0 \\ 3.0 \\ 9.0 \end{bmatrix}$$

(first PC)

$$H2 = (-0.009)\begin{bmatrix} 2.0 \\ 4.0 \\ 4.0 \\ 2.0 \\ 5.0 \end{bmatrix} + (0.692)\begin{bmatrix} 1.0 \\ 2.0 \\ 1.0 \\ 3.0 \\ 1.0 \end{bmatrix} + (0.721)\begin{bmatrix} 3.0 \\ 3.0 \\ 0.0 \\ 3.0 \\ 9.0 \end{bmatrix}$$

(second PC)

$$H3 = (0.748)\begin{bmatrix} 2.0 \\ 4.0 \\ 4.0 \\ 2.0 \\ 5.0 \end{bmatrix} + (0.483)\begin{bmatrix} 1.0 \\ 2.0 \\ 1.0 \\ 3.0 \\ 1.0 \end{bmatrix} + (-0.454)\begin{bmatrix} 3.0 \\ 3.0 \\ 0.0 \\ 3.0 \\ 9.0 \end{bmatrix}$$

(third PC)

Step 6: Determine number of PCs (columns of *H*) to retain
This is determined based on the cumulative percentage of variance accounted for by the PCs. Note: Verify that the first column accounts for 57.36% of the variance and so on.

3.4.4 Weather typing methods

The weather typing approach uses weather types to define links between large-scale circulation and surface weather. GCM variables are classified into a finite number of circulation patterns (CP), which in turn are linked to surface weather. Weather typing methods have a sound physical basis and can be used for multisite downscaling. However, their drawbacks include the fact that GCMs may not simulate the correct frequencies of weather types, and the method is unable to capture the effects of other relevant physical processes. The method also assumes, questionably, that observed relationships between the circulation types and the local climate will be unchanged in the future.

There are two steps in the weather typing approach to downscaling:

1. **Choice of appropriate circulation classification scheme:** The CP classification techniques include subjective and objective methods. In subjective methods such as the Lamb weather types (Jones *et al.*, 1993), the knowledge and experience of meteorologists can be used. However, reproduction of results is not possible and classification is applicable only to specific geographical methods with subjective methods. Objective methods are essentially classification algorithms operating on the selected data set. The objective-classification approach includes methods such as k-means clustering, fuzzy classification, principal component clustering, screening discriminant analysis, and neural network methods. A detailed comparison of the Lamb subjective and objective classification schemes can be found in Jones *et al.* (1993).
2. **Modeling of the circulation–surface climate relationship:** Relationships between the circulation types and surface variables have been modeled using various approaches, including Markov chain models or hidden Markov models, regression analysis, canonical correlation analysis, or sampling from observed current analog data. Fuzzy rule-based approaches and neural networks have also been used.

Weather typing methods have been used for simulating precipitation for a long time. Hay *et al.* (1991) developed a method of precipitation simulation using six weather types. Statistics for the frequency and distribution of precipitation amounts corresponding to each weather type for each month were derived from a 30-year period. These were applied to a Markov model based sequence of weather types to generate precipitation values. Bardossy and Plate (1992) developed a space–time model for daily rainfall, with parameters depending on the atmospheric CP. The model used a power-transformed normal random variable (r.v.) for representing the daily precipitation. The model was applied in the Ruhr catchment, Germany, and reproduced well both the rainfall occurrence probabilities and the amounts. In the methodology used by Bardossy *et al.* (1995), a fuzzy rule-based technique was used for classification of CPs into different states. Each CP type is described by a set of rules and the classification is performed by selecting the CP type for which the degree of fulfillment is highest. Conway and Jones (1998) used three versions of a weather type method for generating daily rainfall series. Two methods used an objective scheme to classify daily circulation types over the British Isles using Lamb's subjective classification. The third method was based on user-defined categories of vorticity. Each method categorized rainfall events according to whether the previous day was wet or dry, and could successfully reproduce the monthly means, persistence, and interannual variability of daily rainfall time series.

The following example discusses an application of weather typing for downscaling daily rainfall.

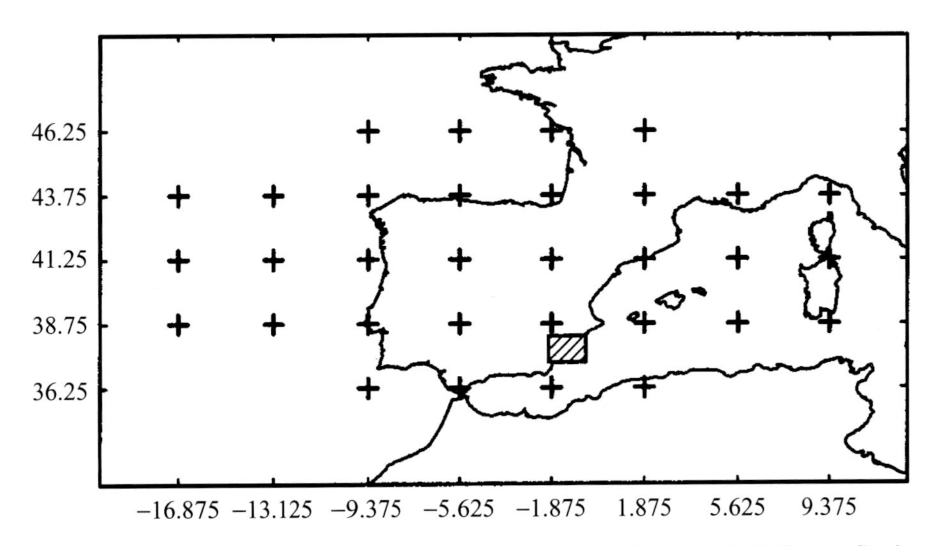

Figure 3.7 Study area for the the automated circulation typing scheme over the Guadalentin Basin (hatched) (*Source*: Goodess and Palutikof, 1998).

Example 3.3 Statistical downscaling: weather typing
An example for statistical downscaling using CPs is presented for the Guadalentin Basin in southeast Spain from a study by Goodess and Palutikof (1998). The large-scale patterns of a predictor variable, gridded sea level pressure, are to be downscaled to local values of a surface climate variable, which is daily rainfall at six stations. Figure 3.7 shows the grid used in downscaling over the basin (shown hatched) in Spain. The model used in the study is the UK Meteorological Office high-resolution GCM (UKTR). Daily output is available from the years 66–75 (10 years) of the 75-year long control and perturbed simulations. In the perturbed simulation, carbon dioxide forcing is increased by 1% per annum and doubles in the year 70. The Lamb weather types, initially developed for the British Isles, are used here for classifying CPs in the Mediterranean climate regime. SLP data for years 1956–89 are interpolated to a 32-point 2.5° latitude by 3.75° longitude grid over the area 36.25° N–46.25° N and 16.88° W–9.38° E, which is the grid spacing for the GCM.

Solution

Step 1: Circulation pattern classification
The large-scale patterns are defined using an automated version of the Lamb weather-type classification scheme (Lamb, 1972), originally developed for the British Isles. The automated scheme is easy to interpret and has a sound physical basis. This scheme categorizes surface flow by direction (with a resolution of 45°) and type (cyclonic/anticyclonic, light/hybrid flow) (Jones *et al.*, 1993). In theory, it can be applied anywhere in Northern Hemisphere mid-latitudes. Eight CP types are identified from sea level pressure data: (i) cyclonic (C), (ii) hybrid-cyclonic (HYC), (iii) unclassified/light flow-cyclonic (UC), (iv) anticyclonic/hybrid-anticyclonic (A/HYA), (v) unclassified/light flow-anticyclonic (UA), (vi) westerly/northwesterly/southwesterly/northerly directional types (W/NW/SW/N), (vii) easterly/northeasterly directional types (E/NE), and (viii) southerly/southeasterly directional types (S/SE).

The mean monthly cycles of the eight circulation-type groups calculated from observed SLP data for the period 1956–89 are shown in Figure 3.8. Mean frequencies simulated by the GCM are also shown. Over the year as a whole, the most frequently observed circulation types are the two unclassified or indeterminate/light-flow types (UC and UA). The UC-type has a very strong seasonal cycle, with a winter minimum and a summer maximum. The UA-type is most frequent in autumn, but does not have a strong seasonal cycle. In contrast, the A/HYA-group has a very strong seasonal cycle, with a pronounced winter maximum. The C- and HYC-types have a late spring/summer maximum and a less pronounced seasonal cycle. The W/NW/SW/N-group has a strong seasonal cycle, which peaks in late autumn/winter. The E/NE-group is one of the least frequent types and does not have a seasonal cycle. The least frequent category is the S/SE-group and is particularly infrequent from May to September. It is verified, on the basis of SLP composite maps, that each of the eight circulation-type groups has a characteristic underlying pressure pattern, which is physically distinct and produces the expected type and direction of surface flow over southeast Spain and the Guadalentin Basin.

Step 2: Project changes in CP frequencies
Seasonal and monthly frequencies of the eight circulation-type groups are calculated using SLP data from the perturbed run. The mean seasonal changes (perturbed-run minus control-run) are shown in Table 3.5. Significant changes are identified using the Mann–Whitney/Wilcoxon rank sum test. It is seen that the largest changes occur in summer. Of the high-rainfall types, there is a significant increase in the frequency of the C- and HYC-types, reflecting the lower mean SLP over the Iberian Peninsula.

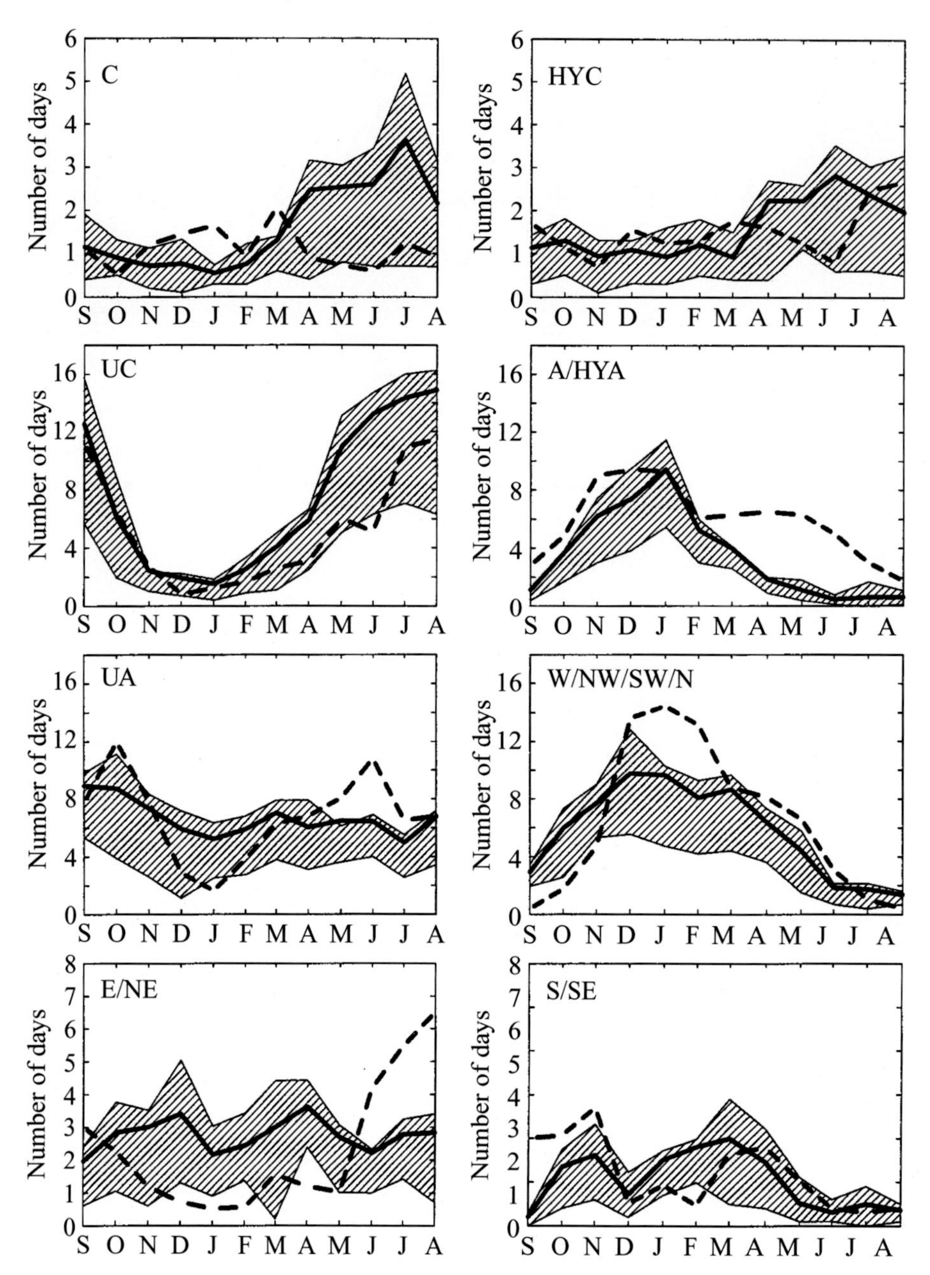

Figure 3.8 Observed and simulated monthly frequencies of the eight circulation types for years 1956–89. Solid line: observed mean frequency. Shaded area: maximum and minimum frequency range observed over any 10-year period. Dashed line: mean frequency calculated from UKTR control-run model output (*Source*: Goodess and Palutikof, 1998).

The frequency of both the E/NE- and S/SE-types increases in summer, but the changes are not significant. Of the low-rainfall types, the A/HYA- and UA-types show a significant decrease in frequency in summer and the W/NW/SW/N group shows a non-significant decrease.

Step 3: Weather generator for rainfall occurrence

A conditional weather generator (CWG) is used in which rainfall occurrence is conditional upon the circulation type of each day, and the transition from one circulation type to another is modeled as a Markov chain process. Rainfall occurrence is defined by two parameters calculated for each of the eight circulation types and for each season: the probability (P_{ct1-8}), expressed as a transition matrix, of the next day being circulation type 1–8, and the mean probability of rain ($PPREC_{ct}$) for each type. On each day, a random number is selected and used to determine the next day's circulation type from the transition matrix. A second random number is selected, and used to determine whether the

Table 3.5 *Mean seasonal changes (perturbed-control run) in the frequency (number of days) of the eight circulation-type groups*

Type	Winter	Spring	Summer	Autumn
C	−0.5	−0.7	+4.4[a]	−0.8
HYC	+0.4	−1.5	+3.1[b]	−0.4
UC	−0.8	+0.3	+3.6	+2.4
A/HYA	0.0	−0.8	−4.3[a]	−2.7
UA	+3.5	+2.3	−7.3[a]	−2.7
W/NW/SW/N	−4.5	+3.1	−1.0	+3.5[b]
E/NE	−0.2	−1.4[b]	+1.1	−0.4
S/SE	+2.1[b]	−1.1	+0.4	+1.2

Source: Goodess and Palutikof (1998)
[a] Changes significant at the 5% level
[b] Changes significant at the 10% level

day is wet or dry. The CWG is used to produce climate-change scenarios based on the assumption that changes in circulation-type frequency will be propagated through to changes in rainfall frequency and amount.

Step 4: Projection of rain-day changes

Three sets of a hundred 30-year-long simulations are performed for the six Guadalentin Basin stations. The sequence of circulation types in each 30-year simulation is dependent on the transition matrix and the random number generator. Thus, the hundred sequences making up each simulation set are different. Output from the following simulation sets is used:

- In the first simulation set, the CWG parameters (transition probabilities P_{ct1-8} and probability of precipitation $PPREC_{ct}$) are calculated from the observed data.
- In the second set of simulations (Cont.), the transition probabilities P_{ct1-8} are calculated from control-run output of the UKTR GCM and $PPREC_{ct}$ is calculated from the observations. These simulations also show the GCM's ability to simulate the frequency of circulation types.
- In the third set of simulations (Pert.), the transition probabilities P_{ct1-8} are calculated from perturbed-run GCM output and probability of precipitation $PPREC_{ct}$ is calculated from the observations.
- The differences between the Pert. and Cont. simulations provide the rain-day climate change scenarios (Table 3.6). These scenarios indicate that the average non-rainy days in the Guadalentin Basin could increase by 10–18% in summer in a future warmer world, and decrease by 5–9% in spring. A very small increase (2–4%) is indicated in winter, and little change in autumn (0–2%).

The next section presents the use of the weather generator approach in statistical downscaling.

Table 3.6 *Mean change (Pert. – Cont.) from the 100 simulations of the mean and SD of the number of rain days simulated by the CWG for six stations in the Guadalentin Basin*

Station ID	Winter	Spring	Summer	Autumn
Mean				
Station 1	**+0.3**	**−0.7**	**+0.3**	0.0
Station 2	**+0.3**	**−0.7**	**+0.7**	**+0.3**
Station 3	**+0.3**	**−1.0**	**+0.6**	+0.1
Station 4	**+0.2**	**−0.5**	**+0.6**	**+0.1**
Station 5	+0.3	−0.8	+0.6	+0.1
Station 6	+0.3	−0.7	−0.4	0.0
SD				
Station 1	**+0.1**	**−0.2**	0.0	0.0
Station 2	**+0.1**	**−0.2**	**+0.1**	**+0.2**
Station 3	0.0	**−0.2**	**+0.2**	0.0
Station 4	0.0	**−0.1**	**+0.2**	0.0
Station 5	0.0	**−0.1**	**+0.2**	**+0.1**
Station 6	**+0.2**	**−0.2**	**+0.1**	0.0

Modified from Goodess and Palutikof (1998)
Significant changes shown in bold

3.4.5 Weather generators

Weather generators (WGs) are statistical models of sequences of local weather variables. Effectually, WG models are random number generators whose output has statistical properties of the weather at a location. Weather generators have been used in a large number of studies on climate change impacts assessment. The standard practice in most WGs is to treat the occurrence and the amount of precipitation separately. Weather generators can be classified as parametric, where data are assumed to follow a specified distribution, and non-parametric or distribution-free. Most parametric WGs follow the Weather Generation Model (WGEN, Richardson and Wright, 1984), where precipitation occurrence is modeled with a Markov chain, and precipitation amounts follow a given distribution function. This is usually taken as an exponential, gamma, or mixed exponential distribution. Parametric WGs have been used for quite a long time in scientific literature (Stern and Coe, 1984; Wilks, 1992). A review of stochastic WG models can be found in Wilks and Wilby (1999), and more recently in Wilks (2010). Models based on kernel-based, multi-variate probability density estimators and k-nearest-neighbor (KNN) bootstrap methods have been used as non-parametric weather generating techniques (e.g., Lall and Sharma, 1996; Rajagopalan and Lall, 1999).

Daily WGs are most common since meteorological data are widely available at this time scale, and most hydrologic impact models are driven by daily data. Daily WGs have two approaches to modeling daily precipitation occurrence: daily precipitation occurrence modeled as a Markov chain conditioned on the previous day's precipitation, or a spell-length approach, where a distribution is fitted to wet and dry spell lengths. Mehrotra and Sharma (2009) present a comparison of three stochastic multi-site WG approaches for downscaling daily precipitation, consisting of a multi-site modified Markov model (MMM), a reordering method for reconstructing space-time variability, and a non-parametric KNN model. Their results indicated that all the approaches could adequately reproduce the observed spatio-temporal pattern of the multi-site daily rainfall, but had varying success for longer time-scale dependences. Mezghani and Hingray (2009) developed a combined downscaling and disaggregation semi-parametric WG for multi-site generation of hourly precipitation and temperature, applied to the Upper Rhone river basin in the Swiss Alps. Daily regional weather variables were first generated from generalized linear models based on daily atmospheric circulation indices from NCEP reanalysis. They were then disaggregated to the required spatial and temporal scales using a KNN approach. The model was found to successfully reproduce statistics for temperature and total and liquid precipitation at 3-hourly scales.

Non-homogeneous hidden Markov models (NHMM) have been used for relating synoptic or large-scale atmospheric patterns to local precipitation (Hughes and Guttorp, 1994; Charles *et al.*, 1999). Hughes and Guttorp (1994) developed a NHMM for relating synoptic or large-scale atmospheric patterns to local precipitation. The model postulated the existence of an unobserved weather state that is conditionally Markov, given atmospheric data. The model was fit to a network of stations in Washington State and was found to reproduce the observed first and second moments closely. Charles *et al.* (1999) extended the NHMM model of Hughes *et al.* (1999) by incorporating precipitation amounts. They used the NHMM to simulate daily precipitation occurrence in southwestern Australia. The conditional distribution of precipitation amount at a site was modeled by regressing transformed amounts on precipitation occurrence at neighboring sites within a given radius. The extended NHMM was found to reproduce frequency characteristics for wet day probabilities, dry and wet spell lengths, amount distributions, and intersite correlations accurately. Charles *et al.* (2004) analyzed the interannual and interdecadal performance of the extended NHMM in southwestern Australia. The model was found to have good overall performance, and reproduced interannual winter precipitation variability well. Mehrotra and Sharma (2006) used a non-parametric non-homogeneous hidden Markov model (NNHMM) to downscale rainfall occurrence at a site using a dynamic weather state indicative of the centroid and average wetness fraction of the rainfall occurrence field. Kilsby *et al.* (2007) used a Neyman–Scott rectangular pulses model as a daily WG in which rainfall is associated with clusters of rain cells, with storm origins arriving as a Poisson process.

Raje and Mujumdar (2009) developed a conditional random field (CRF) model for downscaling daily precipitation. In this method, the predictand sequence and atmospheric variables were represented as a CRF to downscale the predictand in a probabilistic framework. The conditional distribution of the daily precipitation sequence **y**, given the observed daily atmospheric variable sequence **x**, is modeled as a linear chain CRF (Lafferty *et al.*, 2001):

$$p(y/x) = \frac{1}{Z(x)} \exp\left\{\sum_{i=1}^{r}\sum_{k=1}^{K} \lambda_k f_k(y_i, y_{i-1}, x)\right\} \quad (3.7)$$

where $\{\lambda_k\}$ is a parameter vector, and $\{f_k(y, y', x)\}_{k=1}^{K}$ is a set of real-valued feature functions defined on pairs of consecutive precipitation values and the entire sequence of atmospheric data. Some feature functions used include intercept, transition, raw observation, threshold, and difference features to capture multiple characteristics of observed (input) data. Uncertainty in precipitation prediction is addressed through a modified Viterbi algorithm, which predicts the n most likely sequences. Figures 3.9 and 3.10 show results from the CRF-downscaling model applied for downscaling monsoon daily precipitation in the Mahanadi basin in Orissa. Figure 3.9 shows a comparison of kernel density or Parzen window estimates of current and future most likely daily precipitation probability density functions (PDFs) for two locations for two time slices. Comparison of present and future predicted PDFs shows a distinct change in shape, with decreasing frequency of low precipitation and increasing frequency of mid-to-higher precipitation amounts, as projected for the chosen GCM (MIROC3.2) and scenario (A1B) combination.

Figure 3.10 shows projections of wet and dry spell lengths for the years 2045–65 at location 1 from the CRF model. It can be seen that short dry spells of 1–2 days are projected to increase, while mid-length dry spells of 3–9 days are projected to decrease, as a result of climate change. Also, shorter wet spells are likely to decrease, but longer wet spells of above 9 days are likely to increase in future.

Mehrotra and Sharma (2009) presented a comparison of three stochastic multi-site WG approaches for downscaling daily precipitation, consisting of a multi-site modified Markov model, a reordering method for reconstructing space-time variability, and a non-parametric KNN model. Their results indicated that all the approaches could adequately reproduce the observed spatio-temporal pattern of the multi-site daily rainfall, but had varying success for longer time-scale dependences.

The following example explains in detail the steps involved in downscaling using a KNN weather typing approach.

Example 3.4 Statistical downscaling: weather generators

A simple WG using a KNN approach for downscaling daily mean temperature is presented. Climate change impacts on daily temperature in the Punjab region are explored. The GCM used for

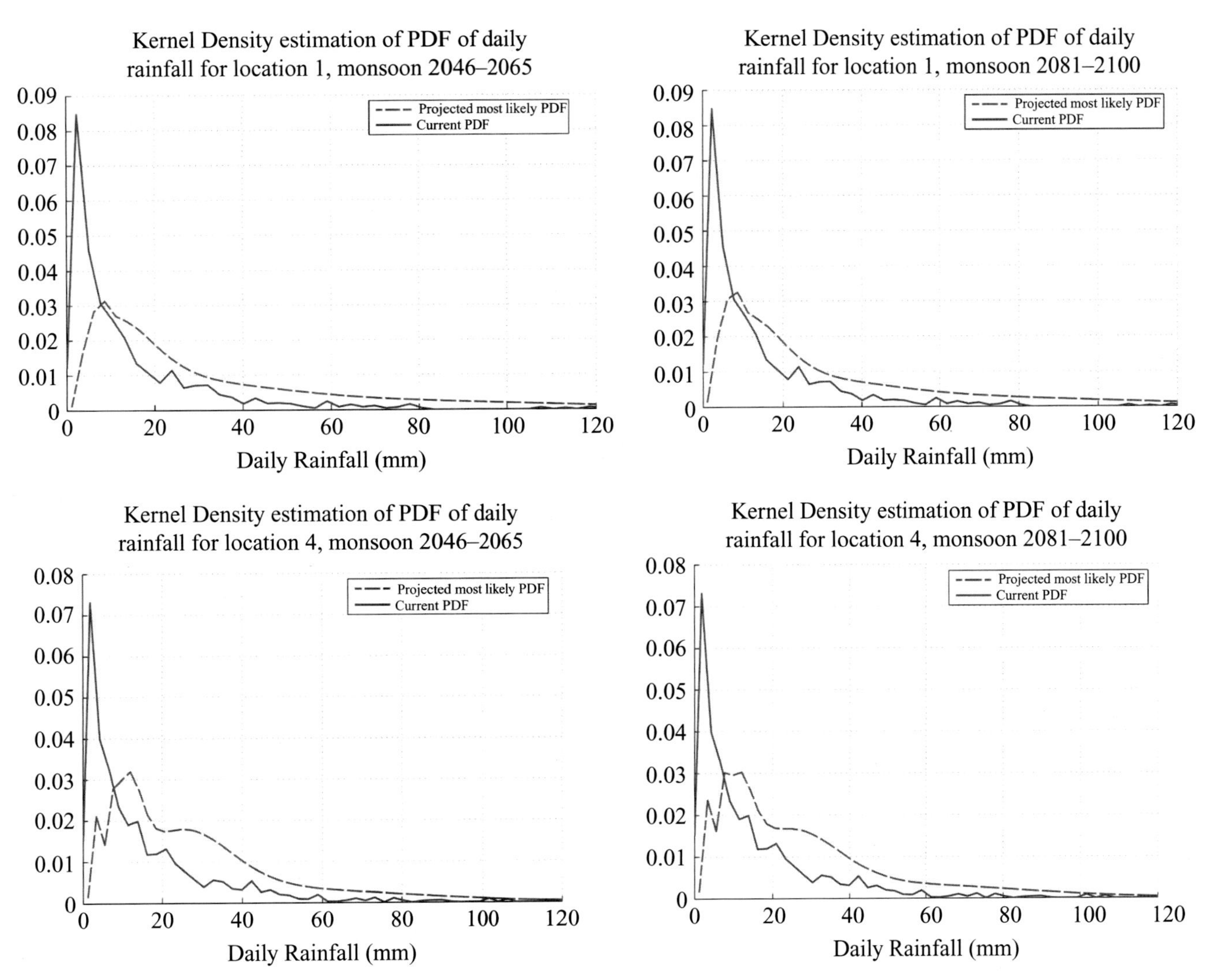

Figure 3.9 Kernel density estimation of probability density of monsoon daily precipitation amount: comparison between current (1951–2004) and future most likely PDFs (*Source*: Raje and Mujumdar, 2009).

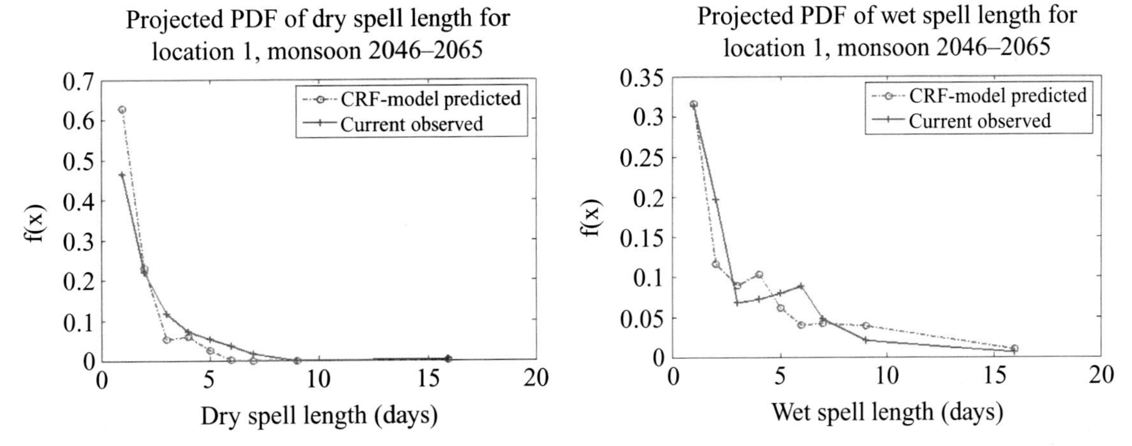

Figure 3.10 Projections for PDFs of monsoon dry and wet spell lengths at location 1 for years 2045–65 (*Source*: Raje and Mujumdar, 2009).

Table 3.7 *Station locations for downscaling temperature*

Station	Latitude (°N)	Longitude (°E)
1	30.5	74.5
2	30.5	75.5
3	30.5	76.5
4	31.5	74.5
5	31.5	75.5
6	31.5	76.5

projecting future climate variables is the Canadian global climate model (CGCM3) developed by the CCCma. The six locations where downscaling is to be performed are shown in Table 3.7. Three scenarios are to be considered for downscaling: A1B, A2, and B1.

Solution

Step 1: Determine predictors for downscaling

The climate variables used as predictors for downscaling daily temperature are determined as follows: mean sea level pressure, surface specific humidity, specific humidity at 850 hPa, surface temperature at 2 m, surface U-wind (eastward) and surface V-wind (northward). These are based on an initial screening of predictor variables through a correlation analysis with daily mean temperature at one of the downscaling locations. Predictor data over an area from 25° N to 34° N and 70° E to 80° E are obtained from the National Center for Environmental Prediction/National Center for Atmospheric Research (NCEP/NCAR) reanalysis data for years 1951–2004 (Kalnay *et al.*, 1996). Gridded daily mean temperature data (at 1 degree by 1 degree resolution, interpolated from station data) from 74.5° E to 76.5° E and from 30.5° N to 31.5° N for years 1951–2004 are obtained from the India Meteorological Department (IMD) and used as predictand for training the models. Daily atmospheric variables from the CGCM3 for the years 2046–65 for three scenarios, A1B, A2, and B1, are used for prediction.

Step 2: Data preprocessing

A linear interpolation is performed to obtain CGCM3 variable values at the respective NCEP grid points. Standardization is then performed to reduce systematic biases in the mean and variances of GCM predictors relative to the observations or NCEP/NCAR data, using the period 1961–90 as a baseline. Principal component analysis is then applied to the predictor variables to reduce and effectively summarize the information from all variables at the 25/30 grid points. The first five PCs accounting for more than 80% of the variance are taken as predictors to train the model. The loadings of the PCA are applied to the standardized CGCM3 data to get their projections in the principal directions.

Step 3: Apply the KNN downscaling model

The KNN algorithm is described for use in a stochastic WG by Lall and Sharma (1996), and Rajagopalan and Lall (1999). The KNN algorithm (Simonovic, 2012) searches for analogs of a feature vector based on similarity criteria in the observed time series. The steps involved in prediction are as follows, slightly modified from the methodology of Gangopadhyay *et al.* (2005):

1. A feature vector is compiled of climate predictor variables for a chosen day. The feature vector ($\vec{F}_f^t$) consists of values for all predictor variables for the feature day t, for which analogs in terms of K nearest neighbors are sought.

$$\vec{F}_f^t = \left[x_1^t \; x_2^t \ldots \; x_k^t\right] \tag{3.8}$$

   where x_i^t is the ith predictor variable on day t and k is the total number of predictors (equal to 5, the number of PCs retained).
2. For each time, the Euclidean distance between the feature vector and the PCs of training data from time $i = 1, 2, \ldots, T$ is computed. The distance d_i of feature day t from training day i is given as

$$d_i = \left[\sum_{j=1}^{5} \left(x_j^t - p_j^i\right)^2\right]^{1/2} \tag{3.9}$$

3. The distances d_i are sorted in ascending order and only the first K neighbors are retained. The choice of K is based on the square root of all possible candidates, i.e., $K = \sqrt{T}$ (Rajagopalan and Lall, 1999; Yates *et al.*, 2003).
4. A weight w_i ($0 < w_i < 1$) is assigned to each of the K neighbors using a bi-square weight function as

$$w_i = \frac{\left[1 - \left(\frac{d_i}{d_K}\right)^2\right]^2}{\sum_{i=1}^{K}\left[1 - \left(\frac{d_i}{d_K}\right)^2\right]^2} \tag{3.10}$$

   where d_K is the sorted distance of neighbor K.
5. A neighbor is selected from the K neighbors as an analog for feature day t. For this, a uniform random number $u \sim U[0,1]$ is first generated. If $u \geq w_i$, then the day corresponding to distance d_1 is selected. If $u \leq w_K$, then the day corresponding to distance d_K is selected. For $w_1 < u < w_K$, the day corresponding to that w_i is selected for which u is closest to w_i.
6. The neighbor day for each feature day obtained using the KNN algorithm is used to select the daily observed precipitation value for that location. Doing this for each future day constitutes the downscaled precipitation series for each of the locations.

Table 3.8 *Performance of KNN downscaling model in independent testing*

Station	Mean 1996–2005 (obs) (°C)	Mean 1996–2005 (computed NCEP) (mm)	Mean 1961–2000 (computed 20C3M) (°C)	SD (obs) (°C)	SD (NCEP) (°C)	Testing (R)
1	23.94	23.70	23.75	7.46	7.34	0.945
2	23.71	23.51	23.57	7.33	7.24	0.945
3	22.37	22.14	22.22	6.69	6.61	0.938
4	23.68	23.39	23.45	7.25	7.30	0.945
5	23.01	22.72	22.80	7.02	7.04	0.943
6	20.50	20.24	20.31	6.50	6.53	0.937

Table 3.9 *Future projections for daily mean temperature at six locations in Punjab*

Station	Current mean (°C)	Mean (computed 20C3M for 1961–2000) (°C)	Mean A1B (2046–65) (°C)	Mean A2 (2046–65) (°C)	Mean B1 (2046–65) (°C)	Mean A1B (2081–2100) (°C)	Mean A2 (2081–2100) (°C)	Mean B1 (2081–2100) (°C)
1	23.77	23.78	24.89	25.05	24.57	25.35	25.90	24.91
2	23.57	23.60	24.68	24.85	24.38	25.14	25.68	24.71
3	22.23	22.25	23.22	23.38	22.94	23.62	24.07	23.25
4	23.47	23.48	24.61	24.75	24.29	25.06	25.62	24.62
5	22.82	22.83	23.91	24.05	23.60	24.34	24.86	23.91
6	20.32	20.34	21.35	21.49	21.05	21.76	22.23	21.35

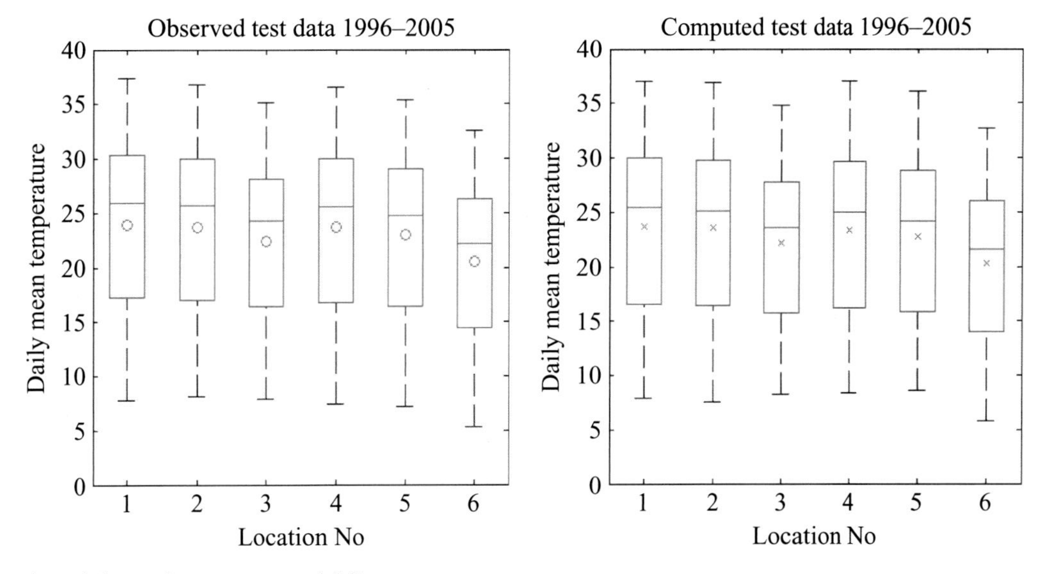

Figure 3.11 Boxplots of observed versus computed daily mean temperatures.

Step 4: Testing the model

The model is trained over years 1969–95 and tested over years 1996–2005. Table 3.8 shows the observed and computed statistics for mean daily temperature for years 1996–2005 and performance in terms of R-value obtained using the KNN model, showing that KNN is able to reproduce both the mean and the standard deviation of daily mean temperature reasonably well, with the mean being slightly underestimated by the model.

Figure 3.11 shows the boxplots of observed versus computed daily mean temperatures for the period 1996–2005 for all downscaling locations. It is seen from Figure 3.11 that the model is able to satisfactorily reproduce the range, median, and means of daily mean temperature in independent testing.

Step 5: Projection of future temperature

Table 3.9 shows projections for future mean daily temperature for two time slices of 2046–65 and 2081–2100 at the six locations,

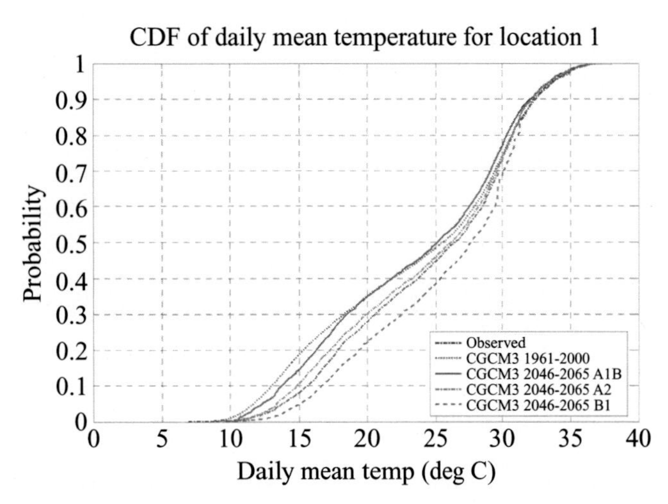

Figure 3.12 CDFs of daily mean temperature at location 1 using KNN method.

under three emissions scenarios, A1B, A2, and B1. Figure 3.12 shows the cumulative distribution functions (CDFs) of current versus projected daily mean temperature under three scenarios for years 2046–65 at location 1. It is seen from the table that the daily mean temperatures are projected to marginally increase compared to that simulated by the model for years 1961–2000, under all three scenarios for the Punjab region.

3.4.6 Transfer functions

Transfer functions are regression-based approaches that derive a direct quantitative relationship between the predictor(s) and the predictand. The methods differ according to the choice of mathematical transfer function, predictor variables, or statistical fitting procedure, and include linear and non-linear regression, artificial neural networks, canonical correlation, and principal component analysis or independent component analysis.

The earliest transfer function models were used for downscaling in the 1990s. Wigley *et al.* (1990) performed regression analyses for temperature, precipitation, mean sea level pressure (MSLP), and 700-mbar geopotential height as the predictors and local temperature and precipitation as the predictands. Using monthly mean data from Oregon, and separate analyses for each month, they reported that the percentage of explained variance ranged from 58 to 87% for temperature and 39 to 76% for precipitation, with large spatial differences in the amount of explained variance. Crane and Hewitson (1998) developed a neural-net-based transfer function to derive daily precipitation from larger-scale geopotential height and specific humidity. The technique was applied in the Susquehanna River basin (USA), where the downscaled precipitation increase was found to be larger than that predicted by the GENESIS GCM. Trigo and Palutikof (1999) examined linear and non-linear transfer functions based on ANNs to downscale precipitation over the Iberian Peninsula. They found that linear or slightly non-linear ANN models were more capable of reproducing the observed precipitation than more complex non-linear ANN models. Dibike and Coulibaly (2005) used a temporal neural network approach for downscaling daily precipitation and temperature in the Saguenay watershed in Canada. When compared to a regression-based model, it was found to better predict daily precipitation extremes and variability. Tatli *et al.* (2005) developed a model to predict precipitation over Turkey using a three-stage process: the potential predictors are preprocessed with PCA to get their significant PCs, then transformed into independent components (ICs), following which a canonical correlation analysis (CCA) is employed between ICs and monthly total precipitation series. Then, a recurrent neural network was trained using the outputs from both a naïve first-order model (using only historical data, assuming trend remains constant over time) and a causal (e.g., regression) linear model.

Ghosh and Mujumdar (2006) examined the future rainfall scenario in Orissa region using a linear regression model with fuzzy clustering. Using MSLP and geopotential height as predictors, they applied fuzzy clustering to the PCs identified using PCA to get fuzzy membership values in each of the clusters. A separate regression relation was developed for each cluster and was modified with a seasonality term. Tripathi *et al.* (2006) used a support vector machine (SVM) for downscaling of precipitation over India. The predictors identified were classified into two groups using cluster analysis representing wet and dry seasons. The SVM regression approach can capture non-linear relationships between variables by the use of kernel functions that implicitly map the data to a higher-dimensional space.

The following example illustrates the transfer function approach through a linear regression downscaling technique.

Example 3.5 Statistical downscaling: transfer functions

A transfer function approach for projection of future monthly precipitation based on fuzzy clustering and linear regression is explained through an example for the Orissa meteorological subdivision, India, with reanalysis data of MSLP as predictor and observed precipitation as predictand, from Ghosh and Mujumdar (2007). Gridded MSLP data used in the downscaling are obtained from NCEP reanalysis (Kalnay *et al.*, 1996). Monthly average MSLP outputs from 1948 to 2002 were obtained for a region spanning 15–25° N in latitude and 80–90° E in longitude that encapsulates the study region. Table 3.10 gives a list of GCMs with available scenarios used in the study.

Solution

The method derives a regression relationship between NCEP reanalysis data for CPs and observed precipitation that is used in

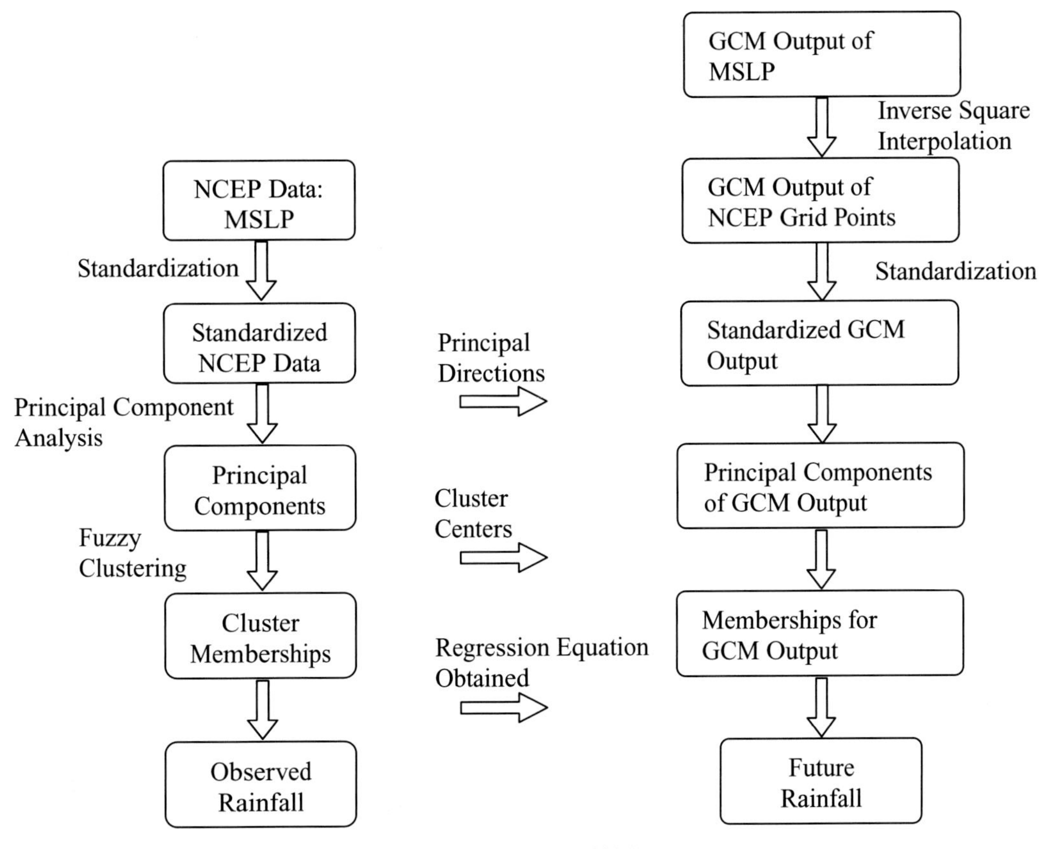

Figure 3.13 Fuzzy clustering based downscaling (*Source*: Ghosh and Mujumdar, 2007).

Table 3.10 *GCMs used and available scenarios*

GCM	Scenarios used in study
CCSR/NIES coupled GCM	A1, A2, B1, B2
Coupled global climate model (CGCM2)	IS92a, A2, B2
HadCM3	IS95a, (GHG + ozone + sulphate), A2
ECHAM4/OPYC3	IS92a, A2, B2
CSIRO-MK2	(IS92a + sulphate), IS92a, A1, A2, B1, B2

Source: Ghosh and Mujumdar (2007)

modeling future precipitation from GCM projections. The training involves three steps: PCA, fuzzy clustering, and linear regression with seasonality terms, as shown in Figure 3.13.

Step 1: Interpolation and standardization

The GCM grid points do not match with NCEP grid points, hence interpolation is performed with a linear inverse square procedure using spherical distances (Willmott *et al.*, 1985) to obtain the GCM output at NCEP grid points. Subsequently, standardization is used to reduce systematic biases in the mean and variances of GCM predictors relative to the observations or NCEP/NCAR data. This involves subtraction of mean and division by standard deviation of the predictor variable for a predefined baseline period of 1960–90, for both NCEP/NCAR and GCM outputs.

Step 2: Preprocessing – PCA

PCA and fuzzy clustering are first applied to the predictor variables before using them for training the downscaling model. PCA is performed on the predictor variables at 25 NCEP grid points to convert them into a set of uncorrelated variables. It is seen that 98.1% of the variability of the original data set is explained by the first ten PCs and therefore only the first ten PCs are used for modeling streamflow.

Step 3: Preprocessing – Fuzzy clustering

Fuzzy clustering is used to classify the PCs into classes or clusters, so that a membership value for each class is assigned to each data point. Fuzzy clustering assigns membership values of the classes to various data points. The important parameters required for the fuzzy clustering algorithm are the number of clusters (c) and the fuzzification parameter (m). The fuzzification parameter controls the degree of the fuzziness of the resulting classification, which is the degree of overlap between clusters. The number of clusters and the fuzzification parameter are determined from cluster validity indices such as fuzziness performance index (FPI) and normalized classification entropy (NCE) (Roubens, 1982). FPI estimates the degree of fuzziness generated by a specified number of classes and is given by

$$FPI = 1 - \frac{cF - 1}{c - 1} \quad (3.11)$$

where

$$F = \frac{1}{T}\sum_{i=1}^{c}\sum_{t=1}^{T}(\mu_{it})^2 \quad (3.12)$$

μ_{it} is the membership in cluster i of the PCs at time t. NCE estimates the degree of disorganization created by a given number of classes as

$$NCE = \frac{H}{\log c} \quad (3.13)$$

where

$$H = \frac{1}{T}\sum_{i=1}^{c}\sum_{t=1}^{T} -\mu_{it} \times \log(\mu_{it}) \quad (3.14)$$

The optimum number of classes/clusters is established on the basis of minimizing these two measures. It is found that an FPI value of 0.25 is achieved for $m = 2.0$ and $c = 2$, and these values are used further in the example.

Step 4: Linear regression

Linear regression is used to model the monthly precipitation with PCs, membership values of the PCs in each of the clusters, and the cross product of membership values and PCs as regressors. An appropriate seasonality term is used to capture the seasonality. The linear regression equation is given by

$$P_t = C + \sum_{i=1}^{I-1}\beta_i \times \mu_{it} + \sum_{k=1}^{K}\gamma_k \times pc_{kt} + \sum_{i=1}^{I-1}\sum_{k=1}^{K}\rho_{ik} \times \mu_{it} \times pc_{kt} \quad (3.15)$$

with

$$C = C^0 + C^1 \times \sin\left(\frac{2\pi p}{12}\right) + C^2 \times \cos\left(\frac{2\pi p}{12}\right) \quad (3.16)$$

$$\beta_i = \beta_i^0 + \beta_i^1 \times \sin\left(\frac{2\pi p}{12}\right) + \beta_i^2 \times \cos\left(\frac{2\pi p}{12}\right) \quad (3.17)$$

$$\gamma_k = \gamma_k^0 + \gamma_k^1 \times \sin\left(\frac{2\pi p}{12}\right) + \gamma_k^2 \times \cos\left(\frac{2\pi p}{12}\right) \quad (3.18)$$

$$\rho_{ik} = \rho_{ik}^0 + \rho_{ik}^1 \times \sin\left(\frac{2\pi p}{12}\right) + \rho_{ik}^2 \times \cos\left(\frac{2\pi p}{12}\right) \quad (3.19)$$

where P_t is the precipitation at time t, pc_{kt} is the kth PC of the CP at time t, and μ_{it} is the membership in cluster i of the PCs at time t; K and I are the number of PCs used and the number of clusters, respectively; β_i, γ_k, and ρ_{ik} are the coefficients of μ_{it}, pc_{kt}, and their product terms, respectively; C is the constant term used in the equation. The membership values μ_{it} in each cluster are assigned to the different points based on fuzzy c-means algorithm. Seasonality is incorporated by Equations 3.16–3.19, where p is the serial number of the month within a year ($p = 1, 2, \ldots, 12$).

Step 5: Validating the regression

The correlation coefficient (r) between the observed and predicted precipitation is considered as the goodness of fit of the regression model. To verify the model, a k-fold cross-validation is performed ($k = 10$), which gives an r value for training of 0.924 and for testing of 0.922. The goodness of fit of the model is also tested with the Nash and Sutcliffe (1970) coefficient. The Nash–Sutcliffe coefficient (E) lies between 0 and 1 and is given by

$$E = 1 - \frac{\sum_t (P_{ot} - \bar{P}_{pt})^2}{\sum_t (P_{ot} - \bar{P}_o)^2} \quad (3.20)$$

where P_{ot} and P_{pt} are the observed and predicted precipitation at time t, respectively, and P_o is the mean observed precipitation. It is obtained as 0.83 for the present model, which is satisfactory.

Step 6: Results for projection

The future projection of precipitation for wet (June, July, August, and September) and dry periods separately for CCSR/NIES GCM with B2 scenario are shown in Figure 3.14.

The downscaling model is seen to significantly underestimate the monsoon interannual variability. This is a limitation often seen in regression models. The results indicate a slight increase in wet-period precipitation and a severe decrease in dry-period precipitation for the particular scenario.

The following section discusses the use of disaggregation models as a tool in hydrologic impacts studies, especially for flood risk projection.

3.5 DISAGGREGATION MODELS

It is often necessary to have rainfall data at shorter time scales than monthly or mean daily precipitation for hydrologic studies. Many hydrologic processes such as infiltration and evaporation are affected not only by the total precipitation but also by the temporal structure, such as storm duration, inter-storm periods and rainfall intensity. River flows during flood events often vary significantly for sub-daily time scales. In order to assess flood risk using derived data for flood peaks and flood volumes, it is important to model flows at hourly or sub-hourly time scales. In practice, only a short record of observed sub-daily flows may be available, but there may be an extended record of observed or modeled daily flows. These daily flows can be disaggregated into sub-daily flows for flood-level simulation. Continuous hydrologic modeling in urban areas also requires short time step (10-minute time step or less) rainfall series, depending on the response time

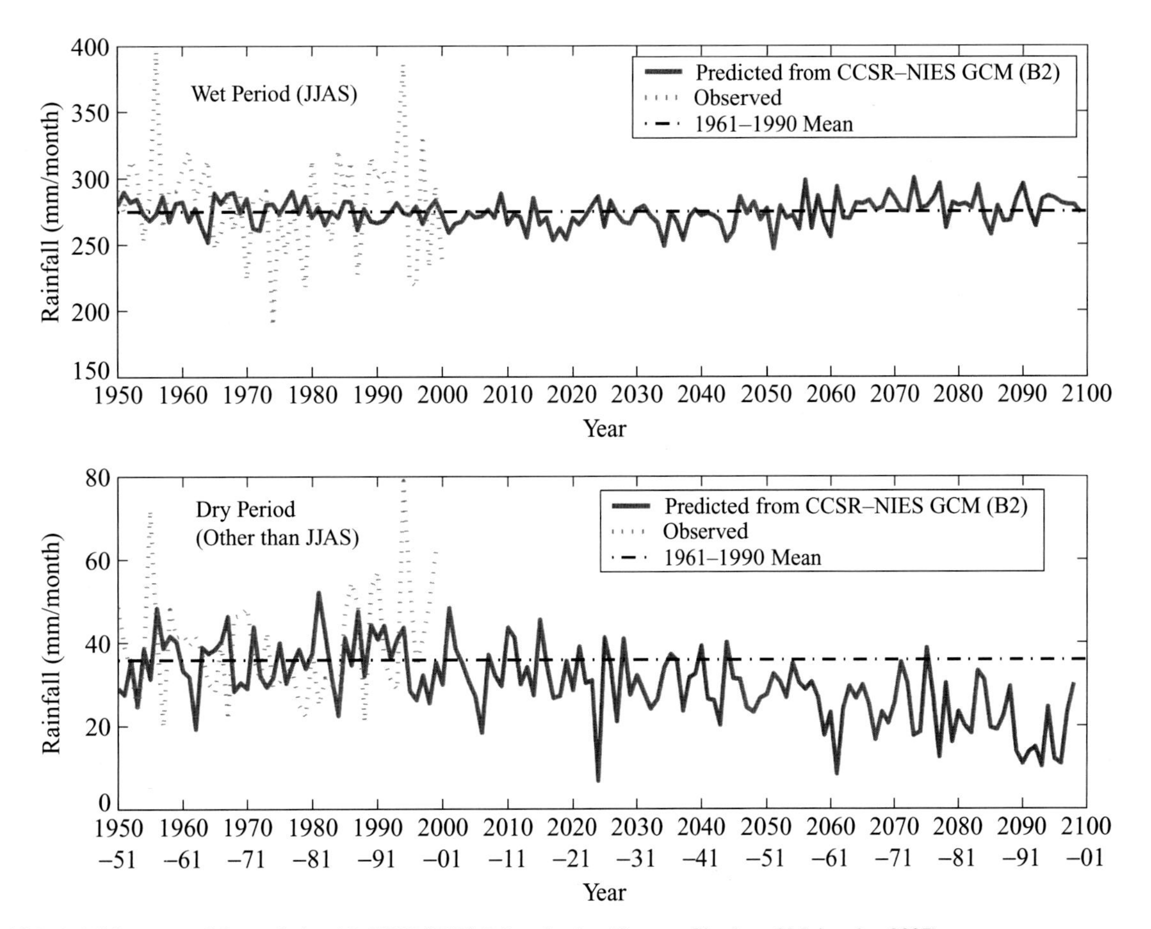

Figure 3.14 Rainfall for wet and dry periods with CCSR/NIES-B2 projection (*Source*: Ghosh and Mujumdar, 2007).

of the watershed. The available rainfall series, however, usually have coarser time steps. In projection of future flood risk, downscaling to sub-daily time scales is also not very prevalent, since GCM output at these scales is not considered reliable. Hence, in order to provide the desired smaller resolution series from coarse resolution time series, an appropriate disaggregation model has to be applied. The basic goal of any disaggregation model is to preserve the statistical properties of the hydrologic time series at more than one level.

Only temporal disaggregation in the context of hydrologic modeling will be dealt with in this chapter. Spatial disaggregation techniques, such as those used to incorporate orographic effects in the disaggregation of rainfall fields from atmospheric to hydrologic models, or for water balance computations in incorporating the spatial variability of rainfall into GCMs, are not discussed here. There are two categories of disaggregation models: deterministic and stochastic, which are discussed in the following sections.

3.5.1 Deterministic disaggregation techniques

The deterministic models are briefly described here as a prelude to stochastic disaggregation. Deterministic models include the linear model (Ormsbee, 1989) and the constant disaggregation model (constant model or CST). In the CST model, the rainfall intensity is assumed to be constant over the whole time period. This model is the simplest possible disaggregation model, where all time steps produced within disaggregated time are wet. The deterministic model is based on a linear interpolation of the disaggregated time step rainfall intensities within the aggregated larger time step. The slopes of the intensity–time linear relationships and the critical time for slope inversion (t^*) are obtained by assuming that the interior temporal rainfall pattern has geometric similarity with the exterior one, as shown in Figure 3.15. The exterior temporal rainfall pattern is defined by the three consecutive hourly rainfall amounts R_i, R_{i-1}, and R_{i+1} of the current, preceding, and following hours, respectively. The intensity time evolution within the rainy hour to disaggregate is then obtained for each of the six typical external patterns (ascending, descending, left or right peak, left or right valley).

The time evolution $g(t)$ of intensity within the rainy hour to disaggregate is (Hingray and Haha, 2005)

$$g(t) = \frac{R_{i-1}}{R_i^*} - \frac{(R_{i-1} - R_i)t}{R_i^* t^*} \quad \text{for} \quad 0 < t < t^* \tag{3.21}$$

$$g(t) = \frac{R_i}{R_i^*} - \frac{(R_i - R_{i+1})(t - t^*)}{R_i^*(T - t^*)} \quad \text{for} \quad t^* < t < T \tag{3.22}$$

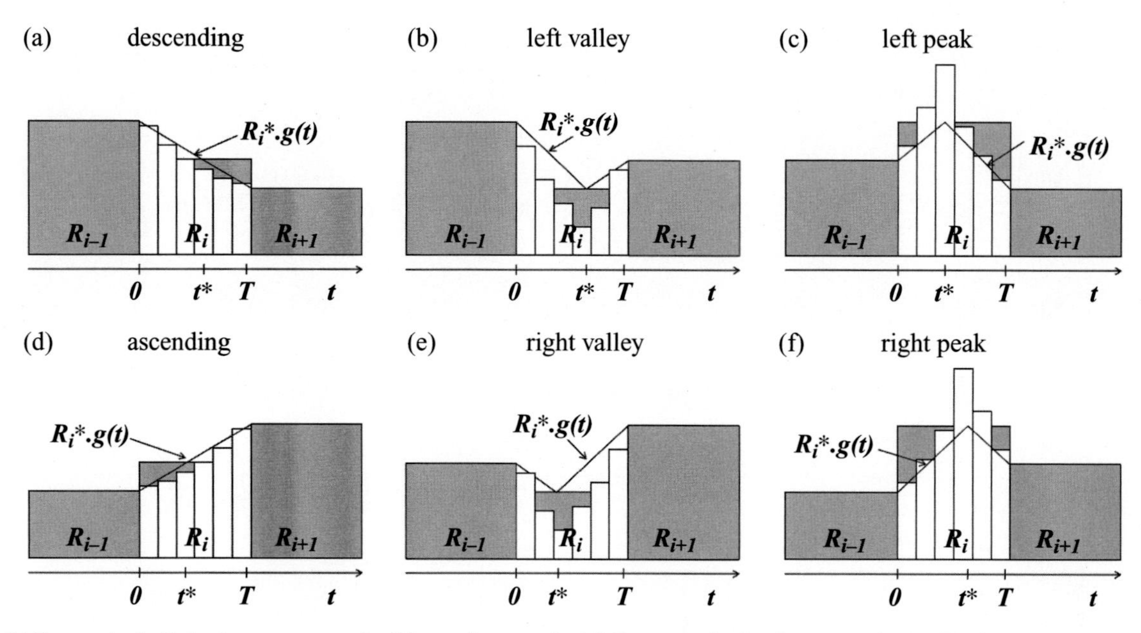

Figure 3.15 Geometric similarity between external and internal temporal rainfall patterns for the six pattern classes. The internal temporal pattern determined by the time evolution $g(t)$ of rainfall intensity is proportional to the piecewise linear function R_i^*. $g(t)$ plotted on the schemes (*Source*: Hingray and Haha, 2005).

where T is the duration of the time period to disaggregate and R_i^* is a normalization factor:

$$R_i^* = \frac{T}{2}(R_i + R_{i+1}) - \frac{t^*}{2}(R_{i+1} - R_{i-1}) \tag{3.23}$$

t^* is the critical time for intensity slope inversion. If the external rainfall temporal pattern is either of ascending or descending type, then

$$t^* = T\frac{R_i - R_{i-1}}{R_{i+1} - R_{i-1}} \tag{3.24}$$

If the external rainfall temporal pattern is either of valley or peak type, then

$$t^* = T\frac{R_i - R_{i-1}}{2R_i - R_{i+1} - R_{i-1}} \tag{3.25}$$

The rainfall amount $R_{i,k}$ assigned to the kth time step is

$$R_{i,k} = G(t_{k+1}) - G(t_k) \tag{3.26}$$

where $G(t)$ is the cumulative rainfall amount from the beginning of the current rainy hour up to time t:

$$G(t) = \int_{\theta=0}^{t} g(\theta)d\theta \tag{3.27}$$

The Ormsbee deterministic disaggregation model also does not produce dry time step rainfall amounts, and does not involve any parameters.

3.5.2 Stochastic disaggregation techniques

Among the earliest disaggregation models, Valencia and Schaake (1973) introduced a flow disaggregation model that became popular in stochastic hydrology. The model preserves the annual and seasonal statistics and has the additive property that the aggregation of the generated seasonal flows yields the generated annual flows. If **X** and **Y** are normalized variables with mean zero, then the basic form of the disaggregation model is (Valencia and Schaake, 1973)

$$\mathbf{Y} = \mathbf{AX} + \mathbf{B} \in \tag{3.28}$$

where **X** is an n-vector of annual values at n sites; **Y** is an $n\omega$ co-vector of seasonal values, in which ω is the number of seasons in the year; **A** is an $n\omega \times n$ parameter matrix; **B** is an $n\omega \times n\omega$ matrix; and $\in$ is an $n\omega$ co-vector of independent standard normal variables. The additivity condition, by which the sum of the seasonal values must add up to the annual values, implies that the columnwise sum of **A** adds to one, and the columnwise sum of **B** adds to zero. The parameters **A** and **B** may be estimated by (Valencia and Schaake, 1973)

$$\mathbf{A} = \mathbf{S_{YX}S_{XX}}^{-1} \tag{3.29}$$

$$\mathbf{BB}^{\mathrm{T}} = \mathbf{S_{YY}} - \mathbf{AS_{XY}} \tag{3.30}$$

where $\mathbf{S_{UV}}$ represents the covariance matrix of random variables **U** and **V**. The Valencia–Schaake model preserves the correlations among the various seasons in the year, but does not preserve the correlation between the last season and the first season of consecutive years. Mejia and Rousselle (1976) attempted to avoid this shortcoming by including an extra term in the original disaggregation model, but their model had some inconsistencies (Lane, 1982). The Valencia–Schaake and Mejia–Rousselle models both also suffered from the limitation that they have a large number of parameters.

Numerous studies have utilized stochastic models for disaggregation of rainfall. An important characteristic for a stochastic rainfall disaggregation model is the preservation of statistical properties of the observed rainfall records, most importantly, rainfall distribution and extreme event characterization (Menabde *et al.*, 1999). The most widely used rainfall simulation models are based on point process theory, such as the Bartlett–Lewis or Neyman–Scott rectangular pulse models (Rodriguez-Iturbe *et al.*, 1987), or scale invariance theory, such as cascade-based models (Olsson, 1998). The multiplicative random cascade model distributes rainfall on successive regular subdivisions with b as the branching number. The ith interval after n levels of subdivision is denoted Δ_n^i (there are $i = 1, \ldots, b_n$ intervals at level n). The dimensionless spatial scale is defined as $\lambda_n = b^{-n}$, i.e., $\lambda_0 = 1$ at the 0th level of subdivision. The distribution of mass occurs via a multiplicative process through all levels n of the cascade, so that the mass in sub-cube Δ_n^i is

$$\mu_n\left(\Delta_n^i\right) = r_o\lambda_n \prod_{j=1}^{n} W_j(i) \quad \text{for} \quad i = 1, 2, \ldots, b^n; \quad n > 0 \tag{3.31}$$

where r_o is an initial rainfall depth at $n = 0$ and W is the cascade generator. Properties of the cascade generator W can be estimated from the moment scaling behavior across scales. In a random cascade, the ensemble moments are shown to be a log–log linear function of the scale of resolution λ_n. The slope of this scaling relationship is known as the Mandelbrot–Kahane–Peyriere (MKP) function (Mandelbrot, 1974). In data analysis, the scaling of the sample moments is used to estimate the slope of the sample moment scaling relationship, and assuming a distribution for the cascade generator, W, the parameters of the cascade model can then be estimated.

Temporal models that represent the rainfall process include the Poisson white noise model, the Poisson rectangular pulses model, the Neyman–Scott clustering model, etc. A modified Bartlett–Lewis rectangular pulses model was used by Bo *et al.* (1994) for disaggregated daily rainfall by estimating six model parameters (from 24- and 48-hour accumulated rainfall data). Salvucci and Song (2000) used a Poisson storm arrival, gamma distributed depth model (PG model) to disaggregate monthly rainfall into daily precipitation events by conditioning the PG model parameters on the total monthly precipitation. Margulis and Entekhabi (2001) developed two simple statistical disaggregation schemes for use with the monthly precipitation estimates provided by satellites. Kumar *et al.* (2000) presented a novel multi-site disaggregation method for disaggregating monthly to daily streamflow at multiple sites. The method used a non-parametric estimation of the conditional probability distribution of disaggregated flows (conditioned on multi-site monthly flows) using k nearest neighbors of the monthly spatial flow pattern. The problem was formulated as a constrained optimization of conditional expectation of daily flows, subject to flow continuity across days and sites, proper summability in space and time, and other constraints to regularize the obtained solution.

The following example provides an illustration of the steps involved in disaggregation of monthly to daily rainfall used by Margulis and Entekhabi (2001).

Example 3.6 Stochastic disaggregation

The use of the Poisson rectangular pulses model (RPM, Rodriguez-Iturbe *et al.*, 1984) is demonstrated through this example for disaggregation of monthly rainfall from Margulis and Entekhabi (2001). The disaggregation technique was tested at the location of Meridian, Missouri, USA, for the month of March. As the temporal and spatial scales of measured rainfall are different from those required for hydrologic models, this model was developed for use with monthly precipitation estimates provided by satellites, for subsequent use with a surface hydrologic model. The scheme is based on a simple statistical model of storm arrival rates and storm structure characteristics. The model is used to generate multiple monthly precipitation time series whose statistical mean is consistent with the aggregated estimate provided by satellites. The ensemble of precipitation realizations can then be used to force hydrologic models to yield statistics of other relevant hydrologic variables, such as soil moisture.

Solution

Step 1: Obtain the parameters of the rainfall process

The RPM model represents the rainfall process with rainfall events described as independent rectangular pulses of duration t_r, with constant intensity i_r, and storm arrivals described by a Poisson process. The Poisson process that describes the occurrence of rainfall events can be characterized by the independent arrival rate $1/E[t_b]$, where $E[t_b]$ is the mean inter-arrival time between storms. For the RPM it is assumed that t_r and i_r follow independent exponential distributions with mean values of $E[t_r]$ and $E[i_r]$. Using the probability distributions of the RPM parameters, theoretical equations for the moments of the integrated rainfall process over disjoint aggregated intervals of length T have been developed (Rodriguez-Iturbe *et al.*, 1984). The theoretical equations for the mean, variance, lag-1 autocorrelation of the integrated precipitation (Y_t), and probability of no rain at 1 hour in terms of the RPM parameters are

$$E[Y_t] = T\frac{E[i_{tr}]E[t_r]}{E[t_b]} \quad (3.32)$$

$$Var[Y_t] = \frac{4E^3[t_r]E^2[i_r]}{E[t_b]}\left\{\frac{T}{E[t_r]} - 1 + \exp\left[-\frac{T}{E[t_r]}\right]\right\} \quad (3.33)$$

$$\rho_{Y_t}(1) = \frac{\left\{1 - \exp\left(-\frac{T}{E[t_r]}\right)\right\}^2}{2\left\{\frac{T}{E[t_r]} - 1 + \exp\left(-\frac{T}{E[t_r]}\right)\right\}} \quad (3.34)$$

$$P[Y_t = 0] \approx \frac{E[t_b] - E[t_r]}{E[t_b]} \quad (3.35)$$

These statistics are functions of the storm structure parameters $E[t_b]$, $E[t_r]$, $E[i_r]$, and the level of aggregation, T. In the present example the parameters determined by Hawk and Eagleson (1992) for each month of the year at many stations across the USA based on historical data are used. In other regions, these parameters would need to be estimated from historical rain gauge data.

Step 2: Disaggregation of rainfall

Two techniques are used for the disaggregation of monthly precipitation, assuming only knowledge of the climatological RPM parameters for the given region and the measured monthly rainfall for a particular month. It is assumed here that below or above average monthly rainfall is due solely to the number of storms that occur during that month and that the expected intensity and duration of each storm remains the same as that given by the $E[i_r]$ and $E[t_r]$ storm structure parameters. The expected number of storms is given by

$$E[N] = \frac{T}{E[t_b]} \quad (3.36)$$

Thus $E[t_b]$ is effectively revised when $E[N]$ is revised, based on the measured rainfall for that month. The expected total monthly rainfall is

$$E[m] = E[N]E[t_r]E[i_r] \quad (3.37)$$

By substituting the measured monthly rainfall ($\hat{m}$) into Equation 3.36, the revised value of $E[t_b]$ from Equations 3.37 and 3.38 is

$$E[t_b]' = \frac{TE[t_r]E[i_{br}]}{\hat{m}} \quad (3.38)$$

The revised parameter $E[t_b]'$, along with $E[t_r]$ and $E[i_r]$, is used in the RPM to produce an ensemble of 100 precipitation time series through a Monte Carlo (MC) technique, which on average preserves the measured monthly rainfall. In another method, the MC samples only those realizations produced by the model whose total monthly accumulation is within a specified tolerance of the total monthly rainfall (till the ensemble size becomes 100).

Step 3: Testing the model

To test the two disaggregation techniques, the RPM is first used to generate a synthetic precipitation time series for Meridian, Missouri, USA, for the month of March, using the appropriate RPM parameters taken from Hawk and Eagleson (1992). For each case, the synthetic time series is treated as the actual precipitation, which is aggregated over the month to be used as the input ($\hat{m}$) to each disaggregation scheme to produce 100 realizations. To measure how well the temporal structure of the realizations matches the actual time series, the mean, variance, lag-1 autocorrelation, and probability of no rain at levels of aggregation (T) that are required for hydrologic applications (up to 24 hours) are compared between the actual time series, those predicted by the theoretical equations using the revised storm parameters, and those averaged over the 100 realizations produced using methods 1 and 2. The results are shown in Table 3.11.

It is seen from Table 3.11 that the theoretical results for the mean rainfall are preserved exactly at all levels of aggregation, which is a direct result of the way in which $E[t_b]$ is revised. It is seen from the results for Method 2 that by constraining outcomes to those within a tolerance of the actual monthly precipitation, not only is the mean preserved, but the predicted variance is also improved. However, the other statistics are not preserved at all levels of aggregation. This occurs in temporal modeling of rainfall due to the dependence of model parameters on the time scale of measurement (Rodriguez-Iturbe *et al.*, 1984). However, in general, the technique is seen to perform relatively well in introducing a reasonable temporal structure into the disaggregated time series.

The following section explains the use of the relatively new MHMs in climate impacts studies in hydrology.

3.6 MACROSCALE HYDROLOGIC MODELS

Macroscale hydrologic modeling is essentially an application of hydrologic models over a large spatial domain. This domain could range from a large river basin, through a continent, to the entire land surface of the globe. The primary reasons for modeling at large scales, much above those at which hydrologic models usually operate, are manifold. A primary application is correction of perceived weaknesses in the representation of hydrologic processes in regional and global atmospheric models. Such models permit an explicit treatment of variability within an atmospheric model grid cell and the routing of water along a river network within and between grid cells. They could also be used in simulation of river flows in large river basins for operational and planning purposes. The use of macroscale models also permits discharge and nutrient load results along the

Table 3.11 *Comparison of temporal structure statistics at different levels of aggregation for Meridian, MS, storm structure parameters. Method 1 and 2: Ensemble mean values are given with the standard deviation in parenthesis*

	T (hours)	Actual	Theoretical	Method 1	Method 2
$E[Y_t]$ (mm)	1	0.14	0.14	0.14(0.10)	0.14(0.00)
	3	0.41	0.41	0.41(0.30)	0.41(0.00)
	6	0.83	0.83	0.82(0.60)	0.83(0.01)
	12	1.65	1.65	1.64(1.20)	1.65(0.01)
	24	3.30	3.30	3.29(2.41)	3.30(0.02)
$Var[Y_t]$ (mm)	1	0.62	0.75	0.74(1.01)	0.59(0.25)
	3	3.93	5.95	5.80(8.38)	4.43(1.63)
	6	13.11	20.03	19.31(28.97)	13.92(5.18)
	12	26.73	59.59	56.27(96.17)	38.24(17.86)
	24	61.76	152.22	141.60(246.22)	90.06(41.39)
$r_{Y_t}(1)$	1	0.71	0.88	0.81(0.12)	0.84(0.07)
	3	0.40	0.68	0.55(0.22)	0.57(0.17)
	6	0.04	0.49	0.3(0.22)	0.35(0.18)
	12	0.08	0.28	0.18(0.19)	0.18(0.16)
	24	−0.01	0.13	0.07(0.18)	0.05(0.18)
$P[Y_t = 0]$	1	0.93	0.95	0.94(0.03)	0.94(0.02)
	3	0.90	–	0.93(0.03)	0.92(0.02)
	6	0.85	–	0.90(0.04)	0.89(0.03)
	12	0.78	–	0.85(0.06)	0.83(0.05)
	24	0.63	–	0.75(0.09)	0.73(0.07)

Source: Margulis and Entekhabi (2001)

coastal boundary to be studied to examine freshwater and nutrient influxes to seas. These models can be also be used for flood forecasting and flood risk assessment under future climate scenarios, since they provide output at the sub-daily and daily scales. The key characteristics of a macromodel are (Vörösmarty *et al.*, 1989):

- The model should be transferable from one geographical location to another, and model parameters should be physically relevant.
- The model should be applied to hydrologic response units (sub-basins) or on a regular grid.
- Runoff routing must be from the point of generation or the grid cell/sub-basin through the spatial domain along the river network.

MHMs simulate water fluxes in two or three spatial dimensions (Vörösmarty *et al.*, 1993) over a large geographical domain. They simulate water exchanges across the land–atmosphere boundary through precipitation, evapotranspiration, and infiltration in a similar manner to soil–vegetation–atmosphere transfer schemes (SVATs). Then, the generated runoff is routed laterally using the simulated topology and stream networks to the basin outlet. The study region has to be divided into suitably sized catchments for which parameters can be determined from available spatial data with little or no calibration to allow application on global scale (Arnell, 1999a). In order to apply the model to study climate or land use change, it has to be based on physical principles. However, the limiting factor is availability of input data, which constrains the parameters and the form of the model. Therefore, most macroscale models are used as conceptual water balance accounting models. It is found that small (or micro) scale processes employed in catchment models are not necessarily dominant at the larger scales, as the process equations are often scale dependent (Dunn, 1998). Two approaches have been used in the development of a macromodel (Xu, 1999a). The "top-down" approach simulates each of the fundamental units (grid cells or sub-basins) as a single lumped catchment, and a simple conceptual hydrologic model is applied within each unit. The "bottom-up" approach identifies representative hydrologic areas and applies detailed physically based hydrologic models that are aggregated to all catchments or fundamental units in a large area (e.g., Kite *et al.*, 1994). Figure 3.16 shows the hierarchical structure of a MHM, with finer-scale, site-specific sub-modules interacting with simulations over broader domains. This structure allows for transmitting impacts of global climate change to regional and local levels, also allowing feedbacks from

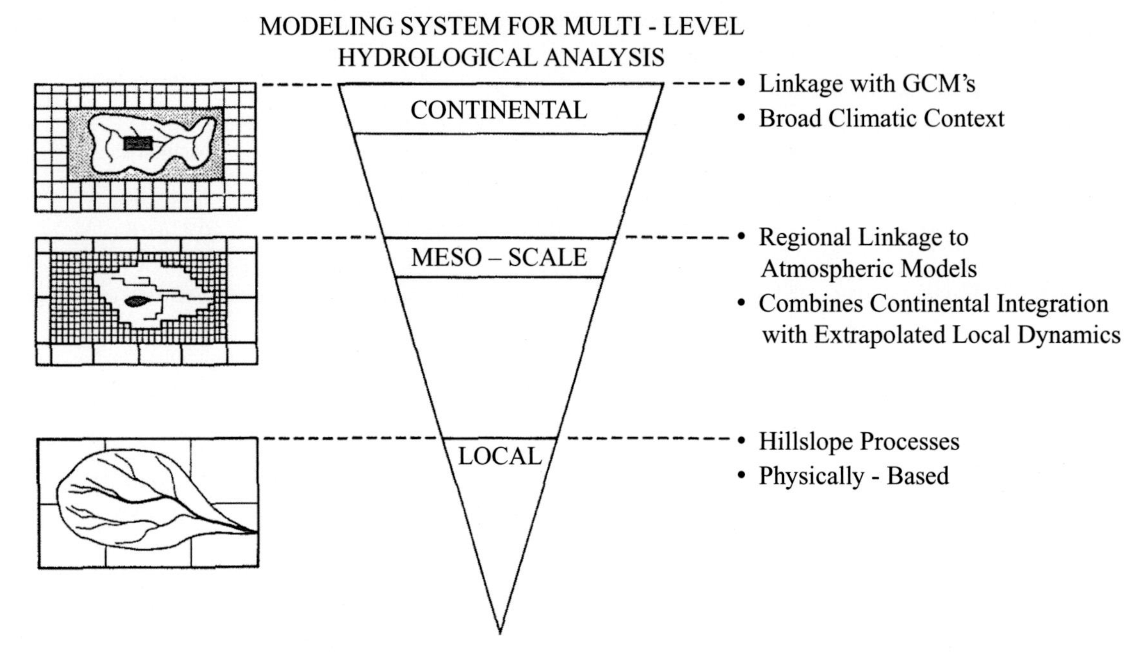

Figure 3.16 Variable resolution strategy for linked atmosphere–macrohydrology models for regional-scale studies (*Source*: Vörösmarty *et al.*, 1993).

physically valid dynamics to reach regional and synoptic scales, as a coupled "top-down" and "bottom-up" approach (Vörösmarty *et al.*, 1993).

The first macroscale model was developed by Vörösmarty *et al.* (1989), who applied a monthly water balance model (WBM) to the Amazon basin. The model has a spatial resolution of 0.5° × 0.5° with parameters derived from gridded soil texture and vegetation databases. In this model, each grid cell was treated as a lumped one-layer soil store with no sub-grid variability of inputs or properties. Excess rainfall over evapotranspiration was routed to subsurface detention pools when field capacity was exceeded. Runoff was released from the rain detention pool in proportion to the current pool storage. Snow accumulation was based on mean monthly temperature, and snowmelt also depends on elevation. The total runoff released from snowmelt and rain pools was passed to a water transport model (WTM), which routed runoff from one cell to another using network topology where channels are treated as linear reservoirs. Vörösmarty *et al.* (1989) concluded that the lack of spatial data sets of sufficient quality and resolution was the main limitation to applying the model in a transient time-varying mode. Yates (1997) developed a simple monthly WBM, which did not include independent data sets such as soil water storage capacity or land cover, but derived the input data from a climate–vegetation classification. The model was found to represent spatially and temporally averaged runoff regimes in large river basins quite well. However, lateral routing was not taken into account in this model, nor in the model of Arnell (1999a,b).

The hydrologic models Macro-PDM of Arnell (1999a,b), the WBM of Vörösmarty *et al.* (1998) and the WaterGap Global Hydrology Model (WGHM, Döll *et al.*, 2003) try to simulate the actual soil moisture dynamics by generating pseudo-daily precipitation from monthly precipitation data and the number of wet days in each month. They use a simple degree–day algorithm to simulate snow, and allow runoff generation to occur even if the soil moisture capacity is not exceeded. The variable infiltration capacity (VIC) model of Liang *et al.* (1994) has distinguishing characteristics such as the representation of sub-grid variability in soil moisture storage capacity as a spatial probability distribution, and the representation of base flow as a non-linear recession. VIC explicitly represents vegetation, and simultaneously solves the surface energy and water balances. Application of the VIC model for nine large river basins across the globe showed that coupling macromodels with a GCM produces a better representation of the recorded flow regime than GCM predictions of runoff for the world's large river basins (Nijssen *et al.*, 2001). Yu *et al.* (2006) developed a new method of coupling coarse-grid regional or global climate models with a much finer grid hydrologic model, designed for interactive climate–hydrologic simulations with explicit changes in individual rivers, lakes, wetlands, and water tables. The relevant quantities (infiltration, runoff) are disaggregated to the finer hydrologic grid based on current near-surface soil moisture in the hydrologic model. Feedbacks on the climate (evaporation, surface heat flux) can be aggregated on the climate grid in the same way. The predicted routing of major rivers, seasonal discharges at outlets of major river basins, and quantities such as continental water-table depths, soil moisture, and groundwater recharges were found to be realistic.

3.7 HYPOTHETICAL SCENARIOS FOR HYDROLOGIC MODELING

Another approach that has been used to assess the hydrologic impacts of climate change is the use of hypothetical climate scenarios. For the initial stage of impact studies, hypothetical scenarios have been used by several researchers (e.g., Xu, 2000). In this approach, present conditions are altered by constructing hypothetical scenarios. Hence, key climatic parameters are changed to cover a reasonable range of possibilities. For example, changes in temperature and precipitation are considered in order to perform a sensitivity analysis of regional water resources under varying climatic conditions. For the evaluation of effects of climate change on flow regime, the climate inputs to a rainfall–runoff model are perturbed and the effect on a statistic of the modeled flows is examined. Generally, hypothetical scenarios assume that air temperature will rise by 0.5–4 °C and precipitation will change by ±10–25% (Xu, 2000; Jiang *et al.*, 2007), and changes in evaporation are also sometimes specified (Arnell, 1992). Several climate scenarios are created by increasing the historical mean temperature by fixed amounts (typically 1–4 °C at a specified step of 0.5 or 1.0 °C) and altering precipitation by fixed percentages (typically ±10–30% at a fixed step of 10%). These hypothetical scenarios are then used to force hydrologic models to evaluate their effect, usually on runoff. This approach avoids the use of results from GCMs altogether, which are generally required for impact studies. However, by definition such studies can only assess the sensitivity of a given basin to climate change.

The general procedure for estimating the impacts of hypothetical climate change on hydrologic behavior is as follows:

1. Determine the parameters of a hydrologic model in the study catchment using current climatic inputs and observed river flows for model validation.
2. Perturb the historical time series of climatic data according to some climate change scenarios (typically, for temperature, by adding $1T = +1, +2, +4$; and for precipitation by multiplying the values by $(1 + 1P/100)$).
3. Simulate the hydrologic characteristics of the catchment under the perturbed climate using the calibrated hydrologic model.
4. Compare the model simulations of the current and possible future hydrologic characteristics.

An important impact of climate change on hydrologic regimes is expected to be on the occurrence of extreme events, such as flood frequencies and magnitudes. The following section explains general procedures used in flood modeling under climate change.

3.8 MODELING OF FLOODS UNDER CLIMATE CHANGE

Floods cause huge loss of life and property, lead to social and environmental damage, and result in economic losses to the tune of US$6 billion annually. Some examples of recent flooding events include the Mississippi River floods in 1993, flooding caused by Hurricane Katrina in 2005, the Danube and Rhine rivers European floods in summer 2002, and Mumbai, India, monsoon flooding in 2005. Climate change is likely to impact climate variability and extremes, and this important aspect has been investigated in several recent studies. Model simulations show that warming may lead to an intensification of the hydrologic cycle and lead to increases in mean and heavy precipitation (Frei *et al.*, 1998). River flooding is usually caused by heavy rainfall, which causes water to overflow into surrounding areas. Flood runoff depends on many factors, including: the amount, intensity, and duration of rainfall; the topography, vegetation and soil characteristics of the catchment; the wetness of the catchment before the storm (antecedent conditions); and evaporation in the catchment. Flooding can also be caused and worsened by snowmelt. Other types of flooding can occur due to inundation by groundwater or high sea levels, or dam-break floods.

Continuous rainfall–runoff simulation at daily, hourly, or even sub-hourly time steps is necessary to model the flood regime of a catchment correctly (Prudhomme *et al.*, 2002), although a design storm method can also be used. However, reliable temporal scales of GCMs are only monthly to seasonal, and hence most current climate impacts studies on floods use monthly GCM outputs in downscaling to station-scale precipitation, which is subsequently used to derive daily rainfall scenarios. In such cases, the modeled flood regimes will vary depending on the assumptions made in the conversion from monthly to daily change projection. Many studies of extremes focus on the 90th or 95th percentile events, since the detection probability of trends decreases for even moderately rare events.

3.8.1 Flood regime description

The flood regime of a catchment is usually described through a flood frequency curve. This curve concentrates only on a selection of the flow series, with the highest flows, or floods. Changes in two flood indicators are usually of interest in flood modeling: the frequency of flood events, described by the change in the return period (i.e., frequency of occurrence) of a flood event of fixed magnitude and the severity of flood events, described by the change in the magnitude of flood events of a fixed return period. The flood frequency curve relates flood magnitude to flood return period.

3.8.2 Underlying assumptions

Flood frequency analysis assumes that the flow regime is stationary, i.e., that the data samples used to construct the flood frequency curves are drawn from a distribution that is not changing. In the context of climate change, this assumption of non-stationarity is difficult to justify, and the use of complex methods is justified (Strupczewski *et al.*, 2001; Prudhomme *et al.*, 2003). However, it may be justifiable to assume stationarity for a time slice (such as 2040s) of interest in the future. Standard probability methodologies will remain valid under this assumption, and thus resulting flood frequency curves are representative of the considered time horizon. Impacts of climate variability patterns, such as PDO or ENSO, on streamflow may also impact the assumptions of stationarity used in flood frequency analysis.

The following section explains the general methodology adopted for flood frequency analysis in hydrology.

3.8.3 Flood frequency analysis

Impact analysis of climate change on floods and low flows is linked to the study of extreme values. Given a data set, such as daily streamflow measurements, drawn from the (unknown) probability distribution of identically distributed random variables, in the classical approach, several statistical functions of the data set, such as the mean, variance, etc., can be calculated. However, for flood analysis, and the statistics of the data set, values lying in the extremes of the data distribution are of interest. Classical distributions of the data can provide an empirical estimation of the frequency of occurrence of events resulting in extreme values. However, in risk analysis, the distribution of extreme values of the data set is used for estimation of frequencies of occurrence of values beyond the observed extremes (Baguis *et al.*, 2008). Flood frequency analysis is the procedure of obtaining the relationship between flood quantiles and their non-exceedance probability using extreme value theory. The generalized extreme value (GEV) (of which extreme value type I (EV I), EV II, and EV III are special cases) and the generalized Pareto distribution (GPD) are frequently used when analyzing data in the context of extreme value theory. The EV I or Gumbel distribution is commonly used for the distribution of annual maxima of streamflows, though the Pearson type 3 or log-normal distribution is also common. The EV III distribution is used for annual low flows. The extreme value distributions have to be calibrated to empirical extremes to be extracted from time series using independence criteria. Most common approaches are the annual maximum series (AMS) method and the peak-over-threshold (POT) (Naden, 1992) or partial duration series (PDS) method. An overview of these methods can be found in Hosking and Wallis (1987), and Madsen *et al.* (1997).

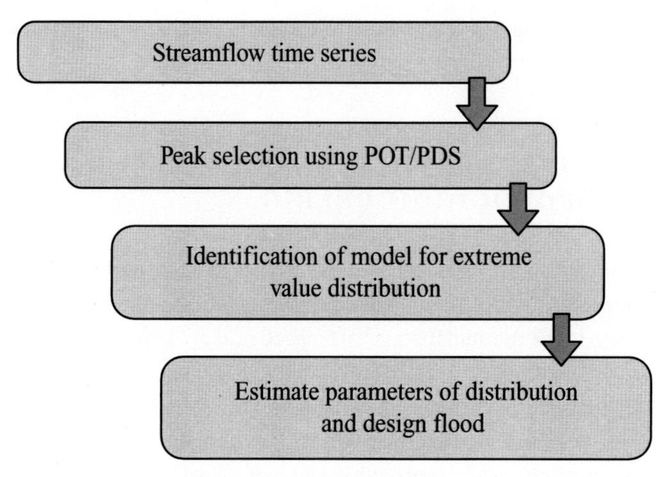

Figure 3.17 Typical procedure for flood analysis.

The magnitude of the *T*-year flood at a site is the amount of streamflow that has a probability $1/T$ of being exceeded in any one year. Figure 3.17 shows the general approach in flood frequency analysis using streamflow series. A selection of values from this series that can be considered peak events is first made (Claps and Laio, 2003), using either the AMS or POT/PDS methods. This results in a transformation of the daily runoff stochastic process into a marked point process having two random variables: n, which is the number of peaks per year, and q, which is the magnitude of the peak event. A model is then fitted to this marked point process, assuming that subsequent peaks are independent, and that the CDF of the flood peaks, $F_Q(q)$, is some known probability distribution. If T is the return period or recurrence interval of a flood q_p with non-exceedance probability p,

$$T = \frac{1}{n\,(1 - p)} \tag{3.39}$$

where n is the average annual number of occurrences of the random variable Q. Here $n = 1$ if we consider an annual maximum data series. Hence, the flood quantile or T-year event is given by

$$q_p = F_Q^{-1}\left(1 - \frac{1}{\lambda T}\right) \tag{3.40}$$

where λ is the expected value of n, i.e., the mean number of events per year. Usually, model selection for the CDF of the flood peaks, $F_Q(q)$, is made using goodness-of-fit criteria such as Akaike information criteria (AIC) or the standard least squares criterion (SLSC). For example, if it is assumed that the number of occurrences, n, per year is Poisson distributed, and that the PDF of the flood peaks, $f_Q(q)$, is exponential or Pareto distributed, then the CDF of the flood peaks is given as (Claps and Laio, 2003)

$$F_Q(q) = 1 - \left(1 - k\frac{(q - q_0)}{\alpha}\right)^{\frac{1}{k}} \quad (3.41)$$

where q_0, α, and k are the location, scale, and shape parameters, respectively. The GPD is a versatile distribution, which reduces to the exponential distribution when $k = 0$. In order to avoid overparameterization, q_0 is usually set as the minimum value of the sample of selected peaks, or directly as the threshold level q_b. The last step in flood frequency analysis is model estimation, i.e., the estimation of λ and the parameters of $F_Q(q)$ based on the available data. Parameter estimation methods commonly used are the method of moments, probability-weighted moments or L-moments, or maximum likelihood.

In the POT method, all values above a specified threshold are used, which is advantageous as compared to the AM method where substantial amounts of potentially useful information is discarded. However, more data points do not guarantee a better estimate of the EV distribution, since this increases the possibility that flood peaks may be interdependent. Some characteristics of the POT method are reviewed by Madsen *et al.* (1997) and by Lang *et al.* (1999). When the extreme value analysis is carried out for a range of aggregation levels or time scales, amplitude/duration/frequency relationships or curves can be derived (intensity/duration/frequency (IDF) for rainfall and discharge/duration/frequency (QDF) for river flow). These are relationships between the intensity or river flow and the frequency or return period for the range of aggregation levels specified (Baguis *et al.*, 2008).

3.8.4 Flood frequency analysis under climate change

For the twentieth century, an increasing risk of 100-year floods was detected in 29 basins larger than 20,000 km^2 by Milly *et al.* (2002). McCabe and Wolock (2002) found a step increase in steamflows predominantly in the eastern part of the USA during the 1970s. Baker *et al.* (2004) also found increasing flows largely in the eastern part of the USA. Many questions arise with respect to impacts of climate change on floods. In addition to the impact on flood magnitudes and frequencies, the timing of flood events could also change with warming. Flood risk would be affected by land use and land cover changes, and changes in temperature, humidity, soil moisture, and groundwater levels would also affect the risk of large floods. Since current extreme value distributions may not hold under global warming, a statistical flood estimation would not be useful under climate change. A more physically based approach is needed, which incorporates meteorological and hydrologic information.

Flood modeling under climate change can use analytical methods or Monte Carlo simulation. For example, IDF curves could be used for deriving future flood frequency distributions analytically (Goel *et al.*, 2000). Synthetic meteorological series input to a rainfall–runoff model could also be used to derive future streamflow time series to which a suitable extreme value distribution could be fitted (Booij, 2005). For example, Bell *et al.* (2007) used a grid-based flow-routing and runoff-production model, with hourly RCM precipitation estimates as input to assess the effects of climate change on river flows in catchments across the UK. Using a resampling method to investigate the robustness of the modeled changes in flood frequency, it was found that changes in flood frequency at higher return periods are generally less robust than at lower return periods. Figure 3.18 shows the derived flood frequency curves for the current and future simulated streamflow series derived using the POT and AM methods for two UK catchments, using the grid-to-grid model from this study. It is seen that for the Blackwater catchment the current and future curves using the two methods are very similar. Both sets of curves cross, indicating a change from slightly increased future peak flows at shorter return periods to decreased future flows at higher return periods. However, for the Taff catchment the curves are different at high return periods, depending on the method used to derive them. The current and future curves diverge for AM data, indicating a greater increase in future flood flows at high return periods. Choice of method is thus an additional source of uncertainty in changing flood risk assessment.

Several other studies have dealt with flood modeling under future climatic conditions. Studies in Britain have examined the spatial pattern of changing runoff for the whole land area of Britain (Arnell, 1992; Pilling and Jones, 1999). Pilling and Jones (1999) used downscaled GCM output for 2050 and for 2065 to drive a hydrologic model and simulate annual and seasonal effective runoff for Britain. Most studies involving model output show increased risk of flooding in the UK under a changed climate (Wilby *et al.*, 2008).

The following example explains the typical steps involved in flood analysis for a catchment, under climate change.

Example 3.7 Modeling of floods

Flood modeling under climate change is studied in a set of five catchments (Figure 3.19) in Great Britain through this example, from Prudhomme *et al.* (2003). Uncertainties in climate change are incorporated using a set of climate scenarios randomly generated by a Monte Carlo simulation, using several GCMs, SRES-98 emissions scenarios, and climate sensitivities. Catchment 25006 is an eastward draining, rural, and impervious catchment in northeast England, while catchment 28039 is highly urbanized. The southern catchment 40005, covered by clay, is impervious, while catchments 55008 and 96001 are both impervious catchments without any urbanized areas, receiving moderate to high precipitation.

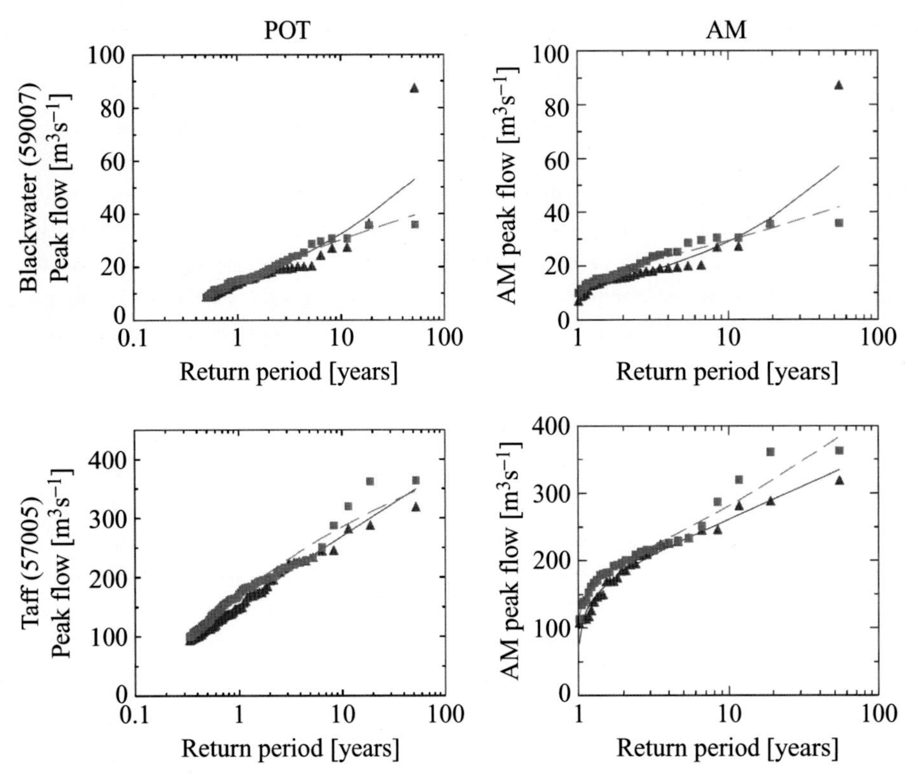

Figure 3.18 Flood frequency curves for current (triangles/solid lines) and future (squares/dashed lines) derived using the POT and AM data series for two UK catchments from simulated flows (*Source*: Bell *et al.*, 2007).

Solution

Step 1: Construct future climatic time series

Results from seven GCM experiments held by the IPCC Data Distribution Centre are considered, with two versions of the Hadley Centre's GCM, each run with four different starting conditions, called "ensemble runs." An overall total of 14 GCM patterns (six from other GCMs, eight from Hadley Centre GCM) were thus considered. Scenarios representing four SRES-98 marker scenarios and nine climate sensitivities (ranging from 1 to 5 °C) were derived from a simple rescaling approach. Monte Carlo simulations, giving equal weight to each individual GCM and each emissions scenario, and different weights to the climate sensitivity parameter, were undertaken to produce a set of 25,000 GCM runs. They include about 25 runs for each of all possible combinations between 14 GCM patterns, four emissions scenarios and nine climate sensitivities.

A proportional change downscaling method is used to construct a future time series of precipitation and potential evaporation for each GCM-scenario–sensitivity combination, as follows:

1. **Spatial downscaling:** A simple spatial downscaling procedure is applied to derive the monthly climate change scenarios for each case study catchment. For every GCM run, absolute differences between the mean monthly estimate representative of the GCM baseline ($1 \times CO_2$ run) and of the time horizon (as modeled by that GCM with a climate change forcing) are calculated for each cell of the GCM. They are then expressed as percentage changes compared to values of the mean monthly 1961–90 climatology derived from observations. This results in a set of monthly percentage changes for each climate variable at the same spatial resolution as the observed climatology, which is finer than that of the GCMs.
2. **Temporal downscaling:** Hourly rainfall series are derived from mean monthly changes using a proportional temporal downscaling technique. The monthly percentage change in precipitation (produced from the spatial downscaling procedure) is applied to each rain event of the same month, so that the same absolute monthly change is achieved with the rainfall increase/decrease for each day changing by the same proportion. The same methodology is applied to potential evaporation (PE).

Thus, the time series of rainfall and PE for three time horizons are constructed from the baseline time series and climate change scenarios, expressed as monthly percentage changes in precipitation and potential evaporation.

Step 2: Simulate streamflow for current and future conditions

For each study catchment, river flow was simulated at an hourly time step using a conceptual rainfall–runoff model, the probability distributed model (Moore, 1985), where the distribution of

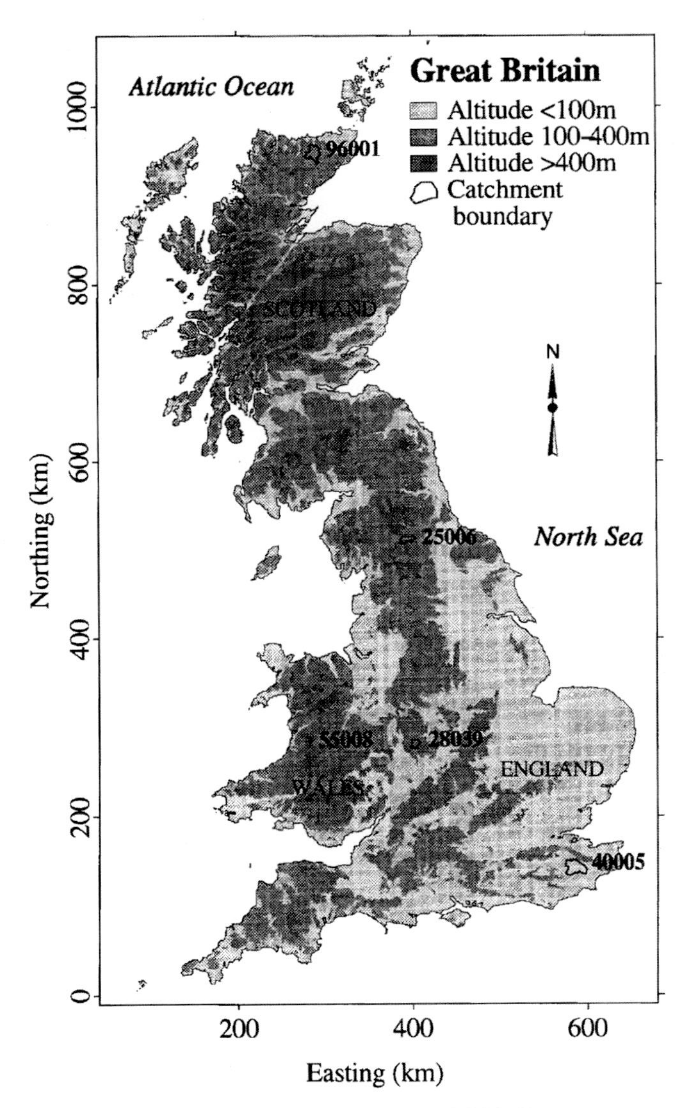

Figure 3.19 Location of the study catchments (25006: Greta at Rutherford Bridge, 28039: Rea at Calthorpe Park, 40005: Beult at Stile Bridge, 55008: Wye at Cefn Brwyn, and 96001: Halladale at Halladale) (*Source*: Prudhomme *et al.*, 2003).

storage capacity over the catchment is described by a specified probability function. The soil moisture store permits runoff to respond non-linearly to rainfall inputs by varying the proportion of the catchment generating fast direct runoff. A semi-automated procedure is used to fit the parameters of the model, by calibrating it for each catchment using historical climatic and hydrologic time series. Streamflow series representative of current and future conditions were simulated using this calibrated model. For example, the rainfall–runoff model (with the fitted parameters) is run using the 2050s time series to simulate flow series assumed representative of the 2050s.

Step 3: Analysis of the flood series
A POT analysis (Naden, 1992) was used. After sampling the $3N$ highest independent flood events, where N is the number of years of the series, a flood frequency distribution was then fitted to the flood sample. The generalized Pareto distribution is chosen because of its flexibility and robustness. Events of 2, 5, 10, and 20-year return periods are analyzed, which describe a range of flood events.

Step 4: Derive confidence intervals associated with flood frequency distributions
Bootstrapping is used to produce a set (199) of randomly sampled flood series, and a flood frequency distribution is fitted to each series. The 95% confidence interval is derived from the ensemble of the resulting 199 flood frequency distributions.

Step 5: Changes in magnitudes of flood quantiles
Table 3.12 shows the return periods associated with the median of results representative of all the runs from individual GCMs, and with the upper and lower quartiles. It is seen that the return period of any chosen flood event is generally shown to decrease for all catchments according to most GCMs, reflecting that flood events of any given magnitude are expected to become more frequent than currently observed on average. Half of the runs of most GCMs are usually within the 95% confidence interval for quantiles of current return period above or equal to five years.

These derived changes in return periods for the median, upper, and lower quartiles can be useful inputs for further analysis and planning of flood protection design measures.

3.9 UNCERTAINTY MODELING

The treatment of uncertainty is relatively underdeveloped in assessing the hydrologic impacts of climate change. However, it is accepted that there are large uncertainties in predictions such as flood risk, arising from various sources. These sources include limitations in scientific knowledge (for example, effect of aerosols) that are differently parameterized in GCMs, randomness, and human actions (such as future GHG emissions), which can be classified as scenario uncertainty. Downscaling of GCM outputs to station-scale hydrologic variables using statistical relationships introduces additional uncertainty. Another source of uncertainty arises in hydrologic modeling. These uncertainties accumulate from the various levels and their propagation up to the regional or local level leads to large uncertainty ranges at such scales (Wilby, 2005; Minville *et al.*, 2008). Several studies explore some component of the uncertainty propagation, such as choice of downscaling model, or choice of emissions and downscaling method combined with hydrologic model parameter uncertainty. But few studies attempt end-to-end uncertainty analysis of hydrologic impacts (Dessai, 2005; Raje and Mujumdar, 2010a,b). It is especially difficult to model the uncertainty in the magnitude–frequency relationship of floods. Non-stationary effects of climate

Table 3.12 *Statistics for new return period of the magnitude of flood quantiles Q(T) (T = return period) associated with individual GCMs for the 2050s*

		96001			25006			28039			55008			40005		
		Scotland			Northern England			Midlands			Wales			Southeast England		
	GCM	Uq	Med	Lq	Uq	Med	Lq	Uq	Med	Lq	Uq	Med	Lq	Uq	Med	Lq
$Q_{(2)}$	CCSR	1.6	1.4	1.2*	1.1	0.9*	0.8*	2.3	2.0	1.7	1.7	1.5	1.4*	3.3	2.7	2.3
	CGCM	1.6	1.4	1.3	1.4	1.2	1.1	1.2*	1.0*	0.9*	1.6	1.5	1.4*	1.3*	1.1*	1.0*
	CSIRO	1.8	1.7	1.5	1.4	1.3	1.2	1.4	1.2	1.1*	1.6	1.5	1.4*	1.4	1.2*	1.1*
	ECHAM4	1.3	1.1*	1.0*	1.3	1.2	1.0*	1.8	1.6	1.4	1.9	1.8	1.7	2.2	2.0	1.8
	GFDL	1.2*	1.1*	0.9*	1.3	1.1	1.0*	1.5	1.3	1.1*	1.7	1.6	1.5	1.6	1.4	1.3*
	HadGGa	1.2*	1.1*	0.9*	1.3	1.1	0.9*	1.5	1.3	1.0*	1.7	1.6	1.4*	1.4	1.3*	1.1*
	HadGGd	1.4	1.2*	1.1*	1.5	1.3	1.1	1.6	1.3	1.1*	1.7	1.5	1.4*	1.5	1.3*	1.1*
	NCAR	1.9	1.7	1.5	1.8	1.6	1.4	2.5	2.1	1.8	2.0	1.9	1.8	1.7	1.5	1.4
	95% CI	3.6		1.2	3.3		1.0	4.5		1.2	3.2		1.4	4.6		1.3
$Q_{(5)}$	CCSR	4.4	3.9	3.4	3.0	2.5	2.2	5.7	4.6	3.9	3.7	3.4	3.1	27.0	14.6	9.6
	CGCM	4.0	3.6	3.3	3.4	3.0	2.7	2.7	2.3	1.9*	3.8	3.5	3.2	3.4	3.0*	2.6*
	CSIRO	4.5	4.1	3.8	3.7	3.3	2.9	3.5	2.8	2.4*	3.8	3.5	3.2	3.7	3.2	2.8*
	ECHAM4	3.5	3.1	2.7*	3.6	3.2	2.8	4.6	3.8	3.2	4.5	4.2	3.9	7.5	6.0	5.1
	GFDL	3.3	2.8	2.5*	3.1	2.7	2.4	3.3	2.8	2.4*	4.0	3.6	3.4	4.5	3.9	3.4
	HadGGa	3.2	2.7	2.3*	3.4	2.8	2.4	3.7	2.9	2.3*	4.1	3.7	3.3	3.7	3.2	2.8*
	HadGGd	3.6	3.2	2.8	3.9	3.2	2.7	4.4	3.4	2.6	4.0	3.6	3.3	4.0	3.2	2.7*
	NCAR	4.6	4.2	3.8	4.5	4.0	3.7	7.5	5.6	4.5	5.1	4.7	4.4	4.4	3.9	3.4
	95% CI	10.2		2.7	10.1		1.4	17.0		2.5	11.9		2.2	22.5		3.0
$Q_{(10)}$	CCSR	9.2	8.3	7.5	6.7	5.6	4.8	11.6	8.9	7.1	6.8	6.0	5.4	227.6	69.2	32.3
	CGCM	8.2	7.5	6.7	6.9	6.2	5.4	5.5	4.2*	3.4*	7.2	6.4	5.8	9.8	7.8	6.4
	CSIRO	9.1	8.4	7.6	7.6	6.7	6.0	7.6	5.6	4.4*	7.2	6.3	5.7	9.8	7.5	5.9
	ECHAM4	7.2	6.4	5.7	7.8	7.0	6.2	9.8	7.3	6.0	8.6	7.9	7.1	28.6	17.0	11.2
	GFDL	6.8	5.9	5.2*	6.0	5.3	4.7	6.4	5.0*	4.1*	7.7	6.8	6.1	12.8	9.5	7.9
	HadGGa	6.6	5.7	4.8*	7.2	6.1	5.2	7.9	5.5	4.1*	8.0	6.9	6.0	9.9	7.9	6.5
	HadGGd	7.6	6.7	6.0	8.3	7.0	5.9	10.9	7.1	5.1	7.9	7.0	6.3	9.1	7.3	5.9
	NCAR	9.3	8.5	7.7	9.6	8.4	7.6	19.2	11.8	8.3	10.6	9.2	8.1	10.3	8.2	7.1
	95% CI	22.1		5.4	24.1		1.7	47.2		5.0	35.0		2.7	162.2		5.2
$Q_{(20)}$	CCSR	19.6	17.4	15.5	15.8	13.1	11.1	23.7	16.3	12.5	12.3	10.4	9.1	1675.9	342.1	98.8
	CGCM	17.0	15.2	13.6	14.4	12.4	11.0	11.8	8.0*	6.1*	13.6	11.4	10.0	53.4	28.3	18.1
	CSIRO	18.5	16.8	15.2	16.4	14.0	12.2	19.0	10.9	7.8*	13.4	11.3	9.8	38.5	19.9	13.2
	ECHAM4	14.4	12.9	11.5	18.1	15.4	13.4	22.2	14.0	10.3	16.9	14.4	12.7	151.9	53.5	25.2
	GFDL	14.3	12.5	11.2	12.1	10.5	9.2	12.7	8.9*	6.8*	15.2	12.6	11.0	52.7	28.2	18.8
	HadGGa	13.4	11.6	9.9*	16.5	13.4	11.2	17.7	10.5	7.4*	15.8	13.1	10.9	40.9	23.8	16.8
	HadGGd	16.2	14.2	12.6	19.0	15.4	12.9	30.4	15.3	9.7	16.1	13.6	11.7	28.2	18.4	13.9
	NCAR	19.0	17.2	15.6	21.9	17.9	15.5	50.7	24.2	15.4	22.9	17.9	14.6	29.0	19.0	14.4
	95% CI	47.1		10.2	57.6		1.9	156.1		9.2	101.4		2.6	5404.5		8.3

Source: Prudhomme *et al.* (2003)

Black boxes show cells with the largest increase; dashed boxes show cells with smallest increases/largest decreases; grey cells show decrease; * cells indicate results outside the 95% confidence intervals.

Uq: Upper quartile, i.e.,75% results below value; Med: median, i.e., 50% results below value; Lq: Lower quartile, i.e., 25% results below value; 95% CI: limits of the 95% confidence intervals.

change, including abrupt change or systematic long-term trends can add to uncertainty in flood risk estimation.

3.9.1 Uncertainty modeling in regional impacts

A commonly used method to study climate impacts on flow regime is to use an ensemble of GCMs, scenarios, and statistical downscaling/regional climate models to provide inputs to a hydrologic model, and study the statistic of interest for the modeled hydrologic variables (e.g., Prudhomme *et al.*, 2003; Minville *et al.*, 2008). The use of several GCMs and scenarios leads to a wide spread in the downscaled hydrologic projection, especially in years far into the future, leading to uncertainties as to which among the several possible predictions should be used in developing responses. Simonovic and Li (2004) have shown the uncertainty lying in climate change impact studies on flood protection resulting from selection of GCMs and scenarios. GCM and scenario uncertainty has been studied in terms of PDFs of a hydrologic drought indicator such as standardized precipitation index (SPI) (Ghosh and Mujumdar, 2007) and using an imprecise probability approach (Ghosh and Mujumdar, 2009).

Dissimilarities between bias-corrected GCM simulations under different scenarios after the year 1990 (end of baseline period) result in different system performance measures that do not validate the assumptions of equi-predictability of GCMs and equipossibility of scenarios. Mujumdar and Ghosh (2008) used possibility theory for deriving a weighted distribution function of monsoon streamflow. Possibility theory is an uncertainty theory essentially used to address partially inconsistent knowledge and linguistic information based on intuition. This intuition about the future hydrologic condition is derived based on the performance of GCMs with associated scenarios in modeling the streamflow of the recent past (1991–2005), when there are signals of climate forcing.

Uncertainty combination has been studied using the Dempster–Shafer theory (Raje and Mujumdar, 2010b) and natural variability linkages have been used for constraining uncertainty in regional impacts (Raje and Mujumdar, 2010a). Prudhomme and Davies (2009) examined uncertainties in climate change impact analyses on the river flow regimes in the UK, using either a statistical or a dynamical downscaling model for downscaling precipitation from an ensemble of GCMs and scenarios, propagated to river flow through a lumped hydrologic model. They showed that uncertainties from downscaling techniques and emissions scenarios are of similar magnitude, and generally smaller than GCM uncertainty. They found that for catchments where hydrologic modeling uncertainty is smaller than GCM variability for baseline flow, this uncertainty can be ignored for future projections, but it might be significant otherwise.

Minville *et al.* (2008) studied the impact of climate change on the hydrology of the Chute-du-Diable watershed in Canada by comparing statistics on current and projected future discharge. They used ten equally weighted climate projections from a combination of five GCMs and two greenhouse gas emissions scenarios (GHGES) to define an uncertainty envelope of future hydrologic variables. Figure 3.20 shows PDFs from the uncertainty analysis of peak discharge, time to occurrence of peak and annual mean discharge for the watershed, for three future time horizons as compared to the control period of 1960–90. They also found that the largest source of uncertainty in their study came from the choice of GCM.

Wilby (2005) studied the relative magnitude of uncertainties in water resource projections arising from the choice of hydrologic model calibration period, model structure, and non-uniqueness of model parameter sets. Using parameter sets of the 100 most skillful model simulations identified by the Monte Carlo sampling, it was found that there is a general increase in the uncertainty bounds of projected river flow changes between the 2020s and 2080s due to equifinality. Figure 3.21 shows projected variations in monthly mean river flows arising because of equifinality, from the study.

Many studies have reported that existing systematic bias in reproducing current climate affects future projections, and must be considered when interpreting results (Prudhomme and Davies, 2009). To account for the bias, weights are assigned to different GCMs based on their bias with respect to the observed data and the convergence of simulated changes across models (Giorgi and Mearns, 2003). Also, most studies cited above have found that uncertainty due to the driving GCM is by far the largest source of uncertainty in hydrologic impacts (Minville *et al.*, 2008; Kay *et al.*, 2009; Prudhomme and Davies, 2009). Some others, however, have found consistent results in hydrologic responses across GCMs, though there are significantly different regional climate responses (Maurer and Duffy, 2005). Many studies have also found that understanding current and future natural variability is important in assessing hydrologic impacts of climate change (Prudhomme *et al.*, 2003; Wilby, 2005; Kay *et al.*, 2009).

The following example demonstrates the uncertainty modeling methodology for GCM and scenario uncertainty in projection of a drought index.

Example 3.8 Uncertainty modeling

The use of non-parametric methods for GCM and scenario uncertainty modeling of future drought is demonstrated, as per Ghosh and Mujumdar (2007). Future PDFs of a drought index are derived using a fuzzy-clustering-based downscaling method (Ghosh and Mujumdar, 2006) for modeling future precipitation (see Example 3.5). The methodology is applied to the case study of the Orissa meteorological subdivision in India to analyze the severity of different degrees of drought in the future.

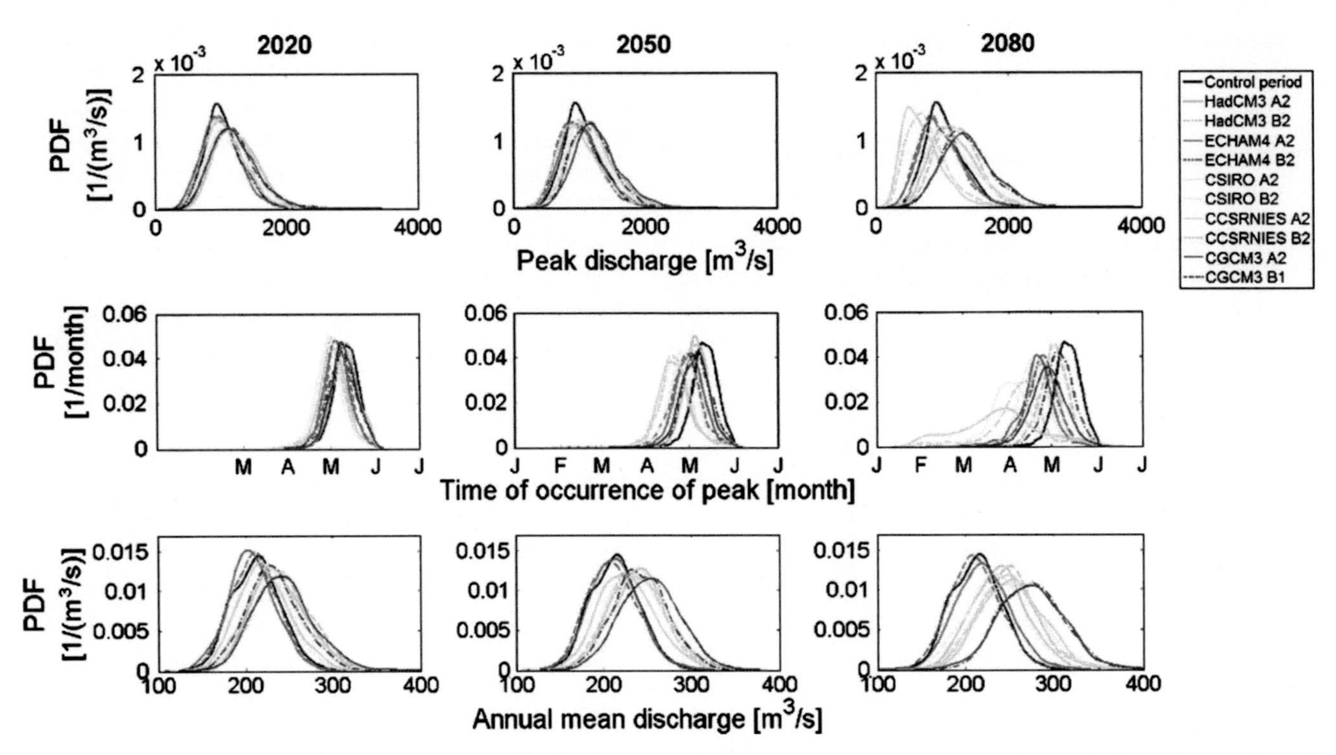

Figure 3.20 Probability density functions of peak discharge (top), time of occurrence of peak (middle), and annual mean discharge (bottom) from the GCM-scenario ensemble for a Nordic watershed for the 2020, 2050, and 2080 time horizons (*Source*: Minville *et al.*, 2008). See also color plates.

Table 3.13 *Drought categories*

Drought category	SPI values
Near normal	0 to −0.99
Mild-to-moderate drought	−1.00 to −1.49
Severe drought	−1.50 to −1.99
Extreme drought	−2.00 or less

Solution

Step 1: Fuzzy-clustering-based downscaling

A statistical relationship based on fuzzy clustering and linear regression is developed between MSLP and precipitation. The methodology and results for the case study are as provided in Example 3.5.

Step 2: Drought indicator (SPI-12) projections

McKee *et al.* (1993) developed the SPI, based on aggregated monthly precipitation, for the purpose of defining and monitoring drought. SPI can be defined by the value of standard normal deviation corresponding to the CDF value of a precipitation event with a known probability distribution. A common procedure adopted for computing SPI is to fit a gamma distribution to the precipitation data, although the Pearson Type III has also been used, and then to transform the data to an equivalent SPI value based on the standard normal distribution. Table 3.13 presents the categories of drought corresponding to their SPI values (McKee *et al.*, 1993).

The parameters required for estimation of SPI, namely, parameters of gamma distribution and non-zero precipitation probability, are estimated based on the observed annual precipitation. Using these parameters, the future annual precipitation (computed from monthly precipitation), downscaled from GCM output, is converted into SPI-12. The SPI-12 is calculated for all GCMs for available scenarios.

Step 3: Uncertainty modeling

The SPI-12 values computed with downscaled outputs from GCMs are considered as the realizations of the random variable SPI-12 where there exists a PDF of SPI-12 in each year. The severity of future drought may be studied by estimating the evolution of the PDF of a drought indicator. Methodologies based on kernel density and orthonormal systems are used to determine the future non-parametric PDF of SPI. Probabilities for different categories of future drought are computed from the estimated PDF.

(a) Normal distribution assumption Here, a normal distribution is assumed for the future SPI-12. The results for each GCM and emissions scenario are taken as the set of independent realizations of SPI-12 and this set is used at each time step to establish the parameters of the normal probability distribution. Figure 3.22(a) shows the average probabilities of drought events for three time slices, years 2000–10, 2040–50, and 2090–2100 using a normal distribution assumption.

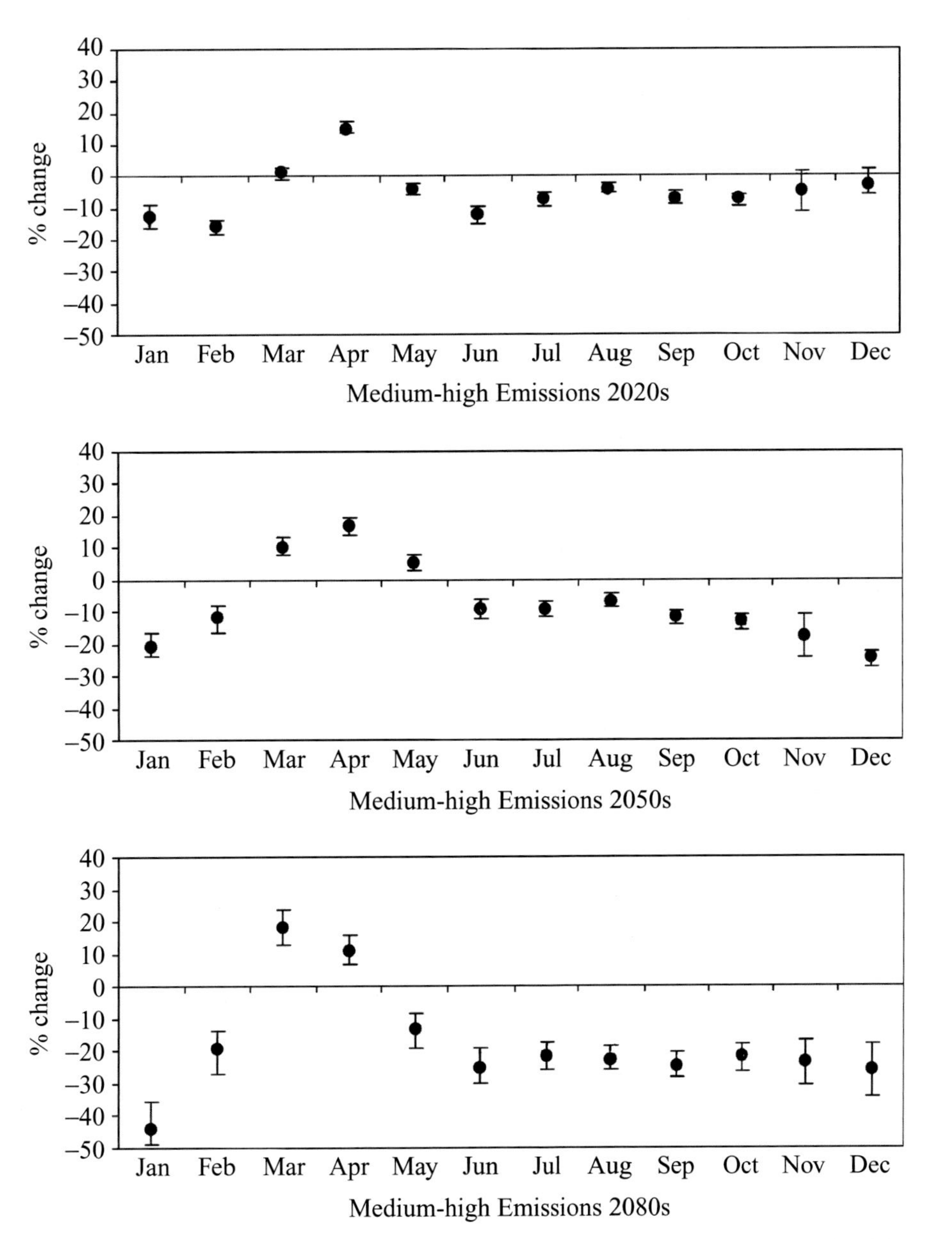

Figure 3.21 Projected changes to monthly mean river flows under medium–high emissions due to non-uniqueness of model parameter sets. Vertical bars show the 95% confidence interval (Modified from Wilby, 2005).

Considerable variations in the probabilities of near-normal conditions and extreme drought are seen from years 2000–10 to 2040–50. The probability of near normal conditions is reduced, and that of extreme drought is increased significantly in the years 2040–50. Variations in the probabilities of different droughts are not significant in the later years, 2040–50 to 2090–2100. This may mean that the assumption of a normal distribution does not result in the correct assessment of drought impacts of climate change for years further in the future. A normal probability plot of SPI-12 for three arbitrarily chosen years also shows that SPI-12 values deviate significantly from the normal distribution.

(b) Kernel density estimation Kernel density estimation entails a weighted moving average of the empirical frequency distribution of the data. Most non-parametric density estimators can be expressed as kernel density estimators, using a kernel function, $K(x)$, defined by a function having the following property:

$$\int_{-\infty}^{\infty} K(x)dx = 1 \tag{3.42}$$

A PDF can therefore be used as a kernel function. A normal kernel (i.e., a Gaussian function with mean 0 and variance 1) is used

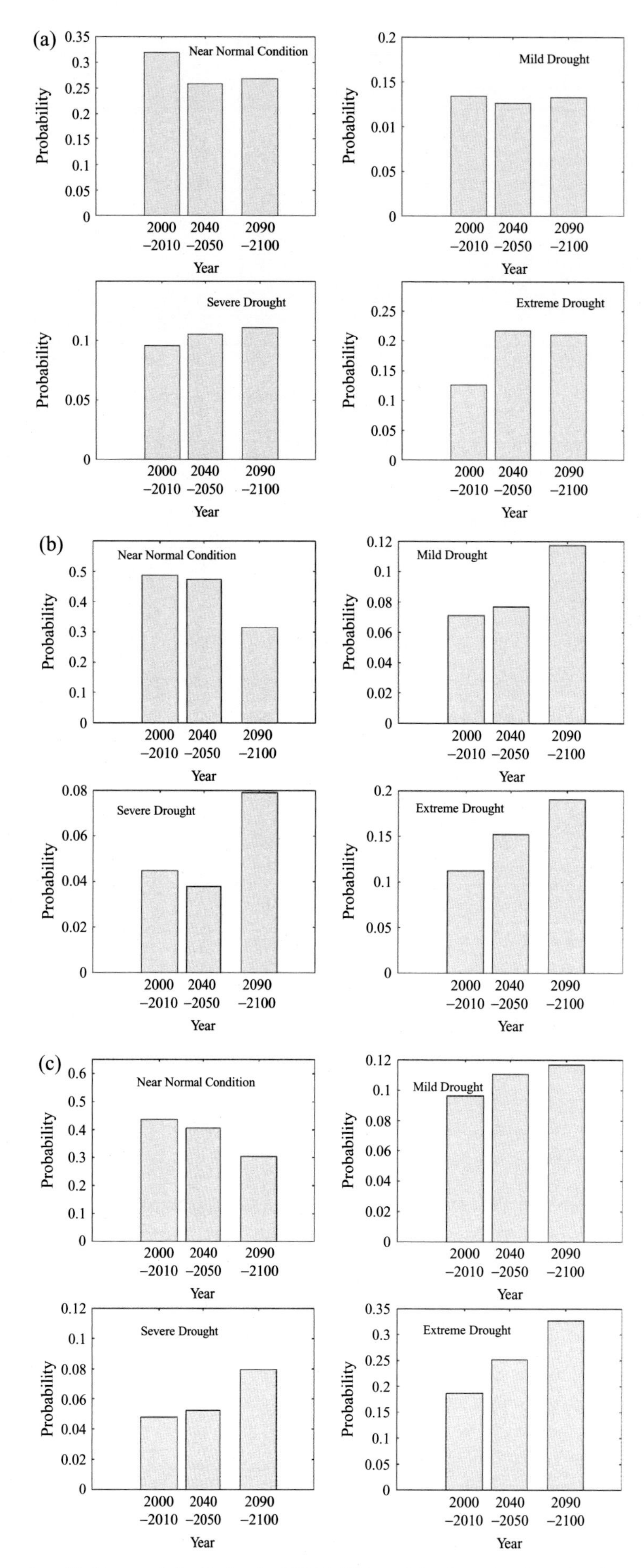

Figure 3.22 Probability of droughts with (a) normal distribution for SPI-12, (b) kernel density estimation, (c) orthonormal series method (*Source*: Ghosh and Mujumdar, 2007).

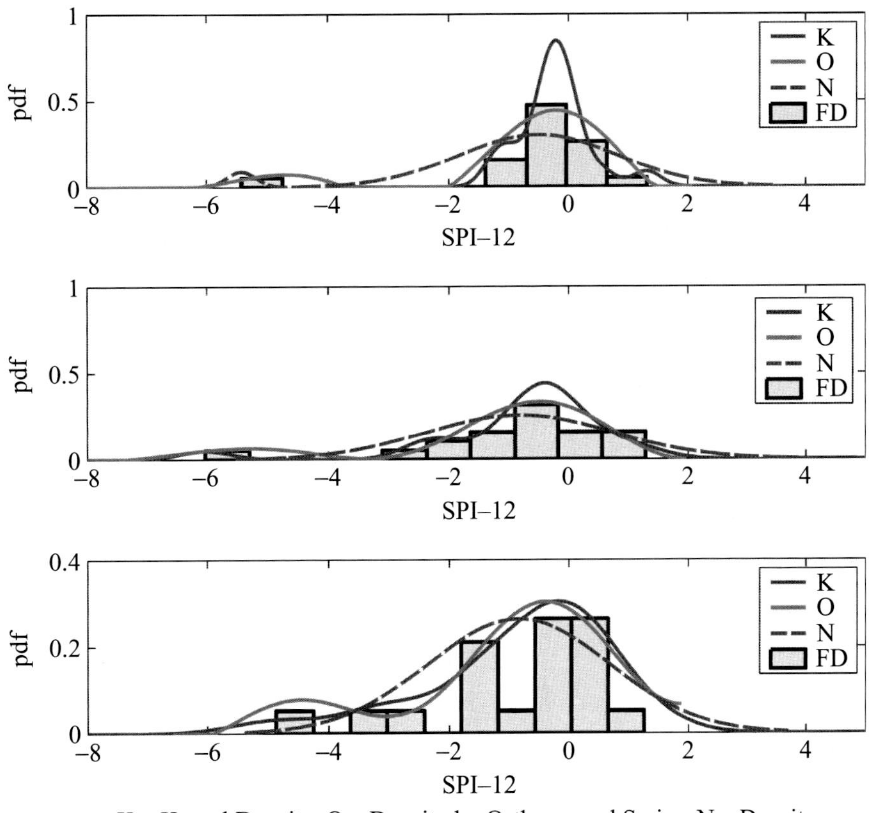

Figure 3.23 Estimation of PDF of SPI-12 for years 2007, 2041, and 2093 (*Source*: Ghosh and Mujumdar, 2007).

here. A kernel density estimator ($\hat{f}(x)$) of a PDF at x is defined by

$$\hat{f}(x) = (nh)^{-1} \sum_{l=1}^{n} K((x - x_l)/h) \tag{3.43}$$

where n is the number of observations (here number of available GCM outputs), x_l is the lth observation (here SPI-12), and h is the smoothing parameter known as bandwidth, which is used for smoothing the shape of the estimated PDF. Figure 3.22(b) presents the probabilities of drought conditions in the years 2000–10, 2040–50, and 2090–2100, as obtained using the kernel density estimation. A significant change is found for the years 2090–2100 from the years 2040–50 in the probabilities of near-normal conditions from the kernel density estimation method, which is absent in the plots obtained from the model based on the assumption of a normal distribution. The probability of extreme drought has a continuously increasing trend in Figure 3.22(b), which is not seen in Figure 3.22(a).

(c) Orthonormal series method A PDF from a small sample can be estimated using the orthonormal series method, which is essentially a series of orthonormal functions obtained from the sample. The summation of the series with coefficients results in the desired PDF. If the orthonormal series as the subset of the Fourier series consisting of cosine functions is selected,

$$\phi_o(x) = 1 \quad \text{and} \quad \phi_j(x) = \sqrt{2}\cos(\pi j x), \quad j = 1, 2, 3, \ldots \tag{3.44}$$

Figure 3.22(c) presents the probabilities of drought conditions in the years 2000–10, 2040–50, and 2090–2100 as obtained using orthonormal series-based density estimation. The results are by and large similar to those of the kernel density estimation except for the probabilities of mild drought. The kernel density estimation procedure projects a sudden increase in the probability of mild drought for the years 2090–2100, whereas such significant change is not observed in the results obtained from the orthonormal series method. From the overall trend in probabilities of all categories of drought, it is seen that the probability of near-normal conditions is likely to decrease, and the probabilities of mild, severe, and extreme droughts are likely to increase over time. The Orissa meteorological subdivision is projected to be more drought-prone in the future.

Step 4: Comparison of results

The PDF of SPI-12 computed using all three methods is presented in Figure 3.23 along with frequency distribution of the sample for

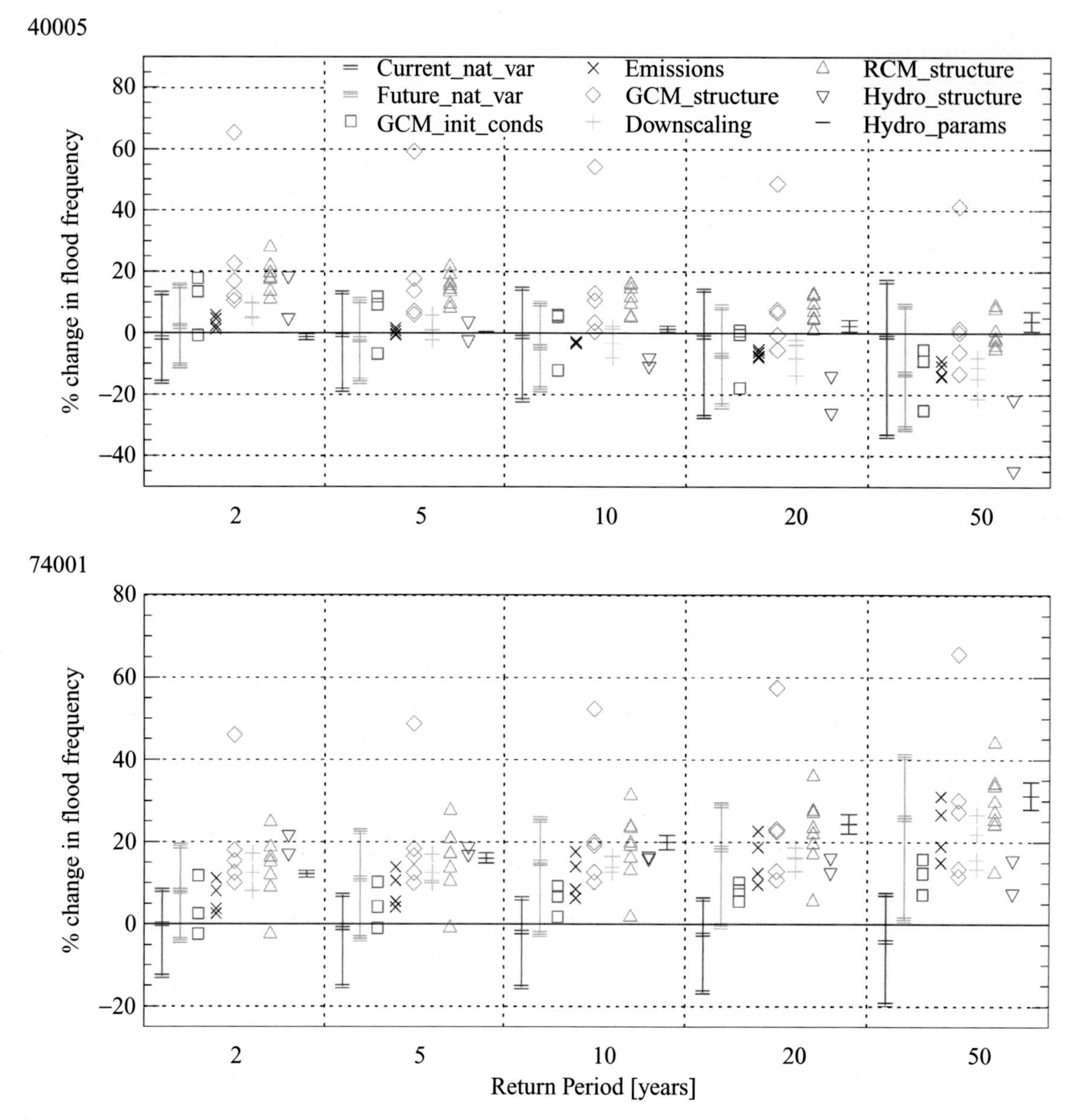

Figure 3.24 Variation in the projected percentage change in flood frequency from the current time to the 2080s, from various sources, for return periods of 2, 5, 10, 20, and 50 years for two catchments. Potential ranges of current and future natural variability from resampling are shown as bars at the median and at the 90% upper and lower bounds (*Source*: Kay *et al.*, 2009).

three arbitrarily chosen years 2007, 2041, and 2093 selected from the three time slices of years 2000–10, 2040–50, and 2090–2100. For all the cases, it is clear from the figure that a normal PDF fails to model the samples of SPI-12, especially the feature of multimodality, in all the three cases. The PDF obtained using orthogonal series closely resembles the shape generated by the frequency distribution. From the overall trend in probabilities of all categories of drought, it may be concluded that the probability of near-normal conditions will decrease, and the probabilities of mild, severe, and extreme droughts will increase over time.

The following section explains uncertainty propagation methods used in climate change impacts studies, which could be used to assess the combined effect of various sources of uncertainty on the modeled hydrologic variables, such as streamflow or soil moisture.

3.9.2 Uncertainty propagation

Prediction uncertainties in climate impacts are interlinked, and propagate through various steps in the climate change impact assessment process. Kay *et al.* (2009) investigated the uncertainty in the impact of climate change on flood frequency in England, through the use of continuous simulation of river flows. Figure 3.24 shows results from the study, which incorporates

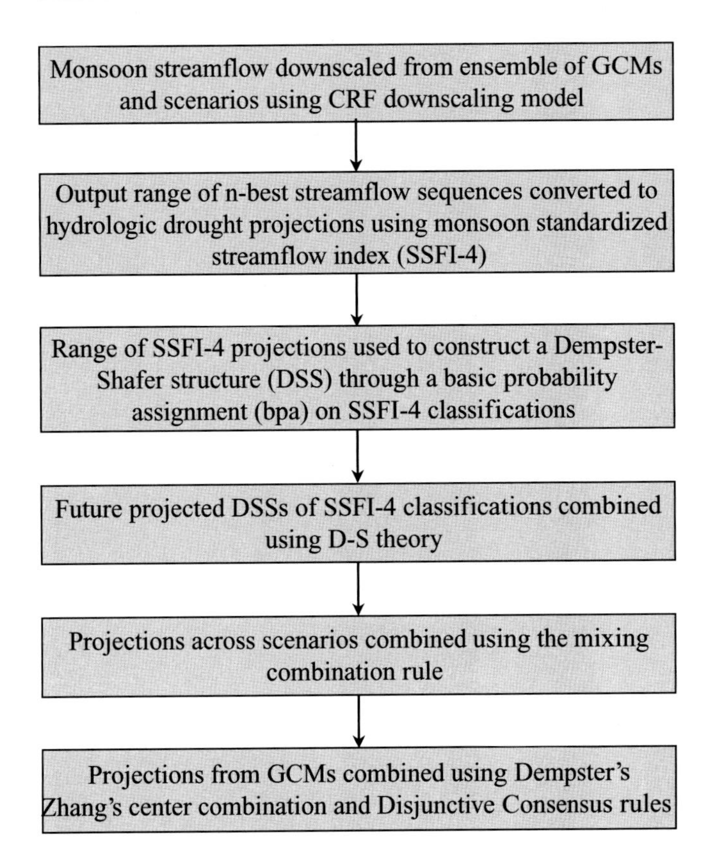

Figure 3.25 Uncertainty combination and propagation using D-S theory.

several sources of uncertainty: future GHG emissions, GCM structure, downscaling methods, RCM structure, hydrologic model structure, hydrologic model parameters, and the internal variability of the climate system. The figure shows the variation in the impact of climate change on flood frequency, from various sources, for five return periods as the percentage change in flood frequency from the current period to the 2080s. It is observed that GCM structure is the dominant source of uncertainty, while that from hydrologic model parameters is the smallest in most cases, although this varies by return period and between the two catchments.

An uncertainty modeling framework was presented by Raje and Mujumdar (2010b), which combines GCM, scenario, and downscaling uncertainty by using a generalized uncertainty measure using the Dempster–Shafer (D-S) evidence theory (Shafer, 1976). The D-S evidence theory, which can be considered a generalized Bayesian theory (Dempster, 1967), is used for representing and combining uncertainty. Unlike a discrete probability distribution on the real line, where the mass is concentrated at distinct points, the focal elements of a D-S structure may overlap one another. The model is hence designed to cope with varying levels of precision regarding the information, and no further assumptions are needed to represent the information. The D-S theory also provides methods to represent and combine weights of evidence (Sentz and Ferson, 2002). Figure 3.25 shows the general uncertainty combination methodology for uncertainty quantification in predicting future monsoon standardized streamflow index (SSFI-4) from monsoon streamflow used in Raje and Mujumdar (2010b). Figure 3.26 shows the results from the case study for the Mahanadi River at the Hirakud Reservoir in Orissa. The mixing combination rule is used first for combining uncertainty across scenarios for each GCM, with equal weights assigned to each scenario projection. Subsequently, combination of GCM uncertainty is performed using the weighted Dempster's rule, the weighted Zhang's center combination rule and the weighted disjunctive consensus rule with equal weights assigned to GCMs.

The difference between plausibility and belief for a given classification in Figure 3.26 shows the associated uncertainty. Since Dempster's rule ignores all conflicting evidence, application of this rule yields the smallest band of uncertainty. Disjunctive consensus based on the union operation does not ignore any evidence and hence shows the largest uncertainty. It is seen that the SSFI-4 projects an increasing probability of drought and decreasing probability of normal and wet conditions in future in Orissa.

Raje and Mujumdar (2010a) presented an uncertainty modeling framework in which, in addition to GCM and scenario uncertainty, uncertainty in the downscaling relationship itself was explored by linking downscaling with changes in frequencies of modes of natural variability. Downscaling relationships were derived for each natural variability cluster and used for projections of hydrologic drought. Each projection was weighted with the future projected frequency of occurrence of that cluster, referred to as "cluster-linking," and scaled by the GCM performance with respect to the associated cluster for the present period, referred to as "frequency scaling." Uncertainty propagation and combination using the D-S theory as used in the work is shown in Figure 3.27. It was shown in this work that a stationary downscaling relationship would either over- or under-predict downscaled hydrologic variable values and associated uncertainty. Results from the work also showed improved agreement between GCM predictions at the regional scale, validated for the twentieth century, implying that frequency scaling and cluster linking may be a useful method for constraining uncertainty.

3.10 SUMMARY

An overview of the background and tools necessary for projecting and evaluating hydrologic impacts due to climate change was presented in this chapter. This subject has a vast scope, and extensive studies have been conducted in the past two decades on topics such as precipitation and streamflow projections for the next century. This precludes an exhaustive coverage of all methodologies

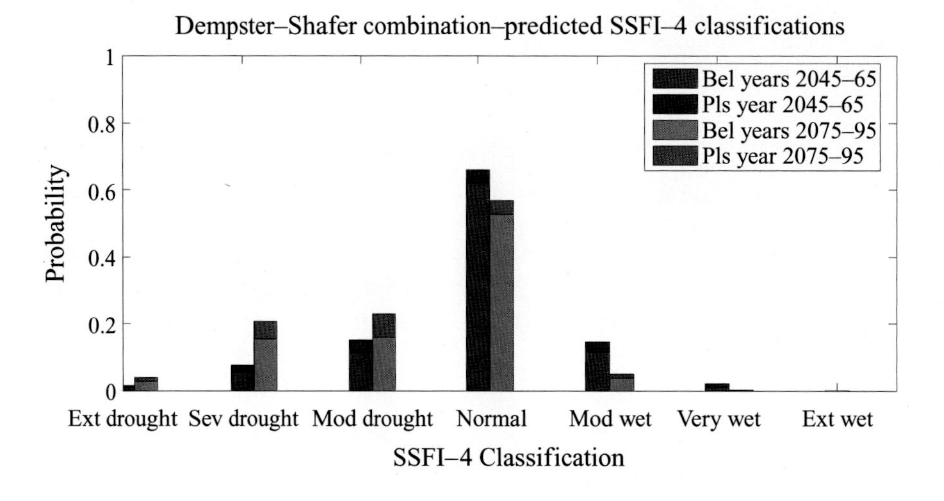

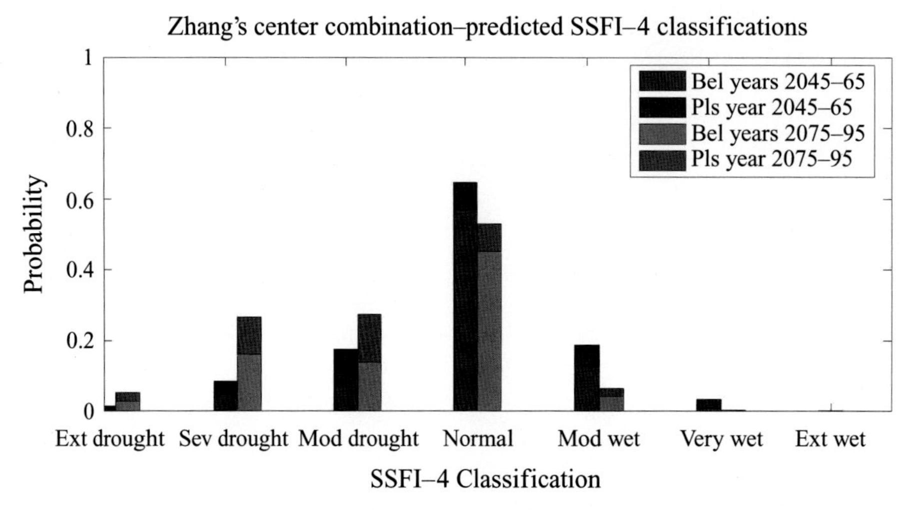

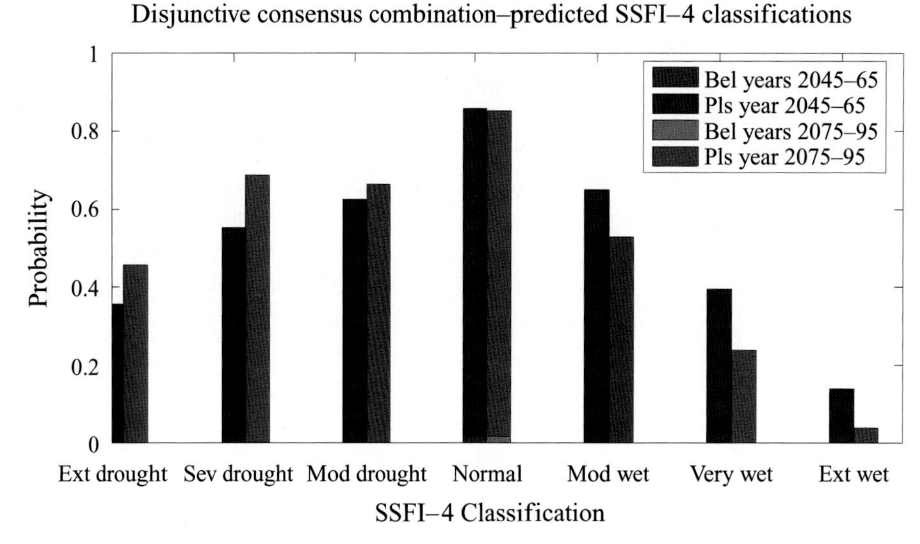

Figure 3.26 Combining uncertainty in terms of belief and plausibility for SSFI-4 classifications for years 2045–65 and 2075–95 using various combination rules (*Source*: Raje and Mujumdar, 2010b).

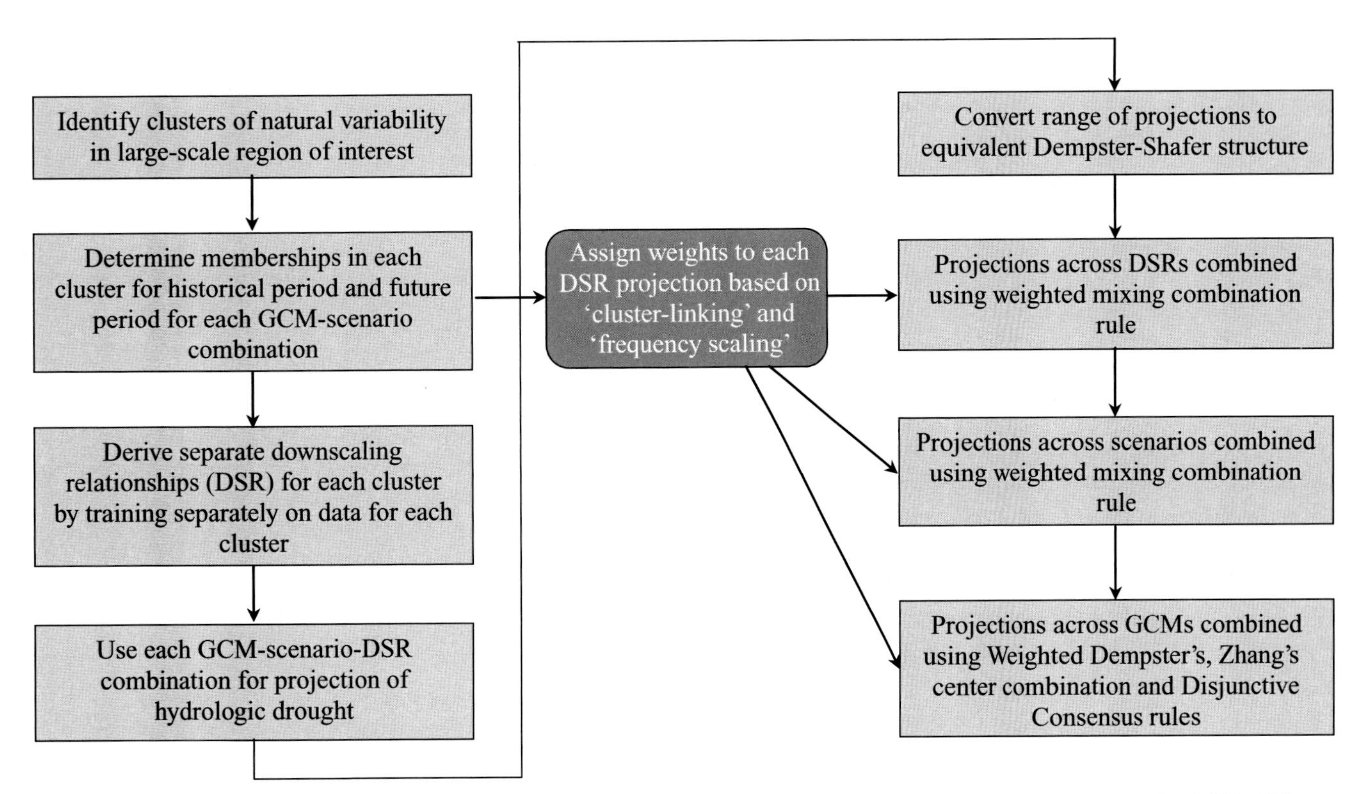

Figure 3.27 Combination of GCM, scenario, downscaling, and downscaling relationship uncertainty using D-S theory and natural variability linkages.

from the literature; however, it is hoped that broad categorizations and distinctions, and significant advances in the field, are covered in this chapter. Statistical downscaling procedures for projection of local hydrometeorological variables from large-scale GCM simulated climate variables are described in detail. Using these variables as inputs along with a suitable hydrologic model will help in evaluation of specific basin-scale impacts of climate change on relevant variables, such as soil moisture or streamflows. Macroscale models presented in the chapter are useful in continental-scale impacts analysis and water balance computations, as well as in adequacy evaluation for GCMs' land surface models and parameterization schemes. Methods presented for temporal disaggregation of seasonal or monthly variables to daily or sub-daily values have direct application in flood frequency analysis. The description of methodologies used for estimating flood frequencies and magnitudes and specific examples are expected to be useful in projection of floods under climate change scenarios. Helping the reader in understanding sources of, and quantifying, uncertainty involved in hydrologic projections under climate change is another goal of the chapter, hoped to have been adequately fulfilled through the discussion of uncertainty propagation techniques and specific examples studied. After reading this chapter, the reader should hence be able to apply the concepts covered for modeling of floods under a changed future climate.

Some broad conclusions can be drawn from the discussion and concepts covered in this chapter. Climate change is likely to lead to changes in the water cycle, which has serious implications for water resources. However, currently available GCMs are unable to model sub-grid-scale hydrologic and convective processes well. In order to study hydrologic impacts of climate change, downscaling methods to predict the regional- or local-scale hydrologic variables from large-scale GCM output have been developed by several researchers. Most impacts studies assume that the basin itself would remain unchanged and ignore changes in vegetation patterns and physiology as well as changes in meteorological parameters, except for precipitation and temperature, due to climate change. However, in practice, climate change will affect not only the precipitation and temperature but also the distribution of glaciers; soil processes, such as drainage capacity and soil quality; land surface processes, such as erosion; and vegetation characteristics, type, and coverage. In addition to downscaling hydrologic variables such as precipitation, temperature and evapotranspiration, it is important to estimate the effect of climate change on flood events, since the timing and magnitude of floods are needed for the design and management of water resources systems. Simple empirical methods have been developed to create daily rainfall scenarios for some catchments, assuming changes in the intensity of the extreme rainfall events. When applied to baseline climatic series, these scenarios show an overall change of the flood regime both in terms of the magnitude and frequency of the extreme events. It is very important to be able to develop realistic scenarios, incorporating the changes in the sub-daily rainfall regime, for flood modeling under climate change.

Current assessments of climate change effects on hydrologic systems are made mainly using offline modeling approaches (Fowler *et al.*, 2007). Outputs from climate model simulations are used to drive hydrologic models and estimate climate change impacts. However, because feedback effects are important, coupled regional modeling systems such as those using MHMs may provide more useful predictions of climate impacts. It should also be noted that hydrologic predictions are burdened with a large amount of uncertainty arising from various sources such as GCM formulations and scenario development. The use of GCM and scenario ensembles is recommended for a realistic assessment of climate change impacts. Cascading uncertainties up to the regional or local level leads to large uncertainty ranges at these scales, and effective uncertainty propagation and combination methods are important for a realistic assessment of hydrologic predictions under climate change. This is especially relevant in flood frequency analysis, where flood risk projections can exhibit significant differences.

EXERCISES

3.1 What are the major factors limiting the direct use of GCMs in hydrologic impact assessment?

3.2 State the broad assumptions of each scenario family as defined in the SRES scenarios (2001). Which scenario has the highest cumulative emissions of carbon dioxide?

3.3 Compare the merits and demerits of dynamical versus statistical downscaling for the purpose of hydrologic impact assessment under climate change.

3.4 What are the assumptions of statistical downscaling?

3.5 What are the three broad approaches used in statistical downscaling? In each of the following cases, examine which approach would be most suitable:

(a) Daily precipitation and temperature scenarios are needed for a river basin, for future flood modeling.
(b) Monthly streamflow inflows for the future are required for reservoir optimization under climate change.
(c) It is necessary to conduct an impact analysis for agricultural and irrigation management in a drought-prone region.
(d) A water supply company needs to project future water balance scenarios under climate change.

3.6 It is necessary to downscale daily precipitation, temperature and evapotranspiration in a river basin. What inter-variable relationships need to be assessed to identify potential predictors for downscaling?

3.7 What other methodologies, apart from those given in Section 3.4.3.2, can be used for bias removal? Is the use of bias removal methodologies justified in downscaling?

3.8 Identify the PCs of the following two-dimensional data. Plot the original data and show the principal directions in this plot. Also plot, separately, the transformed data.

X	Y
2.5	2.4
0.5	0.8
2.3	2.85
3.2	3.1
5.5	5.9
4.7	4.3
1.1	0.9
2.6	2.2

3.9 What are the limitations of PCA when used for dimensionality reduction, prior to statistical downscaling? Which other methodologies can be used for this purpose?

3.10 The figure shows projections of annual maximum discharge for two rivers, under current and future conditions of climate change. Comment on the effect of climate change on the flood regime of each river. Which other data, if available, can be utilized for flood frequency analysis of the rivers?

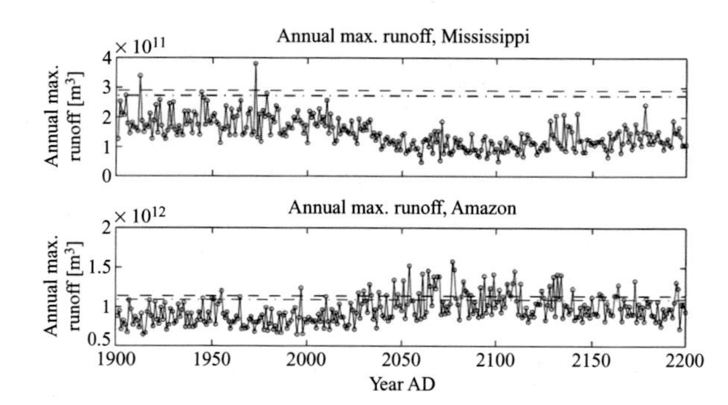

Effect of climate change in two river basins (*Source*: Kleinen and Petschel-Held, 2007)

3.11 What are the main sources of uncertainty in hydrologic prediction under climate change? Which of these sources introduces the largest uncertainty in the process? Which uncertainties need to be factored in each of the cases considered in Exercise 3.5?

3.12 A stream passes through an area considered for residential development. An assessment is needed of the frequency with which the stream would overtop its banks, both under the current climate and under a 2040 climate (the project has a design life of 50 years). There are no streamflow measurements available, but historical monthly rain gauge data are available in the catchment. Explain the steps to assess flood inundation under climate change.

4 Remote sensing for hydrologic modeling

4.1 INTRODUCTION

Remote sensing provides essential inputs required for more effective modeling of different components of the hydrologic cycle. In this chapter, basic concepts of RS that are necessary to understand and analyze images obtained by remote sensing are explained first. Various issues related to satellite RS, such as sensing platforms, passive/active sensors, and spectral reflectance curves are briefly introduced. Important characteristics of digital images, such as spatial, spectral, radiometric, and temporal resolutions, are discussed next. Some useful image processing techniques are explained in three broad categories: image rectification, image enhancement, and information extraction. Extraction of land use/land cover information using different image processing techniques is explained next. Remote sensing inputs that can be obtained for hydrologic modeling, such as rainfall estimation, snow cover/snowmelt estimation, rainfall–runoff modeling, routing of runoff, erosion features, and sediment yield, are discussed and useful techniques are explained in detail. Some of the recent RS missions to obtain detailed information about different components of the hydrologic cycle are presented next. Finally, some of the image processing techniques explained in this chapter are demonstrated using MATLAB.

Remote sensing is the art and science of obtaining information about an object or feature without physically coming into contact with that object or feature. It is the process of inferring surface parameters from measurements of the electromagnetic radiation (EMR) from the Earth's surface. This EMR can be either reflected or emitted radiation from the Earth's surface. In other words, RS is detecting and measuring electromagnetic energy emanating or reflected from distant objects made of various materials, so that we can identify and categorize these objects by class or type, substance, and spatial distribution (American Society of Photogrammetry, 1975).

Remote sensing provides a means of observing hydrologic states or fluxes over large areas. Remote sensing applications in hydrology have primarily focused on developing approaches for estimating hydrometeorological states and fluxes. The primary set of variables includes land use, land cover, land surface temperature, near surface soil moisture, snow cover/water equivalent, water quality, landscape roughness, and vegetation cover. The hydrometeorological fluxes are primarily evaporation and plant transpiration or evapotranspiration and snowmelt runoff.

Basic concepts of RS are introduced in this section.

Electromagnetic energy (E) can be expressed either in terms of frequency (f) or wavelength (λ) of EMR as

$$E = hcf \quad \text{or} \quad hc/\lambda \tag{4.1}$$

where h is the Planck constant (6.626×10^{-34} J s), c is a constant that expresses the celerity or speed of light (3×10^{8} m/s), f is frequency expressed in hertz, and λ is expressed in micrometers (μm; 10^{-6} m). In RS terminology, E is expressed in terms of wavelength, λ. As can be observed from Equation 4.1, shorter wavelengths have higher energy content and longer wavelengths have lower energy content.

The distribution of the continuum of radiant energy can be plotted as a function of wavelength (or frequency) and is known as the EMR spectrum. The EMR spectrum is divided into regions or intervals of different wavelengths and such regions are denoted by different names. The visible region (human eye is sensitive to this region) occupies the range between 0.4 and 0.7 μm. The infrared (IR) region, spanning between 0.7 and 100 μm, has four sub-intervals of special interest for RS: (i) reflected IR (0.7–3.0 μm) and (ii) its film responsive subset, the photographic IR (0.7–0.9 μm); (iii) and (iv) thermal bands at (3–5 μm) and (8–14 μm). Longer wavelength intervals beyond this region are referred to in values ranging from 0.1 to 100 cm. The microwave region spreads across 0.1 to 100 cm, which includes all the intervals used by radar systems. The radar systems generate their own active radiation and direct it towards the targets of interest.

The primary source of energy that illuminates different features on the Earth's surface is the Sun. Solar irradiation (also called insolation) arrives at the Earth at wavelengths determined by the photospheric temperature of the Sun (peaking near 5600 °C). The main wavelength region of solar radiation is from 0.2 to 3.4 μm. The maximum energy (E) is available at 0.48 μm wavelength, which is in the visible green region. As solar energy travels through the atmosphere to reach the Earth, the atmosphere

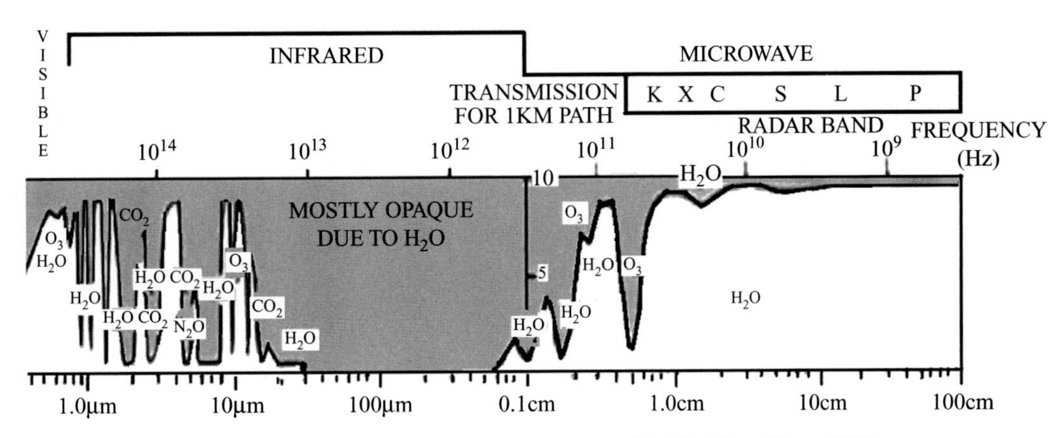

Figure 4.1 Atmospheric windows in the electromagnetic radiation (EMR) spectrum (Modified from Short, 1999).

absorbs or backscatters a fraction of it and transmits only the remainder. Wavelength regions in which most of the energy is transmitted through the atmosphere are referred to as atmospheric windows. Figure 4.1 (Short, 1999) shows the spectrum, identifying different regions with specific names from the visible region to microwave region. In the microwave region, different radar bands are also shown, such as K, X, C, L, and P. In Figure 4.1, shaded zones mark minimal passage of incoming and/or outgoing radiation, whereas white areas denote atmospheric windows. Various constituents of the atmosphere mentioned in Figure 4.1 (such as O_2, CO_2, H_2O) at different wavelengths are mainly responsible for atmospheric absorption or backscatter at those wavelengths.

Most RS instruments on air or space platforms operate in one or more of these windows by making their measurements with detectors tuned to specific wavelengths that pass through the atmosphere. Some of these regions (atmospheric windows) are mentioned in the previous paragraph.

Energy incident on the Earth's surface is absorbed, transmitted, or reflected depending on the wavelength and features of the surface (such as barren soil, vegetation, water body). The amount of energy reflected is measured using different sensors or detectors. The ratio of reflected energy to the incident energy at a specific wavelength is referred to as spectral reflectance, R_λ (expressed as a percentage). Spectral reflectance properties of different features on the Earth's surface vary considerably and are also different at different wavelengths.

4.1.1 Spectral reflectance curves

A spectral reflectance curve of a particular feature is a plot of spectral reflectance of that feature versus wavelength. Typical spectral reflectance curves for vegetation, soil, and water are shown in Figure 4.2 (modified from Lillesand and Kiefer, 2002).

It may be noted that the curves shown in Figure 4.2 represent average reflectance curves of healthy vegetation, dry barren soil, and clear water bodies. Although the reflectance of individual

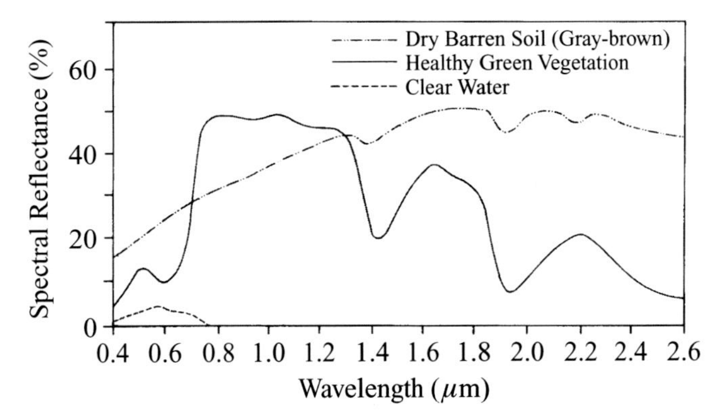

Figure 4.2 Typical spectral reflectance curves for vegetation, soil, and water (Modified from Lillesand and Kiefer, 2002).

features varies considerably above and below the average, these curves demonstrate some fundamental points concerning spectral reflectance. Understanding spectral reflectance curves for different features at different wavelengths is essential to interpret and analyze an image obtained in any one or multiple wavelengths. Some of the salient points are discussed next.

4.1.1.1 SPECTRAL REFLECTANCE FOR VEGETATION

The spectral reflectance curve for healthy green vegetation exhibits the "peak-and-valley" configuration as illustrated in Figure 4.2. The valleys in the visible portion of the spectrum are due to the pigments in plant leaves. Chlorophyll, for example, strongly absorbs energy in the wavelength bands in the range of 0.45 to 0.65 μm within the visible region. Our eyes perceive healthy vegetation as green in color because of the high reflection of green energy within the visible region. If a plant is subject to some form of stress that interrupts its normal growth and productivity, it may decrease or cease chlorophyll production. The result is less chlorophyll absorption in the blue and red bands. Often the red reflectance increases to the point that we see the plant turn yellow (combination of green and red).

In the reflected IR portion of the spectrum, at 0.7 μm, the reflectance of healthy vegetation increases dramatically. In the range from 0.7 to 1.3 μm, a plant leaf reflects about 50% of the energy incident upon it. Most of the remaining energy is transmitted, since absorption in this spectral region is minimal. Plant reflectance in the 0.7 to 1.3 μm range results primarily from the internal structure of plant leaves. As the leaf structure is highly variable between plant species, reflectance measurements in this range often permit discrimination between species, even if they look the same in visible wavelengths. Likewise, many plant stresses alter the reflectance in this region and sensors operating in this range are often used for vegetation stress detection.

Beyond 1.3 μm, energy incident upon vegetation is essentially absorbed or reflected, with little to no transmittance of energy. Dips in reflectance occur at 1.4, 1.9, and 2.7 μm as water in the leaf strongly absorbs the energy at these wavelengths. So, wavelengths in these spectral regions are referred to as water absorption bands. Reflectance peaks occur at 1.6 and 2.2 μm, between the absorption bands. At wavelengths beyond 1.3 μm, leaf reflectance is approximately inversely related to the total water present in a leaf. This total water is a function of both the moisture content and the thickness of a leaf.

4.1.1.2 SPECTRAL REFLECTANCE FOR SOIL

The spectral reflectance curve for soil in Figure 4.2 shows considerably less peak-and-valley variation in reflectance compared to that for vegetation, i.e., the factors that influence soil reflectance act over less specific spectral bands. Some of the factors affecting soil reflectance are moisture content, soil texture (proportion of sand, silt, and clay), surface roughness, presence of iron oxide, and organic matter content. These factors are complex, variable, and interrelated. For example, the presence of moisture in soil decreases its reflectance. As with vegetation, this effect is greatest in the water absorption bands at 1.4, 1.9, and 2.7 μm. Clay soils also have hydroxyl ion absorption bands at 1.4 and 2.2 μm. Soil moisture content is strongly related to the soil texture. For example, coarse, sandy soils are usually well drained, resulting in low moisture content and relatively high reflectance. On the other hand, poorly drained fine-textured soils generally have lower reflectance. In the absence of water, however, the soil itself exhibits the reverse tendency, i.e., coarse-textured soils appear darker than fine-textured soils. Thus, the reflectance properties of soil depend on a number of factors. Two other factors that reduce soil reflectance are surface roughness and the content of organic matter. The presence of iron oxide in a soil also significantly decreases reflectance, at least in the visible region.

4.1.1.3 SPECTRAL REFLECTANCE FOR WATER

The spectral reflectance curve for water clearly shows the energy absorption at reflected IR wavelengths. Locating and delineating water bodies with RS data is done more easily in reflected IR wavelengths because of this absorption property. However, various conditions of water bodies manifest themselves primarily in visible wavelengths. The energy/matter interactions at these wavelengths are very complex and depend on a number of interrelated factors. For example, the reflectance from a water body can stem from an interaction with the water's surface (specular reflection), with material suspended in the water, or with the bottom surface of the water body. Even in deep water, where bottom effects are negligible, the reflectance properties of a water body are not only a function of the water per se but also of the material in the water.

Clear water absorbs relatively little energy of wavelengths less than 0.6 μm. High transmittance typifies these wavelengths with a maximum in the blue-green portion of the spectrum. However, as the turbidity of water changes (because of the presence of organic or inorganic materials), transmittance and therefore reflectance change dramatically. For example, water bodies containing large quantities of suspended sediments normally have much higher visible reflectance than clear water. Likewise, the reflectance of water changes with the chlorophyll concentration involved. Increases in chlorophyll concentration tend to decrease reflectance in blue wavelengths and increase it in the green wavelengths. These changes have been used to monitor the presence and estimate the concentration of algae based on RS data. Reflectance data have also been used to determine the presence or absence of tannin dyes from bog vegetation in lowland areas, and to detect a number of pollutants, such as oil and certain industrial wastes.

Many important characteristics of water, such as dissolved oxygen concentration, pH, and salt concentration, cannot be observed directly through changes in water reflectance. However, such parameters sometimes correlate with observed reflectance. Thus, there are many complex interrelationships between the spectral reflectance of water and particular characteristics. Further details on the spectral characteristics of vegetation, soil, and water can be found in Swain and Davis (1978).

4.1.2 Passive/active remote sensing

Depending on the source of energy for EMR, RS can be classified as passive or active. In the case of passive RS, the source of energy is that available naturally, such as from the Sun. Most RS systems work in the passive mode using solar energy as the source of EMR. Any object that is at a temperature above 0 K (kelvin) will emit some radiation. However, the energy content will be approximately proportional to the fourth power of the temperature. Thus, the Earth will also emit some radiation since its ambient temperature is about 300 K. Passive sensors can also be used to measure the Earth's radiance, but they are not very popular as the energy content is very low. In the case of active RS, the source of energy is generated and emitted from a sensing platform. Most microwave RS is done through active RS. As a

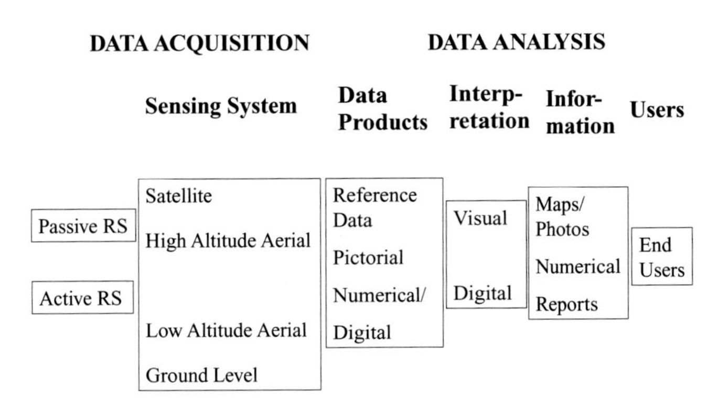

Figure 4.3 Schematic showing different RS platforms for data acquisition and various components of data analysis.

simple analogy, passive RS is similar to taking a picture with an ordinary camera whereas active RS is analogous to taking a picture with camera having built-in flash.

4.1.3 Remote sensing platforms

Remote sensing platforms can be classified as follows based on the elevation from the Earth's surface at which these platforms are placed (Figure 4.3):

- ground-level remote sensors are very close to the ground and they are basically used to develop and calibrate sensors for different features on the Earth's surface;
- low-altitude aerial RS;
- high-altitude aerial RS;
- space shuttles;
- polar-orbiting satellites;
- geostationary satellites.

From each of these platforms, RS can be done in either passive or active mode.

Geostationary or geosynchronous satellites are used for communication and meteorological purposes. Such a satellite is stationary with respect to a point on the equator, i.e., the satellite must be geosynchronous with an orbital period of 24 hours. Its sense of direction is the same as that of the rotation of the Earth on its axis, i.e., west to east. It is placed at a high altitude of about 36,000 km. It is on the equatorial plane with a heavily inclined orbit of 180° (angle of inclination of orbital plane with respect to equator is measured clockwise). It can yield large-area coverage of 45% to 50% of the total globe (footprint) at regular intervals of time.

Polar-orbiting or sun-synchronous satellites are essentially used for RS. The polar orbit is used to take advantage of the Earth's rotation on its axis so that newer segments (or sections) of the Earth come under the view of the satellite, provided the orbital period of the satellite is shorter than the rotational period of the Earth (24 hours). Typically the orbital period for a RS satellite is 103 minutes with its orbital plane inclined at 99°. Nadir is the point of intersection on the surface of the Earth of the radial line between the center of the Earth and the satellite. This is the point of shortest distance from the satellite. The circle on the surface of the Earth described by the nadir point as the satellite revolves is called the ground track. Swath refers to the width of the image perpendicular to the ground track. It is expressed in kilometers on the ground or number of pixels per row in an image.

Remote sensing satellites are placed in near-polar, near-circular, inclined, medium-period, and sun-synchronous orbits for the following objectives:

- near-polar – for global coverage;
- near-circular – for uniform swath;
- inclined – for differences in gravitational pull between equator and poles;
- medium period – for global coverage;
- Sun-synchronous – for constant angle between the aspects of incident solar radiation and viewing by the satellite.

A schematic diagram showing different RS platforms for data acquisition and various components of data analysis is presented in Figure 4.3.

4.2 DIGITAL IMAGES

Based on the type of recording of remotely sensed data, images may be in photographic or digital form. Photographic RS is restricted to aerial RS. Photographic RS can be in the following forms:

- Panchromatic – black and white (B&W): records EMR in the visible ranges, 0.4–0.7 μm.
- Photographic IR (B&W): records EMR in the range 0.76–0.88 μm, which is useful for delineating land–water interfaces, drainage systems with and without water, and vegetation surveying.
- Natural color photography: human visual system can differentiate thousands of colors but only few grey levels. Colors are produced according to human perception.

Although images can be produced using film, they can also be recorded electronically using charge coupled devices (CCDs). Such images are referred to as digital images. A digital image is a regular grid array of squares (pixels) where each grid cell is assigned a digital number (DN).

Low DN indicates low reflectance and high DN indicates high reflectance. In a 7-bit system, DN values range from 0 to 127, whereas in an 8-bit system, DN values range from 0 to 255. Digital images offer the following advantages:

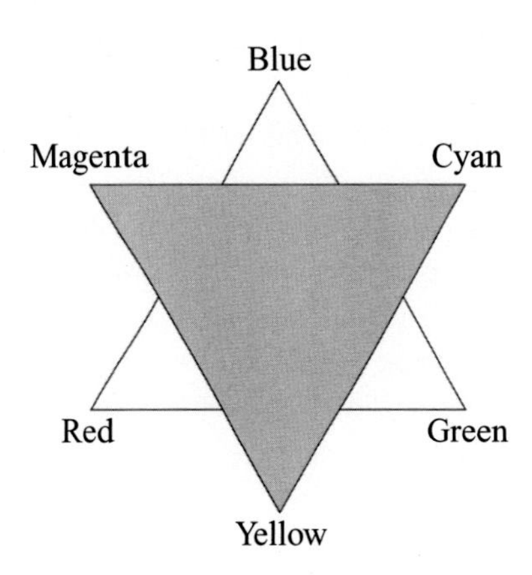

Figure 4.4 Primary and complementary color triangles.

- Photographic film cannot record EMR beyond 1 μm, whereas digital images can record far beyond 1 μm wavelength.
- It is not possible to produce and store large number of films on board a satellite.
- Solid-state electronic devices, CCDs, are very reliable, use little power, and are small and light.
- Data that are obtained digitally can be transmitted easily without any degradation.
- Digital data are in a form that can be readily processed on computers.
- Digital image processing techniques are very powerful for image analysis.

4.2.1 Color composites

The primary additive colors are red, green, and blue and their respective complementary colors are cyan, magenta, and yellow. Color triangles are presented in Figure 4.4, showing the primary and complementary colors as the opposite vertices of a triangle and its superimposed inverse.

If white light is passed through a color filter it is possible to subtract one of the primary colors. For example, a yellow filter subtracts blue light but passes green and red. When two primary colors of light of equal intensity are added, the complementary color of the third primary color will be produced. For example, if red and green are added, yellow will be produced, which is the complementary color of blue.

When the three primary colors are superimposed in equal amounts, then grey tones ranging from black to white are produced. On the other hand, if the three primary colors are superimposed in unequal amounts, then a number of colors are produced.

A good knowledge of the concept of color mixing is essential to prepare color composites from multispectral images. Multispectral images are produced by simultaneously obtaining images of the same scene at different wavelengths. For example, the blue, green, and red wavelengths in the visible region and near-infrared (NIR) bands can produce four different spectral images of a region. These multispectral images can be analyzed individually or by combining any three of them to produce a color composite. Since there are only three primary colors, any three spectral images can be combined to produce a color composite. As already mentioned, if three primary colors are superimposed (mixed) in unequal amounts, then a number of colors are produced. If the multispectral images are from a 7-bit system, DN values in each image range from 0 to 127. Thus, for a pixel, when color mixing takes place for the three spectral bands considered, a large number of color tones can be produced (128^3 in case of a 7-bit system).

When multispectral images in the blue, green, and red wavelengths are given the three primary colors blue, green, and red, respectively, to prepare a color composite, the resulting composite is referred to as a true color composite (TCC) or natural color composite. On the other hand, when multispectral images in blue, green, and red wavelengths are given the three primary colors red, green, and blue, respectively, to prepare a color composite, the resulting composite is referred to as a false color composite (FCC). False color composites are essential in multispectral image analysis as the images that are not in the visible region are given one of the primary colors to produce a color composite. For example, a NIR band image can be given a red color, a red band image can be given a green color, and a blue band image can be given a blue color to produce a false color composite. Thus, by choosing any three of the four spectral images available, ${}^4C_3$, a number of color composites can be produced. Of them, one color composite will be a TCC and all others will be FCCs. Among the number of color composites that can be produced, the FCC of a particular combination, namely, NIR band – red color, red band – green color and green band – blue color, is referred as the standard FCC, as this particular combination is expected to contain more information when compared to all the other FCCs. A typical standard FCC is shown in Figure 4.5. On a standard FCC, vegetation will appear in red (since it has high reflectance in the NIR band, which is given the red color), while a deep clear water body will appear black. The standard FCC showing a portion of Uttara Kannada district, Karnataka, India, presented in Figure 4.5, was prepared using IRS (Indian Remote Sensing) satellite 1C LISS III data. Some of the features that can be noticed in Figure 4.5 are:

- the Arabian Sea on the left (appearing in blue tones in shallow coastal regions and in black for the deeper portion);
- west-flowing river in top half of the image;
- dense vegetation (in red tones);
- small white patches associated with their shadows in black are clouds.

Based on the spectral reflectance curves and color mixing (refer to color triangles) any natural feature can be identified with a specific color in a FCC. It may be noted that if the same band

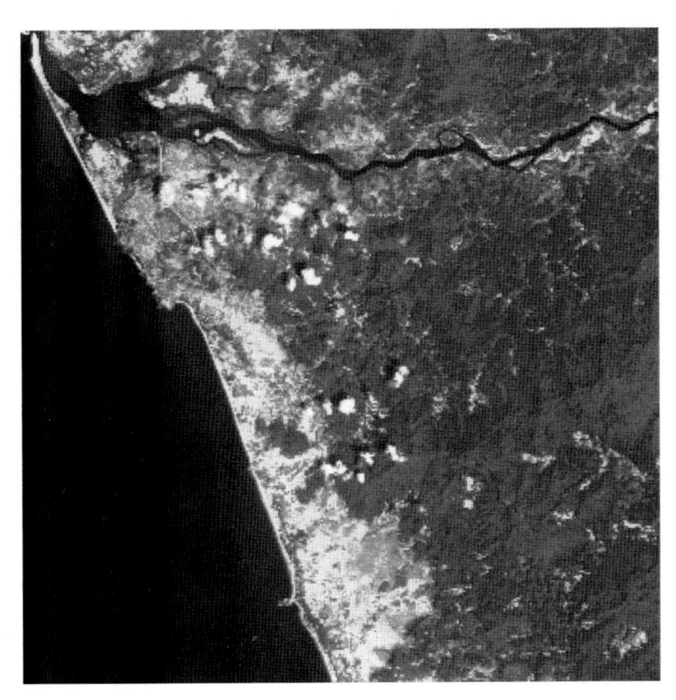

Figure 4.5 Standard false color composite (FCC) prepared using IRS–1C LISS III data. See also color plates.

data (say NIR) are given all the three primary colors to produce a composite, it will appear as a black and white image since all the three primary colors are mixed in equal proportion for any pixel.

4.2.2 Image characteristics

4.2.2.1 SPATIAL RESOLUTION

This is a measure of the area or size of the smallest dimensions (pixel) on the Earth's surface over which an independent measurement can be made by the sensor. It is expressed by the size of the pixel on the ground in meters (pixel – picture element – is square). For example, the NOAA AVHRR (National Oceanic and Atmospheric Administration, Advanced Very High Resolution Radiometer) sensor has a very coarse spatial resolution of 1.1 km, whereas sensors on the Quickbird satellite offer a fine spatial resolution of less than 1 m.

4.2.2.2 SPECTRAL RESOLUTION

The spectral resolution of a sensor characterizes the ability of the sensor to resolve the energy received in a spectral bandwidth to identify different constituents of the Earth's surface. Spectral resolution is defined as the spectral bandwidth of the sensor/detector over which it is sensitive. Spectral resolution is expressed as width of the band in μm. Two different surfaces (A and B) may be indistinguishable in a single band but can be differentiated if the same bandwidth is divided into two or more spectral bands. For example, if the entire spectral visible region (0.4 to 0.7 μm) is considered as one band, the corresponding sensor is known as a panchromatic sensor. The spectral resolution of a panchromatic sensor is approximately 0.3 μm. Alternatively, the visible region can be sliced into three spectral bands, namely, 0.4–0.5, 0.5–0.6, and 0.6–0.7 μm, producing three images within the visible region. The image produced from multiple spectral bands of the same region is called a multispectral image. For a given spatial resolution, many features on the Earth's surface can be better distinguished using an image of higher spectral resolution, compared to an image with coarser spectral resolution.

Hyperspectral data A significant advance in sensor technology stemmed from subdividing spectral ranges of radiation into spectral bands (intervals of continuous wavelengths), allowing sensors in several bands to form multispectral images of the same region under investigation. For example, the visible region (0.4 to 0.7 μm) can be divided into three spectral bands say 0.4–0.5, 0.5–0.6, and 0.6–0.7 μm. These three spectral bands approximately represent the blue, green, and red portions of the visible region. Thus, most of the atmospheric windows can be divided into a number of spectral bands. A typical RS satellite has four to seven spectral bands, mainly in the visible and IR regions. For example, Landsat has seven spectral bands with three in the visible region and four in the IR region, whereas an IRS satellite has four spectral bands with three in the visible region and one in the NIR region.

Recently, sensors that are sensitive to very narrow spectral regions have been developed. With the advances in development of sensors, atmospheric windows could be divided into a large number of spectral bands to obtain images of the same region in hundreds of spectral bands. Such images are referred to as hyperspectral images or data. For example, NASA EO-1 (National Aeronautics and Space Administration, Earth Observation) Hyperion provides high-resolution hyperspectral images capable of resolving radiation from 0.4 to 2.5 μm into 220 spectral bands.

4.2.2.3 RADIOMETRIC RESOLUTION

Radiometric resolution of a sensor is the measure of the number of grey levels measured between pure black (no reflectance) and pure white (high reflectance). Since the reflectance is recorded by a CCD in the form of DNs, radiometric resolution is expressed in bits. In a single-bit system, reflectance is recorded in only two levels, namely, 0 for no reflectance and 1 for reflectance. In a 2-bit system, reflectance can be distinguished in four levels. In a typical RS satellite sensor, radiometric resolution will be 7 bits (128 levels) or 8 bits (256 levels). However, some sensors have very high radiometric resolution (VHRR). For example, NOAA AVHRR has a radiometric resolution of 11 bits (2048 levels).

4.2.2.4 TEMPORAL RESOLUTION

Temporal resolution of a RS system is the measure of the interval of time in which the data are obtained for the same

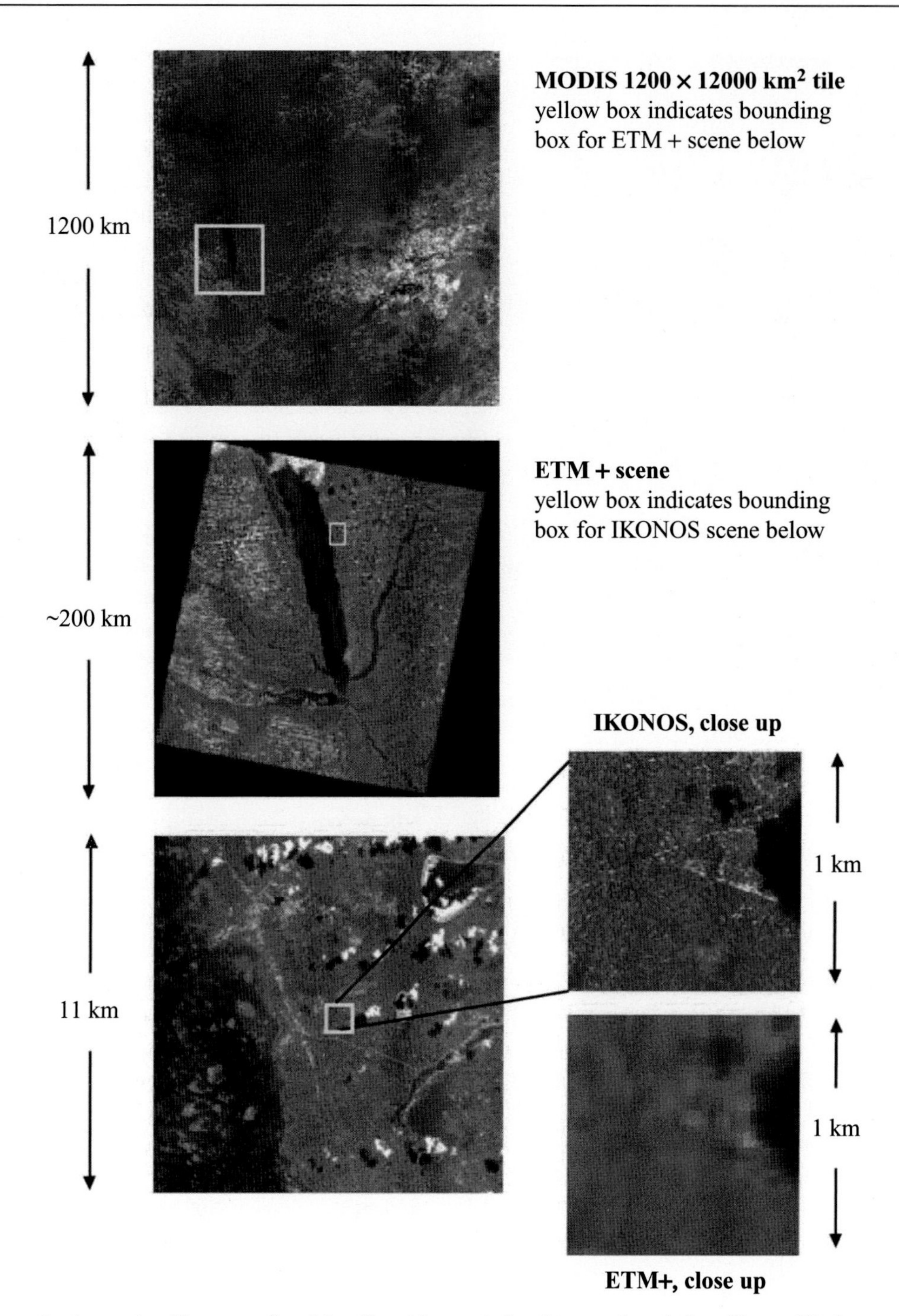

Figure 4.6 Remote sensing images in wide ranges of spatial, radiometric, spectral, and temporal resolutions (*Source*: Morisette *et al.*, 2002). See also color plates.

region. This is applicable only to RS satellites, as they revisit the same region at regular intervals of time depending on their orbital cycle and global coverage time. Temporal resolution varies from less than 1 hour to approximately 30 days. For example, temporal resolution of NOAA AVHRR is 1 day, while it is 17 days in the case of Landsat TM (Thematic Mapper). Repeated coverage of a given region at regular intervals of time enables monitoring of changes in land use/land cover, rate of urbanization, and dynamic events such as cyclone, flood, volcano, or earthquake.

Remote sensing images covering the Mongu Core Site, Zambia, in wide ranges of spatial, radiometric, spectral, and temporal resolutions obtained from different satellite platforms are shown in Figure 4.6 (Morisette *et al.*, 2002).

4.2.2.5 EFFECTS OF CHANGES IN RESOLUTION

Increasing the spectral, spatial, and radiometric resolutions of a system may decrease the signal to noise ratio (SNR) to such an extent that the reflectance data received by the sensor may not be reliable. SNR also depends on the strength of the signal available. For example, to maintain a uniform SNR in IRS, the first three bands have 0.1 μm bandwidth (radiometric resolution) while the fourth band has a bandwidth of 0.3 μm. Depending on the specific feature to be studied, one (or more) of the spectral, spatial, radiometric, and temporal resolutions of a system may have to be compromised. For example, NOAA AVHRR has a very high radiometric resolution (11 bits) and temporal resolution (1 day) but has a very coarse spatial resolution of 1 km. In spite of this, NOAA AVHRR data are very useful for vegetation condition monitoring at regional scale. Effects of changes in resolution are also shown in Figure 4.6.

Digital image processing plays a vital role in the analysis and interpretation of remotely sensed data (Moik, 1980). In particular, data obtained from satellite RS, which is in digital form, can be best utilized with the help of digital image processing. Digital image processing involves substituting the DN associated with a particular pixel by another DN, in order to enhance a particular feature. Image processing can be broadly classified into three categories, namely, image rectification, image enhancement, and information extraction. Image rectification involves applying geometric and other corrections to restore from some errors that may have occurred in acquiring the image through satellite RS. Image enhancement and information extraction are two important components of digital image processing. Image enhancement techniques help in improving the visibility of any portion or feature of the image whilst suppressing the information of other portions or features. Information extraction techniques help in obtaining statistical information about any particular feature or portion of the image. These techniques are discussed in detail and illustrated in the next section.

4.3 IMAGE RECTIFICATION

Image rectification or restoration, which may also be called preprocessing of images, involves applying various corrections to the image. They include geometric corrections, atmospheric corrections, solar illumination corrections, etc. Geometric corrections are performed to rectify an image from geometric distortions that may result due to Earth's rotation, small variations in the satellite's orbit and attitude (pitch, roll, and yaw), etc.

4.3.1 Co-registration of images/maps

Image restoration/rectification involves co-registration of an uncorrected image to a geo-referenced data set or image with the help of ground control points (GCPs). A GCP is a point whose position can be determined on the uncorrected image (row and column) and also on the geo-referenced data set or image (latitude, longitude, or grid coordinates). The unrectified image is tied down to a map or rectified image using these GCPs and the polynomials calculated. The unrectified image is then transformed using the polynomial equations to obtain a geometrically corrected image. For proper rectification of an image, it should be ensured that the GCPs are reliable and evenly distributed throughout the image.

4.4 IMAGE ENHANCEMENT

Image enhancement deals with enhancing certain features or components in an image by applying various image processing techniques. The processing helps in maximizing clarity, sharpness, and details of features of interest towards information extraction and further analysis. There is no single enhancement procedure that is best suited for all purposes. The one that best displays the features of interest to the analyst has to be chosen.

4.4.1 Contrast stretching

The sensitivity range of any RS detector is designed to record a wide range of terrain brightness from black basalt plateaus to sand deserts under a wide range of illuminating conditions. Very few individual scenes have a brightness range that utilizes the full sensitivity range of these detectors. This fact can be observed by plotting the histogram of an image. The histogram can be obtained by plotting the number of pixels for each DN versus DN. A typical histogram is shown Figure 4.17 in Section 4.8.1.2. Most of the DNs of an image will be distributed in a narrow range rather than the total range of DNs, thus reducing the contrast in an image. To produce an image with the optimum contrast ratio, it is important to utilize the entire brightness range of the display medium. Techniques for improving image contrast are among the most widely used enhancement procedures.

4.4.1.1 LINEAR CONTRAST STRETCH

In linear contrast stretching, new DNs are assigned to an output image by assigning the lowest and highest DN in the input image values (say in the range 0 to 255 for an 8-bit system) for the output image and stretching all intervening DNs accordingly:

$$DN_{\text{new}} = \frac{(DN - DN_{\text{min}})}{(DN_{\text{max}} - DN_{\text{min}})} \tag{4.2}$$

Thus, a DN value in the low end of the original histogram is assigned to extreme black and a value at the high end is assigned to extreme white. The improved contrast ratio of the image with linear contrast stretch will enhance different features in the image. Most of the image processing software displays an image only after linear stretching by default. For color images, the individual bands are stretched before being combined in a color image.

4.4.1.2 NON-LINEAR CONTRAST STRETCH

Non-linear contrast enhancement is done in different ways. A uniform distribution stretch (or histogram equalization) redistributes the original histogram to produce a uniform population density of pixels along the horizontal DN axis. This stretch applies the greatest contrast enhancement to the most populated range or brightness values in the original image. The middle range of brightness values are preferentially stretched, which results in maximum contrast. The uniform distribution stretch strongly saturates brightness values at the sparsely populated light and dark tails of the original histogram. The resulting loss of contrast in the light and dark ranges is similar to that in the linear contrast stretch but not as severe. A typical histogram equalized image is shown in Figure 4.18 in Section 4.8.1.2.

A Gaussian stretch is a non-linear stretch that enhances contrast within the tails of the histogram. This stretch fits the original histogram to a normal distribution curve between the 0 and 255 limits, which improves contrast in the light and dark ranges of the image. This enhancement occurs at the expense of contrast in the middle grey range.

Piecewise linear stretch, logarithmic stretch, and power-law stretch are some other contrast stretching methods (Sabbins, 1986).

4.4.2 Density slicing

Density slicing converts the continuous grey tone of an image into a series of density intervals, each corresponding to a specified digital range. Slices may be displayed as areas bounded by contour lines or each slice (digital range) can be assigned a different color, thus converting a black and white single-band image into a color image. This technique emphasizes subtle grey-scale differences that may be imperceptible to the viewer. A density-sliced color image shows much more clarity than the black and white image.

4.4.3 Thresholding

Threshold can be applied to an image to isolate a feature represented by a specific range of DNs. To calculate the area of lakes, DNs not representing water bodies are a distraction. If the highest DN for water is 35 (in NIR image) and is used as the threshold, all DNs greater than 35 are assigned 255 (saturated to white); DNs less than or equal to 35 are assigned zero (black). Lakes will be much more prominent in the image after such thresholding.

This approach can also be adapted to black-out portions of the image, which may not be of interest for a specific region analysis. For example, by overlaying a watershed boundary on an image, the pixels outside the watershed boundary can be assigned a DN value of zero. Thus the analyst can focus only on the region of interest by using this technique.

4.4.4 Filtering techniques

If a vertical or horizontal section is taken across a digital image and the DNs are plotted against distance, a complex curve is produced. An examination of this curve would show sections where the gradients are low, corresponding to smooth tonal variations on the image, and sections where the gradients are high and the DN values change by large amounts over short distances. Filtering is a process that selectively enhances or suppresses particular wavelengths within an image. Two approaches are used to digitally filter images, namely, convolution filtering in the spatial domain and Fourier analysis in the frequency domain. Convolution filtering is the most frequently used approach and is explained next.

A filter is a regular array or matrix of numbers that, using simple arithmetic operations, allows the formation of a new image by assigning new pixel values depending on the results of the arithmetic operations. The convolution matrix is typically a square matrix, for example 3×3, although rectangular matrices can also be used, which must contain odd number of rows and columns. The matrix (filter) is initially positioned in the top left corner of the image and a new DN is calculated for the pixel covered by the central matrix cell (row 2, column 2). The matrix (also called the digital filter or kernel) is then passed through the image data set by shifting one column, and the same calculation is performed and a new DN applied to the pixel in row 2, column 3. Based on the elements used in the matrix and the procedure used for calculating the new DN, different digital filters are developed for different purposes. A typical procedure would be to multiply the matrix (filter) element with the corresponding DN value of the pixel just below it and all such products. Thus, for a 3×3 filter, there will be nine products (multiplications) that should be added to obtain the new filtered value, which could be used to replace the central pixel value in the original image for the current location of the filter. In some cases, the filtered value could be added or subtracted to the central pixel value to obtain the new value. For example, an averaging digital filter comprises a 3×3 matrix in which all the elements are 1. The filtered value obtained is divided by 9 and is used to replace the central pixel value of the image. When such a filter is applied on an image, it will replace every pixel with the average of the surrounding DN values. Thus, an averaging digital filter reduces the effects of the noise component of an image.

It may be noted that filtering decreases the size of the original image depending on the kernel size. In the case of a 3×3 digital filter, there will not be any filtered values for the pixels in the first row, last row, first column, and last column, as these pixels cannot become central pixels for any placement of a 3×3 digital filter. Thus, the filtered image from a 3×3 kernel will not contain the first row, last row, first column, and last column of the original image.

4.4.5 Edge enhancement

Most interpreters are concerned with recognizing linear features in images, such as joints and lineaments. Edges are generally formed by long linear features such as ridges, rivers, roads, railways, canals, folds, and faults. Such linear features (edges) are important to geologists and civil engineers. Geographers map manmade linear features such as highways and canals. Some linear features occur as narrow lines against a background of contrasting brightness and others as the linear contrast between adjacent areas of different brightness. In all cases, linear features are formed by edges. Some edges are marked by pronounced differences that may be difficult to recognize. Contrast enhancement may emphasize brightness differences associated with some linear features. This procedure, however, is not specific for linear features because all elements (not just the linear elements) of the scene are enhanced equally. Digital filters have been developed specifically to enhance edges in images and they fall into two categories: directional and non-directional. Directional filters will enhance linear features that have specified orientation (say those oriented to 30° N), whereas non-directional filters will enhance linear features in almost all orientations. Some of the non-directional filters are Prewitt gradient, Sobel, Canny, and Laplacian filters. A typical edge-enhanced image is shown in Figure 4.24 in Section 4.8.6.4.

4.4.5.1 LAPLACIAN FILTER

The Laplacian filter (Sabbins, 1986) is a 3×3 kernel with a high central value, 0 at each corner and -1 at the center of the edge (Figure 4.7). The filter is placed over a 3×3 array of original pixels and each pixel is multiplied by the corresponding value in the kernel. The nine resulting values are summed (four of them are zeros) and the resulting kernel value is combined with the central pixel of the 3×3 array; this number replaces the original DN of the central pixel and the process is repeated.

0	-1	0
-1	0	-1
0	-1	0

Figure 4.7 Laplacian filter.

The Laplacian filter will enhance edges in all directions except those in the direction of the movement of the filter (i.e., linear features with east–west orientation will not be enhanced). A weighting factor (<1 to diminish, >1 to accentuate the effect of the filter) can be used to further enhance or diminish the edges in images. The resulting kernel value is multiplied by the weighting factor before adding it to the central pixel.

4.4.5.2 DIRECTIONAL FILTER

The directional filter will enhance linear features with a specific orientation (direction). The direction in which the edges are to be enhanced is specified in degrees with respect to north. Angles within the first quadrant are considered with negative sign and those falling in the fourth quadrant are considered with positive sign. This filter consists of two 3×3 kernels, as shown in Figure 4.8(a), which are referred to as left and right kernels. The right kernel is placed over the array of original pixels and each

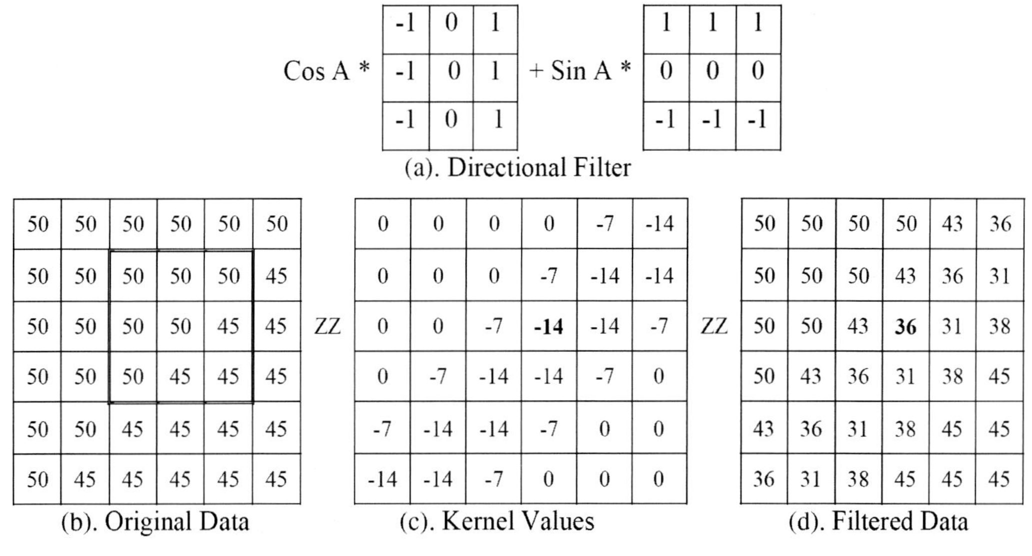

Figure 4.8 Edge enhancement using directional filter.

pixel is multiplied by the corresponding kernel value. The resulting sum of nine values is multiplied by sin(A), where A is the angle specified. Similarly, the left kernel is applied on the same array of original pixels and the resulting sum of nine values is multiplied by cos(A). The resulting two kernel values are summed to obtain the directional filter value, which is then added to the central pixel.

The directional filter can be demonstrated by applying it to a data set (Figure 4.8(b)) in which a brighter area (DN = 50) is separated from a darker area (DN = 45) along a total lineament that trends northeast (A = −45°). The profile along ZZ in Figure 4.8(b) shows the brightness difference (DN = 5) across the lineament.

The filter is demonstrated by applying it to the array of nine pixels shown by the box over the original data set (Figure 4.8(b)) and the resulting kernel and filtered values are shown in bold in Figure 4.8(c) and (d). The sequence of operations is as follows:

1. Place the right filter kernel over the array of original pixels and multiply each pixel by the corresponding filter value. When these nine resulting values are summed up, the resulting value is 10.
2. Determine the sine of the angle (sin(−45°) = −0.71)) and multiply this by the value obtained in step 1 (−0.71 × 10) to give a filter value of −7.
3. Place the left filter kernel over the pixel array and repeat the process; the resulting value is −7.
4. Sum the two filtered values (−14). This value replaces the central pixel in the array of the original data. When steps 1 through 4 are applied to the entire pixel array, the resulting kernel values are shown in the array in Figure 4.8(c).
5. The filtered values for each pixel are then combined with the original value of the pixel to produce the array as shown in Figure 4.8(d).

The contrast ratio of the lineament in the original data set (50/45 = 1.11) is increased in the final data set to (50/31 = 1.61), which is a contrast enhancement of 45% [100 × (1.61 − 1.11)/1.11 = 45]. As shown in profile ZZ, the original lineament occurs across an interface only one pixel wide, whereas the enhanced lineament is four pixels wide.

An alternate way to express this operation is to say that the filter has passed through the data in a direction normal to the specified lineament direction. In this example the filter passing through the normal direction is N45°W. In addition to enhancing features oriented normal to this direction of movement, the filter also enhances linear features that trend obliquely to the direction of filter movement. As a result, many additional edges of diverse orientations get enhanced. The directionality of the filter may also be demonstrated by passing the filter in a direction parallel with a linear trend. As a result, these fractures appear subdued, whereas north and northeast trending features are strongly enhanced.

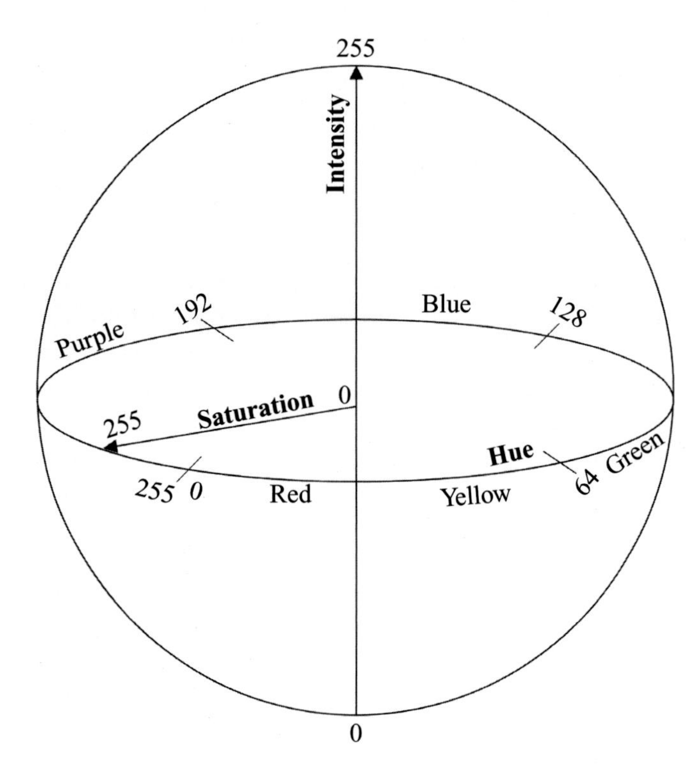

Figure 4.9 IHS sphere (Modified from Sabbins, 1986).

4.4.6 Intensity, hue, saturation images

The additive system of primary colors (red, green, and blue, or RGB system) is well established. An alternate approach to color is the intensity, hue, and saturation system (IHS), which is useful as it presents colors more nearly as a human observer perceives them. The IHS system is based on the color sphere (Figure 4.9) in which the vertical axis represents the intensity, the radius represents the saturation, and the circumference represents the hue. The intensity (I) axis represents brightness variations ranging from black (0) to white (255) and no color is associated with this axis. Hue (H) represents the dominant wavelength of color. Hue values commence with 0 at the midpoint of red tones and increase counterclockwise around the circumference of the sphere to conclude with 255 adjacent to 0. Saturation (S) represents the purity of color and ranges from 0 at the center of the color sphere to 255 at the circumference. A saturation of 0 represents a completely impure color, in which all wavelengths are equally represented and which the eye will perceive as a shade of grey that ranges from white to black depending on intensity. Intermediate values of saturation represent pastel shades, whereas high values represent purer and more intense colors. The range from 0 to 255 is used here to be consistent with the 8-bit scale, any other range of values (0 to 100, for example) could equally well be used as IHS coordinates. Buchanan (1979) described the IHS system in more detail.

When any three spectral bands of a sensor data are combined in the RGB system, the resulting color images typically lack saturation, even though the bands have been contrast stretched. The under-saturation is due to the high degree of correlation between spectral bands. High reflectance values in the green band, for example, are accompanied by high values in the blue and red bands and so pure colors are not produced.

To overcome this problem, a method of enhancing saturation was developed (Sabbins, 1986) that consists of the following steps:

- Transform any three bands of data from the RGB system into the IHS system in which the three component images represent intensity, hue, and saturation. The equations for making this transformation are

$$I = R + G + B; \quad H = \frac{G - B}{I - 3B}; \quad S = \frac{I - 3B}{I} \tag{4.3}$$

  In Equation (4.3), R, G, and B refer to corresponding DN values of red, green, and blue components of the FCC under consideration. Equation (4.3) is used to compute I, H, and S values for every pixel in the image, thus obtaining I, H, and S images.
- The intensity image is dominated by albedo and topography. Sunlit slopes have high intensity values (bright tones), water has low values, and vegetation has intermediate intensity values, as do most rocks. In the hue image, low DNs are used to identify red hues in the sphere (Figure 4.9); vegetation has intermediate to light grey values assigned to the green hue. The lack of a blue hue is shown by the absence of very bright tones. The original saturation image is very dark because of the lack of saturation in the original data. Only the shadows and rivers are bright, indicating high saturation for these features.
- Apply a linear contrast stretch to the original saturation image. This results in an overall increase in brightness of the image. It also improves discrimination between terrain types.
- Transform the intensity, hue, and enhanced saturation images from the IHS system back into three images of the RGB system. These enhanced RGB images are used to prepare the new color composite image.

The IHS transformed image will have a significant improvement over the original image due to wide range of colors and improved discrimination between colors.

4.4.6.1 COMBINING OPTICAL AND MICROWAVE DATA

Images obtained in the visible and IR ranges are also referred as optical data, while those obtained in the microwave region are referred to as microwave data. IHS transformation and its inverse

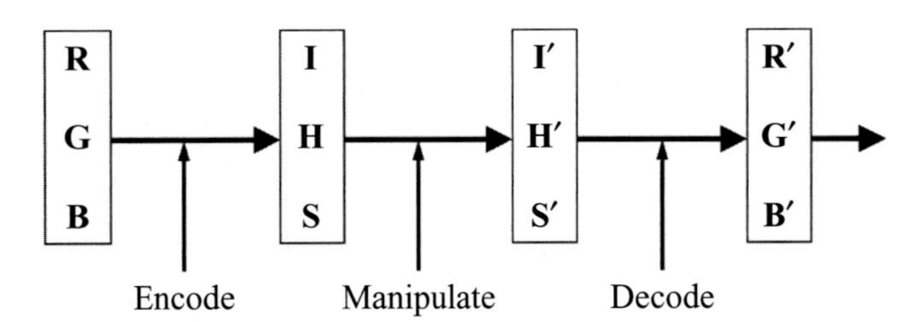

Figure 4.10 Schematic for combining multispectral images with PAN or microwave images.

are also useful for combining images obtained in optical and microwave ranges. For example, a digital radar image could be geometrically registered with IRS or Landsat optical data. After the optical bands have been transformed into IHS values, the radar image data may be substituted for the intensity image. The new combination (radar, hue, and enhanced saturation) can then be transformed back into an RGB image that incorporates both radar and optical data. This procedure is useful to combine radar data (which has cloud penetration capability) with optical data and thus provide information on regions even below the clouds. A similar procedure can be used to combine images acquired with different spectral and spatial resolution. For example, panchromatic (PAN) data with high spatial resolution and coarse spectral resolution can be merged with IRS LISS data (RGB) by substituting an intensity image of IHS-transformed RGB with PAN data. A typical procedure for combining such images is schematically shown in Figure 4.10.

Similarly, TM data can be merged with SPOT PAN data. This transformation significantly enhances the image, providing more detailed spectral and spatial discrimination when compared to either PAN data or standard FCC. Landsat MSS (Multispectral Scanner System) data merged with PAN data using IHS transformation are shown in Figure 4.11. In this figure a river and its flood plains can be very clearly seen.

4.4.7 Time-composite images

When images containing partial cloud cover can be acquired every day, as in the case of NOAA AVHRR, time-composite imagery can be produced without cloud cover by using the following steps:

- Co-register images acquired over number of days (say 15 days).
- Area with cloud cover is identified from the first imagery and is replaced by the next imagery of the same area.
- Cloud cover (if any) from this composite imagery is replaced with the third imagery.
- This procedure is repeated 15 times (say over 15 days of imagery).

Figure 4.11 Synergetic image prepared using IRS–1C LISS III and PAN data. See also color plates.

- Composite imagery is used for further analysis, assuming that the minor changes taking place within the 15 days are negligible.

The NRSC (National Remote Sensing Center), Hyderabad, has used such time-composited imagery of NOAA AVHRR over 15 days for agricultural drought assessment and analysis.

4.4.8 Synergetic images

In synergetic images, images with different spatial, spectral, and radiometric resolutions are combined along with ancillary data, such as elevation information and location information, to further enhance the image. Basically, this procedure is used to transfer data from different systems into a single and portable data set. For this, it is important that separate bands are co-registered with each other and that they contain the same number of rows and columns. FCC can be produced by considering any three bands (may be of different spectral or spatial resolution). Non-remote sensing data, such as topographic data, elevation information, etc., may also be merged through digital elevation modeling. Other non-remote sensing data, such as names of locations, villages, towns, roads, and rivers, can also be merged.

4.4.9 Digital mosaics

Typically, satellite RS imagery will be rectangular in shape, covering a certain region as per the path and row of the satellite's navigation. Digital mosaics are prepared by matching and splicing together individual images to get an entire image for any political, geological, or hydrologic boundary. Digital mosaics are generally prepared by combining different images acquired from the same sensor. Typical steps involved in the preparation of digital mosaics are as follows:

- Geometrically register adjacent images with the help of GCPs in the regions of overlap.
- Eliminate duplicate pixels.
- Merge the images to prepare a composite image.
- Contrast stretch the composite image to have uniform appearance.

Differences in contrast and tone between adjacent images may cause a checkerboard pattern that is common on many mosaics. This can be avoided by contrast stretching the composite image, instead of contrast stretching each component image separately. Digital mosaics thus obtained will have the entire region of interest in one image with all the features in it uniformly enhanced, and the image can be used for further analysis. A typical digital

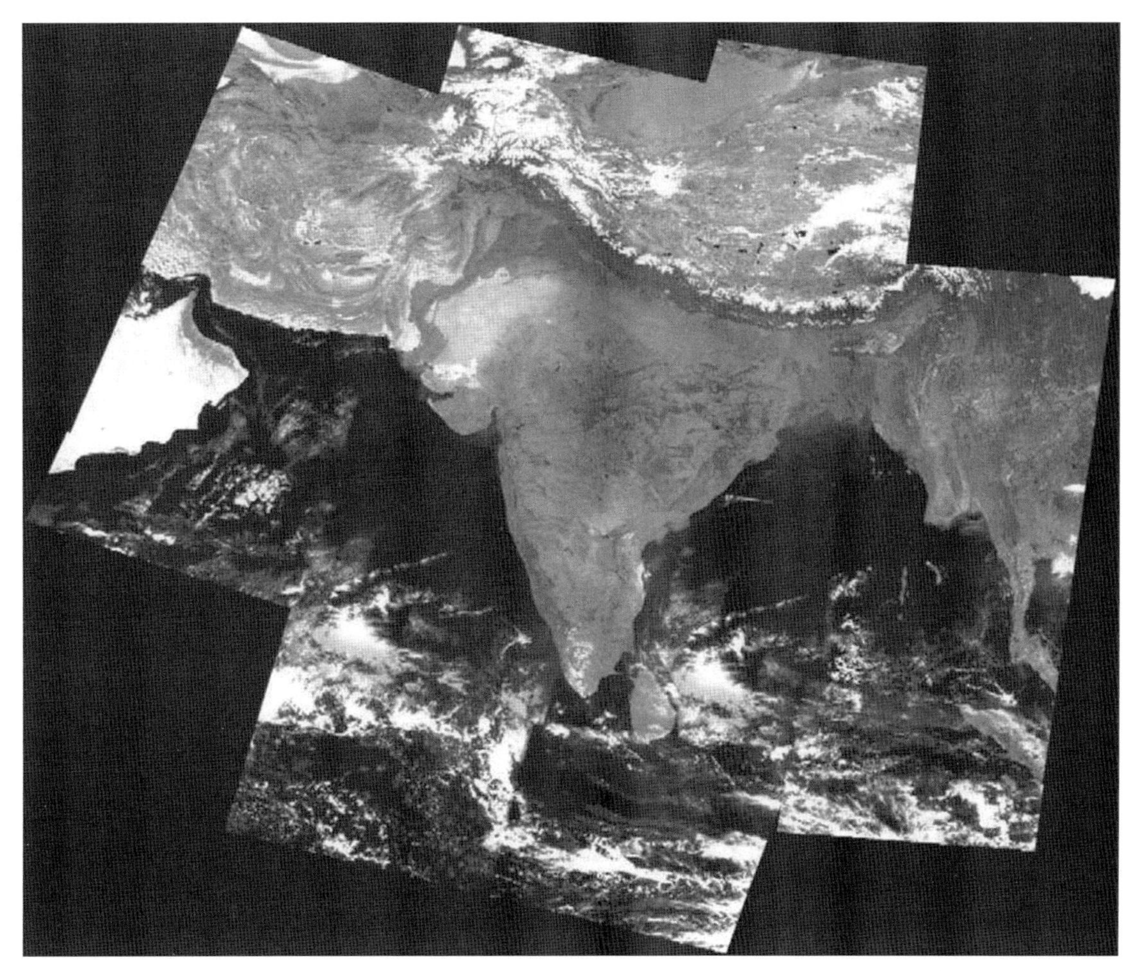

Figure 4.12 Digital mosaic of images obtained using the Ocean Color Monitor (OCM) sensor onboard the IRS-P4 (Oceansat) satellite (*Source*: http://nrsc.gov.in). See also color plates.

mosaic (FCC) of images obtained using the Ocean Color Monitor (OCM) sensor onboard the IRS-P4 (Oceansat) satellite is shown in Figure 4.12. In this mosaic, the whole of India and the surrounding regions can be seen.

4.5 IMAGE INFORMATION EXTRACTION

Image restoration and enhancement processes utilize image processing techniques to provide corrected and improved images for further analysis and study by human interpreters. The computer makes no decisions in these procedures. However, processes that identify and extract information from images are available which can be studied under the category of image information extraction techniques. It may be noted that these techniques should be applied only after image restoration and enhancement procedures. Some of the important image information extraction procedures are explained next.

4.5.1 Principal component images

In any multispectral image, the DN values are typically highly correlated from band to band. This can be noticed schematically from the scatter plots of DNs for pixels in different spectral bands (say bands 1 and 2 of Landsat TM). An elongated distribution pattern of the data points in a scatter plot indicates that as brightness in one band increases, brightness in the other band also increases. A three-dimensional plot of three spectral bands, say Landsat TM bands 1, 2, and 3, would show the data points in an elongated ellipsoid, indicating correlation of the three bands. This correlation means that if the reflectance of a pixel in one band (Landsat band 2, for example) is known, the reflectance in adjacent bands (Landsat bands 1 and 3) can be predicted. The correlation also means that there is much redundancy in a multispectral data set. If this redundancy could be reduced, the amount of data required to describe a multispectral image could be compressed.

The principal components transformation, originally known as the Karhunen–Loeve transformation (Loeve, 1955), is used to compress multispectral data sets by calculating a new coordinate

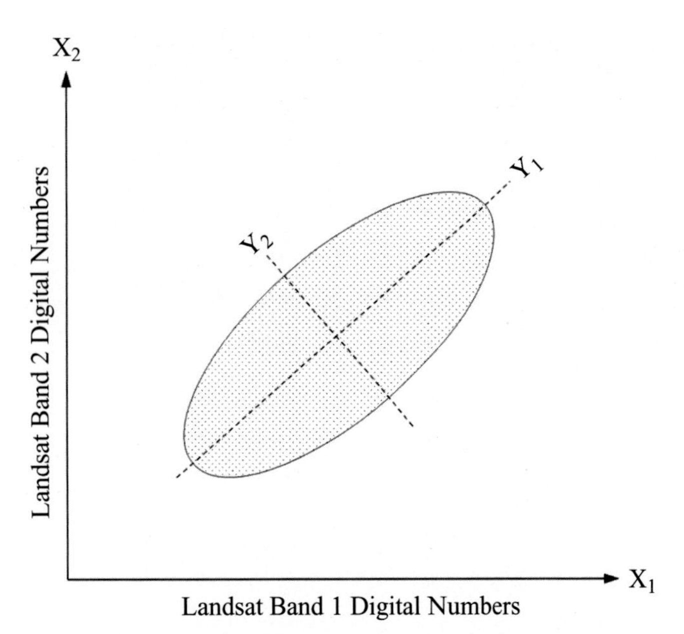

Figure 4.13 Scatter plot of Landsat band 1 (X_1 axis) and band 2 (X_2 axis) showing correlation between these bands. A PC transformation was used to generate a new coordinate system (Y_1, Y_2).

system. For the two bands of data in Figure 4.13, the transformation defines a new axis (y_1) oriented in the long dimension of the distribution and a second axis (y_2) perpendicular to y_1.

The mathematical operation results in a linear combination of pixel values in the original coordinate system that results in pixel values in the new coordinate system:

$$y_1 = \alpha_{11}x_1 + \alpha_{12}x_2; \quad y_2 = \alpha_{21}x_1 + \alpha_{22}x_2 \tag{4.4}$$

where (x_1, x_2) are the pixel coordinates in the original system; (y_1, y_2) are the coordinates in the new system; and α_{11}, α_{12}, α_{21}, and α_{22} are constants. In Figure 4.13, note that the range of pixel values for y_1 is greater than the ranges for either of the original coordinates, x_1 or x_2, and that the range of values for y_2 is relatively small. The same PC transformation may be carried out for multispectral data sets consisting of any number of bands. Additional coordinate directions are defined sequentially. Each new coordinate is oriented perpendicular to all the previously defined directions and in the direction of the remaining maximum density of pixel data points. For each pixel, new DNs are determined relative to each of the new coordinate axes. A set of DN values is determined for each pixel relative to the first PC. These DNs are then used to generate an image of the first PC. A similar procedure is used to produce images for the remaining PCs. The preceding description of the PC transformation is summarized from Swain and Davis (1978) and additional information is given in Moik (1980).

When this technique is applied to data from n spectral bands of the same region, n PC images are produced. However, the first few PC images will contain most of the information present in all the n spectral bands.

Visualization of the PCs is simple, but to create the PC axes it is necessary to calculate the length of the PC axes and their direction. These are computed by determining the eigenvalues (length) and eigenvectors (direction) from the variance–covariance matrix between the DNs in the different spectral bands under consideration. Principal component analysis and the procedure to compute the eigenvalues and eigenvectors is explained in Chapter 5. In a typical PCA of Landsat TM bands 1, 2, 3, 4, 5, and 7 (band 6 will not be considered as it has different spatial resolution), the first PC contains over 65% of the total variance in all the six bands, while PC4, PC5, and PC6 combined will have a little over 2% (i.e., the first three components explain 98% of the total variance). Thus, the information in six spectral bands is compressed into three PC images (50% compression) with a negligible loss of about 2% information. Noise in multispectral images is relegated to PC5 and PC6. This technique can be used for information extraction as well as compression of multispectral images (thus reducing the storage requirements significantly). In the case of hyperspectral data with more than 200 spectral bands, PCA can be used to extract most of the information available in the 200 spectral bands into a few PC images, thus reducing the storage requirement significantly.

Any three PC images can be combined to create a color image by assigning the data that make up each image to a separate primary color (RGB). FCCs produced from the first three PC images will contain more information than any FCC produced from three spectral bands. Summarizing, the PC transformation has the following advantages:

- Most of the variations in a multispectral data set are compressed into the first few PC images.
- Noise is relegated to the last few PC images.
- Spectral differences between materials will be more apparent in PC images than in individual bands.

4.5.2 Ratio images

Ratio images are produced by dividing the DN in one band by the corresponding DN in another band for each pixel, stretching the resulting value to the original 7- or 8-bit system, and plotting the new values as an image. A total of 15 ratio images plus an equal number of inverse ratios (reciprocals of the first 15 ratios) may be prepared, using different combinations from the six original bands. In a ratio image, black and white extremes of the grey scale represent pixels with the greatest difference in reflectivity between the two spectral bands. The darkest signatures are areas where the denominator of the ratio is greater than the numerator. Conversely the numerator is greater than the denominator for the brightest signatures. If denominator and numerator are the same, there is no difference between the two bands. For example, the spectral reflectance curve shows that the maximum reflectance of vegetation occurs in IRS band 4 (reflected IR) and that reflectance is considerably lower in band 2 (green). The ratio image 4/2 results when the DNs in band 4 are divided by the DNs in band 2. The brightest signatures in this image correlate with the vegetation.

Like PC images, any three ratio images may be combined to produce a color image by assigning each image to a separate primary color (RGB). The color variations of the ratio color image express more geologic information and have greater contrast between units than the conventional FCC. Ratio images emphasize differences in slopes of spectral reflectance curves between the two bands of the ratio. In the visible and reflected IR regions, the major spectral differences of materials are expressed in the slopes of the curves. Thus, individual ratio images and ratio color images enable one to extract reflectance variations. However, a disadvantage of ratio images is that they suppress differences in albedo. Materials that have different albedos but similar slopes of their spectral curves may be indistinguishable in ratio images. Ratio images also minimize differences in illumination conditions, suppressing the expression of topography. Thus, ratio images can also be used to reduce the variation in DN values of the same feature located in sunlit and shadow regions of the image.

In addition to ratios of individual bands, a number of other ratios may be computed. An individual band may be divided by the average for all the bands, resulting in normalized ratios. Another combination is to divide the difference between two bands by their sum. For example, (band 4 – band 3)/(band 4 + band 3) of Landsat TM. Ratios of this type are used to study crop conditions and crop yield estimation.

4.5.3 Vegetation indices

It can be noticed from the spectral reflectance curve for vegetation that reflectance is very high for vegetation in the NIR band while it is very low in the red (R) band (chlorophyll absorption band). This contrast is unique for vegetation. Thus, ratio images in the NIR and R bands can be used to emphasize the vegetation component in any image. Several ratio images have been developed using images in the R, NIR, and shortwave infrared (SWIR) regions, which are referred to as vegetation indices. Images obtained by using vegetation indices can be used to identify different vegetation species, health and stage monitoring, and yield estimation. Some of the important vegetation indices (Huete *et al.*, 1997) are explained next in three categories, namely, intrinsic indices, soil-line related indices and atmospherically corrected indices.

4.5.3.1 INTRINSIC INDICES

- Ratio vegetation index (RVI)

$$RVI = \frac{NIR}{R} \tag{4.5}$$

- Normalized difference vegetation index (NDVI)

$$NDVI = \frac{NIR - R}{NIR + R} \tag{4.6}$$

- Normalized difference wetness index (NDWI)

$$NDWI = \frac{SWIR - MIR}{SWIR + MIR} \tag{4.7}$$

- Green vegetation index (GVI)

$$GVI = \frac{NIR + SWIR}{R + MIR} \tag{4.8}$$

where *MIR* represents middle infrared.

4.5.3.2 SOIL-LINE RELATED INDICES

It may be noticed that the maximum reflectance for vegetation in the NIR band is approximately 50–60% of the incident energy. The remaining energy is essentially transmitted by the vegetation. That portion of the energy will be reflected based on the background to the vegetation, which could be soil or more layers of vegetation. Soil-line related indices are developed to apply corrections for a soil background. Initially, by using the DN values in the R and NIR bands of those pixels that contain only barren soil (such DN values are referred to as R_{soil} and NIR_{soil} here), a regression equation of the following form is developed to obtain a and b:

$$NIR_{soil} = a R_{soil} + b \tag{4.9}$$

Then, the following indices can be obtained:

- Perpendicular vegetation index (PVI)

$$PVI = \frac{NIR - aR - b}{\sqrt{1 + a^2}} \tag{4.10}$$

- Weighted difference vegetation index (WDVI)

$$\alpha = NIR_{soil}/R_{soil} \tag{4.11}$$

$$WDVI = NIR - \alpha R \tag{4.12}$$

- Soil adjusted vegetation index (SAVI)

$$SAVI = \frac{(1 + L)(NIR - R)}{NIR + R + L} \tag{4.13}$$

where L depends on the background to vegetation and is typically taken as 0.5.

4.5.3.3 ATMOSPHERICALLY CORRECTED INDICES

Reflectance in any spectral band depends on the amount of incident energy. Some of the incident energy is attenuated by the atmosphere, especially in shorter wavelength regions. This attenuation will result in unwarranted variation in vegetation indices especially when vegetation index images of two different dates are compared. To overcome this problem, atmospherically corrected indices have been developed (Huete *et al.*, 1997). Two of them are given below.

- Atmospherically resistant vegetation index (ARVI) (γ depends on aerosol type)

$$RB = R - \gamma(B - R) \tag{4.14}$$

$$ARVI = \frac{(NIR - RB)}{(NIR + RB)} \tag{4.15}$$

- Soil adjusted atmospherically resistant vegetation index (SARVI)

$$SARVI = \frac{(1 + L)(NIR - RB)}{NIR + RB + L} \tag{4.16}$$

4.5.4 Change detection analysis

Change detection images provide information about seasonal or other changes. The information is obtained by comparing two or more images of an area that were acquired at different times. The first important step for such a procedure is to register the images acquired at different times for the same region by using corresponding ground control points (GCPs). After registration, the DNs of one image are subtracted from those of an image acquired earlier or later. The resulting values for each pixel will be positive, negative, or zero, with zero indicating no change. The next step is to plot these values as an image in which a neutral grey tone represents zero. Black and white tones represent the maximum negative and positive differences, respectively. Contrast stretching is employed on the difference image for scaling to the original 7- or 8-bit system to emphasize the differences more clearly. The agricultural practice of seasonal variation between cultivated and fallow fields can be clearly shown by the light and dark tones in the difference image of images acquired during different seasons. Change detection processing is useful for producing difference images for other RS data, such as between night-time and daytime thermal IR images. The change detection procedure can also be used to monitor urban sprawl and land use/land cover changes on seasonal, annual, or decadal scales.

4.5.5 Multispectral classification

For each pixel in a multispectral image, such as a Landsat TM image, the spectral brightness is recorded for seven different wavelength bands. A pixel may be characterized by its spectral signature, which is determined by the relative reflectance in the different wavelength bands. Multispectral classification is an information extraction process that analyzes these spectral signatures and then assigns pixels to categories based on similar signatures.

By plotting the DN values of each mutispectral image, a cluster diagram can be generated. The reflectance ranges of each band form the axes of a multidimensional coordinate system. Plotting additional pixels of the different terrain types produces multidimensional clusters or ellipsoids. The surface of the ellipsoid forms a decision boundary, which encloses all pixels for that terrain category. The volume inside the decision boundary is called the decision space. Classification programs differ in their criteria for defining the decision boundaries. In many programs the analyst is able to modify the boundaries to achieve optimum results. For the sake of simplicity the cluster diagram is explained with only three axes. In actual practice the computer employs a separate axis for each spectral band of data (say seven for Landsat TM).

Once the boundaries for each cluster, or spectral class, are defined, the computer retrieves the spectral values for each pixel and determines its position in the classification space. Should the pixel fall within one of the clusters, it is classified accordingly. Pixels that do not fall within a cluster are considered unclassified. In practice, the computer calculates the mathematical probability that a pixel belongs to a class. If the probability exceeds a designated threshold (represented spatially by the decision boundary), the pixel is assigned to that class. In general, a more accurate classification result is obtained if a greater number of spectral bands are used. However, increasing the number of bands used in classification greatly increases the computing time. This problem can be overcome by using PC images.

There are two major approaches to multispectral classification: supervised classification and unsupervised classification.

4.5.5.1 UNSUPERVISED CLASSIFICATION

In unsupervised classification, the computer separates the pixels into classes with no directions from the analyst by discretizing the spectral ranges of multispectral images into a pre-specified number of classes. Thus, it is a technique that groups the pixels into clusters based upon the distribution of the DNs in the image. An unsupervised classification program, such as ISODATA clustering, requires the following information:

- maximum number of classes,
- maximum number of iterations,
- threshold value.

Unsupervised classification operates in an iterative fashion. Initially it assigns arbitrary means to the classes and allocates each pixel in the image to the class mean to which it is closest. New class means are then calculated and each pixel is again compared to the new class means. This procedure is repeated over a number of iterations. Pixels move between clusters after each iteration until a threshold is reached. A threshold of 0.98 implies that the program terminates when less than 2% of the pixels move between adjacent iterations. The classes produced from unsupervised classification are spectral classes and may not correlate exactly with "information classes" as determined by supervised classification.

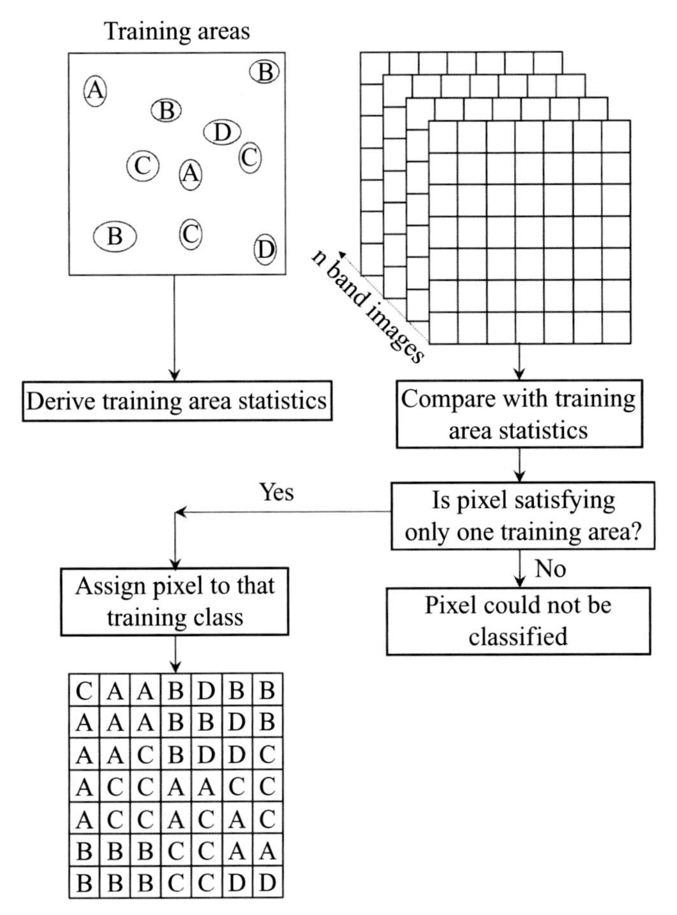

Figure 4.14 Generalized sequence of steps in a supervised classification process (Modified from Gibson and Power, 2000).

4.5.5.2 SUPERVISED CLASSIFICATION

In supervised classification, the analyst defines on the image a small area, called a training site, that is representative of each terrain category, or class. Spectral values for each pixel in a training site are used to define the decision space for that class. After the clusters for each training site have been defined, the computer then classifies all the remaining pixels in the scene. Various classification methods may be used to assign an unknown pixel to one of the classes. The choice of a particular classifier or decision rule depends on the nature of the input data and desired output. The parametric classification algorithm assumes that the observed measurement vectors, X_c, obtained for each class in each spectral band during the training phase of the supervised classification are Gaussian in nature. The non-parametric classification algorithm makes no such assumption. The generalized sequence of steps in a supervised classification process (Gibson and Power, 2000) is presented in Figure 4.14.

It is instructive to review the logic of several of the classifiers. The parallelepiped, minimum distance, and maximum likelihood classification algorithms are the most frequently used supervised classification algorithms (Sabbins, 1986), and they are explained next.

Parallelepiped (or box) classification algorithm This is a widely used decision rule as a first approximation for classification as it is computationally the least intensive compared to other algorithms. It is based on simple Boolean logic. Training data in n spectral bands are used in performing the classification. Brightness values from each pixel of the multispectral imagery are used to produce an n-dimensional mean vector, M_c, as follows:

$$M_c = \mu_{c1}, \mu_{c2}, \mu_{c3}, \ldots, \mu_{cn} \tag{4.17}$$

where μ_{cl} is the mean value of the training data obtained for class c in band l out of m possible classes as previously defined. S_{cl} is the standard deviation of the training data class c in band l out of m possible classes. Using a one standard deviation threshold, a parallelepiped algorithm assigns the pixel in class c if, and only if,

$$\mu_{cl} - S_{cl} \leq X_{i,j,l} \leq \mu_{cl} + S_{cl} \tag{4.18}$$

where $c = 1, 2, 3, \ldots, m$, number of classes and
$l = 1, 2, 3, \ldots, n$, number of bands.

Therefore, the low and high decision boundaries are defined as

$$LOW_{cl} = \mu_{cl} - S_{cl} \quad \text{and} \quad HIGH_{cl} = \mu_{cl} + S_{cl} \tag{4.19}$$

Thus, the parallelepiped algorithm becomes

$$LOW_{cl} \leq X_{i,j,l} \leq HIGH_{cl} \tag{4.20}$$

The decision boundaries form an n-dimensional parallelepiped in feature space. Any pixel that does not satisfy any of the Boolean logic criteria is assigned to an unclassified category. The parallelepiped algorithm is a computationally efficient method of classifying RS data. Unfortunately, in some cases parallelepipeds overlap each other and hence an unknown candidate pixel might satisfy the criteria of being in more than one class, leading to ambiguity.

Minimum distance to means classification algorithm This procedure is also computationally simple and commonly used. It can result in better classification accuracies, compared to other more computationally intensive algorithms such as the maximum likelihood classifier. For minimum distance classification, distance to each mean vector for each unknown pixel is to be evaluated. Many formulae are available to calculate these distances. Among them, the most commonly used formulae are Euclidian distance (EUCD) and round-the-block distance (RTBD) as given below:

$$\text{EUCD} = \sqrt{\sum (X_{i,j,l} - \mu_{cl})^2} \quad \text{RTBD} = \sum |X_{i,j,l} - \mu_{cl}| \tag{4.21}$$

where μ_{cl} represents the mean vectors for class c measured in band l. Any unknown pixel is assigned to one of the training classes that has least distance. Thus, none of the pixels will be left unclassified.

Maximum likelihood classification algorithm This decision rule assigns each pixel having features X to the class c whose units are most probable or likely to have given rise to feature vector X. It assumes that training data statistics for each class in each band are normally distributed. Maximum likelihood classification makes use of mean measurement vector M_c for each class and co-variance matrix V_c of class c and band l. The decision rule applied to unknown measurement vector X is decided in a given class if, and only if,

$$P_c \geq P_i \quad (4.22)$$

where $i = 1, 2, 3, \ldots, m$ with m representing the number of classes.

The pixel is assigned to class c only if the probability for class c is higher than for any other class, and P_c is computed as

$$P_c = [-0.5 \log_e(Det(V_c))] - \left[0.5(X - M_c)^T \left(V_c^{-1}\right)(X - M_c)\right] \quad (4.23)$$

Probability contours are created around each training area and a pixel is assigned to a class depending upon the value of the probability contours that encompass it. The maximum likelihood classifier is generally considered to be the most powerful, but is also considered the most computationally intensive.

The three supervised classification algorithms explained are schematically described (Gibson and Power, 2000) in Figure 4.15. Many more supervised classification algorithms have also been developed using advances in statistical tools and clustering algorithms.

4.5.5.3 GROUND TRUTH COLLECTION/VERIFICATION

Training areas are selected by visiting the regions in the image and collecting sample areas that are uniquely representative of a particular class. This procedure is referred to as ground truth location. Sometimes, to exactly identify the corresponding sample area from an image on the ground, a global positioning system (GPS) receiver can be used.

A sufficient number of pixels for each surface class must be delineated in order to ensure that a representative sample is obtained for each class. The training areas for any one class should not be concentrated in one part of the image but should encompass the entire scene. The histograms for training areas should be unimodal and conform to a normal distribution (Campbell, 1996; Chen and Stow, 2002). The training areas should be as separate and uniquely representative as possible, otherwise a substantial overlap between classes may occur and pixels will be misclassified. Sometimes it may not be possible to ensure that classes are discrete because they may have similar reflectance characteristics in the bands that are being classified. In such a situation it may be preferable to merge the training sites and consider them as a single class. It may also be preferable to isolate an individual class, and this is simply achieved by assigning a value of zero to all other classes.

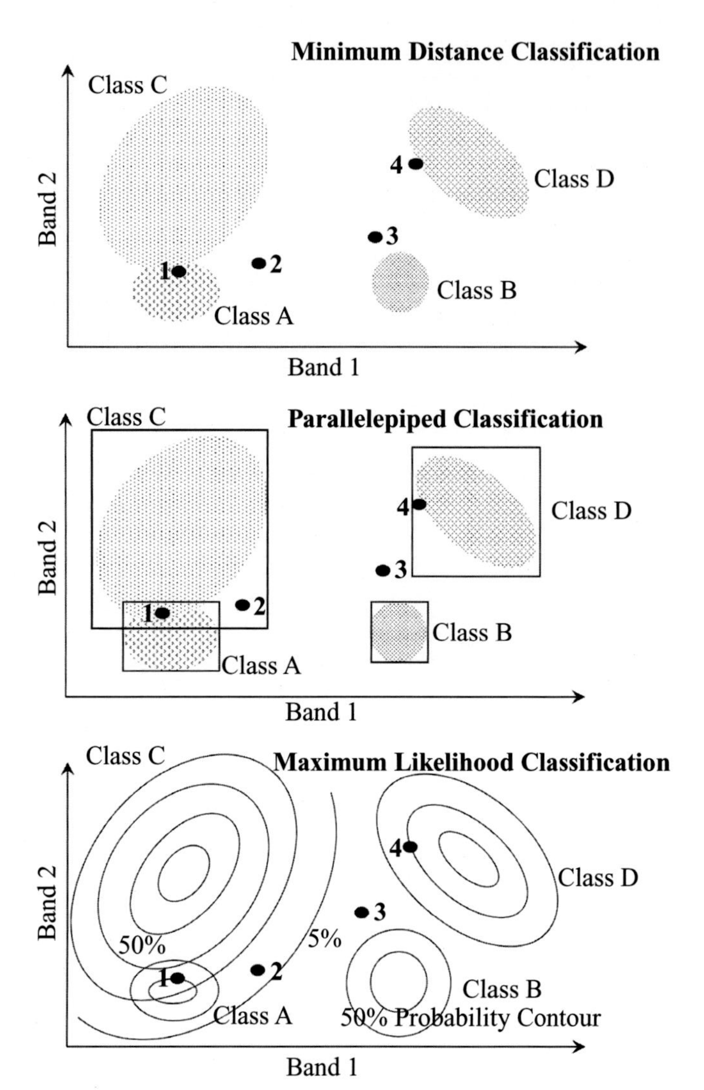

Figure 4.15 Schematic representation of supervised classification algorithms (Modified from Gibson and Power, 2000).

Classification accuracy Typically, a portion (say 70%) of the ground truth data collected for each class will be used as training data for supervised classification of the image and the remaining portion of the ground truth data is used for testing the accuracy classification. If a significant portion of the pixels in the testing data are classified accurately, then the classification procedure can be considered as good and the results can be utilized. However, if most of the testing data are misclassified, then the procedure should be repeated by refining the training areas or re-sampling them.

Confusion matrix A confusion matrix (Kohavi and Provost, 1998) contains information about actual and predicted classifications done by a classification system. Performance of such systems is commonly evaluated using the data in the matrix. Each row of the matrix represents the instances in a predicted class, while each column represents the instances in an actual class. One benefit of a confusion matrix is that it is easy to see if the system is confusing two classes (i.e., commonly mislabeling one as another). Thus, a confusion matrix is an array showing relationships between true and predicted classes. Entries on the diagonal of the matrix show the number of pixels correctly classified. Entries off the diagonal show the misclassifications. The totals for each row will equal the number of pixels for which the ground truth data are available for the testing phase. A confusion matrix can be analyzed to assess the classification accuracy. This approach will be more relevant if the amount of training and testing data available for each class are considerably different.

4.6 LAND USE/LAND COVER INFORMATION

Some of the most vital information required for hydrologic modeling that can be obtained from satellite RS is land use/land cover information. Most of the image enhancement and information extraction techniques explained in Sections 4.4 and 4.5 are useful for this purpose. Some of the useful techniques for obtaining land use/land cover information from RS imagery are reiterated here.

- Panchromatic images (with high spatial resolution) can be merged with multispectral images (with high spectral resolution) using IHS transformation. Resulting images enable discrimination of spatial features of even smaller magnitude (lesser than 5 m in spatial resolution). In addition, when these images can be obtained at very frequent intervals (high temporal resolution), changes in land use/land cover can be monitored more effectively.
- Although FCC is very useful for interpreting land use/land cover information, it facilitates use of only three spectral band images at a time. More useful information can be obtained if more spectral images are used simultaneously. This can be achieved by producing PCA images of multispectral images. FCC thus obtained from the first three PCA images is more effective in obtaining land use/land cover information.
- Multispectral clustering algorithms discussed in Section 4.5.5 and their advancements such as fuzzy clustering algorithms are also very useful in obtaining land use/land cover information.

4.7 UTILITY OF REMOTE SENSING FOR HYDROLOGIC MODELING

Remote sensing can play a useful role in harnessing the wealth of available water resources (Peck *et al.*, 1981). Remote sensing techniques indirectly measure hydrologic variables. So the electromagnetic variables measured by RS techniques have to be related to the hydrologic variables empirically or with transfer functions. There are several areas in the field of water resources wherein RS proves useful for effective applications – particularly in surveying and inventorying. The International Satellite Land-Surface Climatology Project (ISLSCP) encourages scientists to use RS data to better understand natural processes on the global land surface (e.g., Sellers *et al.*, 1995). Large-scale field experiments have been conducted to validate RS algorithms for estimating surface parameters and fluxes such as evapotranspiration (WMO, 1988). There is ample scope for the application of RS in the assessment of various components of the hydrologic cycle, quantification of these components in various environs and the fluxes of water through these environs. Some of the fields include hydrologic studies, river morphology, reservoir dynamics and sedimentation, watershed conservation, command area planning and management, flood estimation and forecasting, groundwater studies, water quality and environmental protection. Remote sensing applications in hydrology are briefly reviewed by Schultz (1988). A review of RS applications to groundwater studies is presented by Meijerink (1996). An exhaustive list of various applications to water resources, wherein RS may substitute or complement or supplement the conventional methods, is given by Balakrishnan (1987). A state-of-the-art review on satellite RS applications in irrigation management is given by Bastiaanssen (1998). This section presents an overview of the applications of RS to various aspects of assessment and management of water resources.

Hydrologic processes are highly dynamic phenomena, which vary both in time and in space. Conventional measurements of these hydrologic processes are accomplished by *in-situ* or point measurements. These are then interpolated or extrapolated to get the aerial estimates. The advent of RS technology has opened new vistas for the study of various components of the hydrologic cycle. There are three broad categories for using RS in hydrologic studies (Salmonson, 1983).

- Simple qualitative observations/assessments are made. For example, visual observation of a photo showing that water from industrial effluent in a stream has a different color than the stream water, suggesting a site for collection of a sample.
- Information on geometric form, dimensions, patterns, geographic location, and distribution is derived for features such as land cover categories that influence runoff, evapotranspiration, and soil moisture.
- Development of correlation between the remotely sensed observations and the corresponding point measurements on the ground for estimation of a hydrologic parameter. Examples include the estimation of rainfall, soil moisture, snow depth, sediment load, etc.

Some examples of parameters used in hydrologic modeling, which have been derived from satellite data with sample references, are presented in Table 4.1.

Table 4.1 *Parameters in hydrology and water resources currently obtainable from satellite RS data*

Variable	Satellite/sensor	Wavelength or frequency	Spatial resolution	Coverage	Sample reference
Snow-covered area	NOAA	0.62,10.8 mm Bands 1 & 4	1 km	2 per day	Kite, 1989; Carroll and Baglio, 1989
Snow depth	GOES	0.65 mm	2 km	2 per hour	Donald *et al.*, 1990
	Nimbus 7	37 GHz	30 km		Chang *et al.*, 1982
Snow-water equivalent	SSM/I	19.3, 37 GHz	25 km	2 per day	Slough and Kite, 1992
	MOS-1-MSR	23, 31 GHz	23–32 km		Walker *et al.*, 1990
Surface temperature	NOAA	10.80 mm (Band 4)	1 km		Dousset, 1989
Evapotranspiration	NOAA	0.62,0.91, 10.8,12.0 mm			Price, 1980
	GOES	0.64,11.5 mm	2–8 km	2 per hour	Bussieres *et al.*, 1989
Precipitation	Meteosat	0.65 mm	3 km		Pietroniro *et al.*, 1989
Land cover/land use	Landsat5-MSS	0.55,0.65, 0.75,0.95 μm	80 m	8–16 days	Whiting, 1990; Kite and Kouwen, 1992
Vegetation	NOAA	0.62,0.91 μm	1 km	2 per day	Allison *et al.*, 1989
	IRS-1C/1D (WiFS)	0.62,0.67 μm	188 m	5 days	Krishna Prasad *et al.*, 1999
Suspended sediment/algal growth	Landsat5- MSS IRS-P4 OCM	0.55,0.65, 0.75,0.95 μm	80 m	8–16 days	Ritchie and Schiebe, 1986 Kaur and Rabindranathan, 1999
Spring runoff	Nimbus 5	19 GHz	30 km	2 per hour	Wankiewicz, 1989
Temporal changes in snowmelt	ERS-1	C-band (5.3 GHz) SAR	30 m	35 days	Donald *et al.*, 1992
Temporal changes in soil moisture	JERS-1	L-band (1 GHz) SAR	30 m	35 days	Pietroniro *et al.*, 1993
Groundwater	Landsat	0.95 μm	80 m	8–16 days	Bobba *et al.*, 1992
Water depth	Landsat	0.48,0.56, 0.66 μm	30 m	8–16 days	Hallada, 1984
Oceanography	IRS-P4 OCM MSMR	412,443,490, 520,555,670,765 nm 6.6,10.65,18, 21 GHz	360 m	2 days	Kalyanaraman, 1999

Modified from Kite and Pietroniro (1996)

4.7.1 Rainfall estimation

The conventional means of rainfall monitoring using a network of rain gauges has limitations due to their sparse density, more so in remote areas and over oceans. The wavelengths most commonly used for rainfall studies are (Barret and Martin, 1981):

Visible (VIS): 0.5–0.7 μm
Infrared (IR): 3.5–4.2 μm and 10.5–12.5 μm
Microwave (MW): 0.81–1.55 cm

4.7.1.1 DELINEATING THE BOUNDARIES OF AREAS LIKELY TO GET RAIN

Most of the meteorological satellites currently used for precipitation estimation are either geostationary or polar-orbiting satellites. Geostationary satellites (e.g., NOAA, GOES, GMS, Meteosat, INSAT) typically carry IR and VIS imagers with spatial resolution from 1 to 4 km. A geostationary satellite positioned over the equator can provide high-frequency (hourly or more frequent) images of a portion of the tropics and middle latitudes, while a polar-orbiting satellite provides roughly twice-daily coverage of the entire globe. Polar orbiters also fly in a low Earth orbit, which is more suitable for the deployment of MW imagers on account of their coarse resolution. High-quality MW imagery was made available after the launch of SSM/I (Special Sensor Microwave Imager) with spatial resolution on the order of 10 km. SSM/I data are found more reliable for rainfall estimation. However, they suffer from two limitations, namely, coarse temporal resolution and coarse spatial resolution. The geometrical effects of three-dimensional rain clouds on radiances observed by an oblique-viewing radiometer such as SSM/I have been examined and found to be considerable (Haferman *et al.*, 1994).

A good number of satellite-based rainfall estimation algorithms are documented in the literature. For example, 55 algorithms were evaluated in the Third Algorithm Intercomparison Project (AIP-3) of the Global Precipitation Climatology Project (GPCP) including 16 VIS/IR, 29 MW and 10 mixed IR/MW algorithms. Petty and Krajewski (1996) presented a review of some of the important algorithms.

The cloud indexing method (Follansbee, 1973) is normally used to estimate daily rainfall over an area through statistical averaging of cloud–rainfall relationships. In this approach, the three most important types of clouds – cumulonimbus, cumulocongestus, and nimbostratus – are considered. This method is widely used in support of broad-scale hydrology.

To meet the growing need for stably calibrated, long time-series data sets for global climate change, a number of organized scientific programs or projects are actively promoting the development, validation, and/or access to the user community of satellite-derived precipitation products. Pathfinder precipitation products from SSM/I data using the GSCAT algorithm are available freely from the Distributed Active Archive Center (DAAC) of NASA, USA (http://daac.gsfc.nasa.gov). The National Climate Data Center (NCDC) is currently producing and distributing rainfall products for both 1° and 2.5° grids (http://www.ncdc.noaa.gov) from a combination of archived MW and IR satellite data, surface rain gauge data, and numerical forecast model output.

4.7.1.2 RADAR DATA FOR AREAL MEASUREMENT OF RAINFALL

Radar is an active microwave RS system operating in the 1 mm to 1 m region of the electromagnetic spectrum. In a radar system, a pulse of electromagnetic energy is transmitted as a beam that is partially reflected by cloud or rainfall, back to the radar. Radar has the unique capability to observe the areal distributions of rainfall and provide real-time estimates of rainfall intensities. Operational use of ground-based radar for rainfall monitoring is limited due to its smaller range and high cost. With the advent of satellites, it has become possible to obtain spatially continuous and homogeneous data over large areas, including oceans, in real time. The use of satellite data for estimating rainfall is mainly based on relating brightness of clouds observed in imagery to rainfall intensities (Hall, 1996).

4.7.2 Snow-cover/snowmelt estimation

Snow in hydrology is a renewable resource and is one of the most complicated parameters to be measured. In India, during the summer months, rivers rising in the Himalayas are substantially fed by snowmelt runoff. Periodic snow-cover monitoring is essential for assessing the snowmelt runoff likely to occur. Conventional methods have serious limitations in the study of dynamic processes such as snow, and its monitoring is complicated by inaccessibility. Considering the vastness of snow-clad watersheds, aerial surveys are expensive and of little use. Satellite RS has a vital role to play in obtaining near real-time snow-cover area (SCA) maps with good accuracy. Depletion curves of SCAs indicate the gradual areal diminishment of the seasonal snow cover during the snowmelt season. The typical shape of depletion curves can be approximated by the equation (Balakrishnan, 1987)

$$S = 100/(1 + e^{bn}) \tag{4.24}$$

where S is SCA in percent obtained from satellite imagery, b is a coefficient, and n is the number of days before ($-$) and after ($+$) the date at which $S = 50\%$.

The snow extent over the globe has become available on a daily basis since 1972 from the NOAA AVHRR satellite, assuming no cloud interference at 1.1 km spatial resolution. Passive microwave observations of snow cover extent are assured every 1–2 days from SSM/I due to its cloud penetration capability, but the spatial resolution is coarse (about 25 km).

The US National Weather Service (NWS) distributes products of periodic river-basin snow-cover extent maps (http://www.nohrsc.nws.gov) from NOAA AVHRR and Geostationary Operational Environmental Satellite (GOES). Some good examples of operational applications of such snow extent data are already available in India. Kumar *et al.* (1991) obtained operational forecasts of daily and weekly snowmelt runoff in the Beas (5,144 km^2) and Prabati (1,154 km^2) river basins in India using satellite products. Another type of application to operational snow hydrology is production of monthly hemispherical snow maps from GOES and NOAA AVHRR data for climatological applications. The EOS Multifrequency Passive Microwave Radiometer (MPMR) can obtain 8 km resolution by using a much larger antenna than currently used for SSM/I. A detailed review of applications of space-borne RS for snow hydrology is given by Rango (1996).

4.7.3 Rainfall–runoff modeling

Runoff in the form of volume of water flowing through a river cross section during a specified time interval cannot be measured on the basis of RS data alone. This may evolve in future, if it becomes possible to measure the width and depth of a river cross section with the aid of RS data and flow velocities by ultrasonic or laser methods. At present RS data are indirectly used to determine runoff with the aid of hydrologic models (Schultz, 1996). Remote sensing data are used either as model input or for the determination of model parameters or both. Most existing models do not incorporate the use of RS capabilities. Therefore it is necessary to develop structures of hydrologic models that are amenable to the spatial and temporal resolution provided by RS data. Areal distribution of information from RS data can be better utilized by the application of distributed deterministic models. The spatial resolution of a model structure and the spatial resolution of its input data should have some correlation. For example, it makes little sense to structure a model in space according to the IRS 1-C PAN pixel size of 5 m and use it as an input to analyze precipitation data from one gauge for an area of several thousand square kilometers. Further, there has to be a reasonable correlation between the applied resolution to time and to space. For example, if only monthly runoff values are being used, there is no need to seek high spatial resolution data (say IRS LISS II with a spatial resolution of 36.25 m). Peck *et al.* (1981) identified 13 variables that can be obtained using RS data with some degree of success. They also presented a review of the existing hydrologic models with regard to their adaptability to RS data. Similar studies were also presented in Balakrishnan (1987) and Schultz (1996).

Non-linear regression models based on RS and ground truth data were used for runoff modeling. A detailed sensitivity analysis for river basins in tropical West Africa was carried out by Papadakis *et al.* (1993) to obtain information on the resolution required for the modeling effort in time, space, and spectral channels. Meteosat data from European geostationary satellites with a high temporal resolution of 30 min and spatial resolution of 5 km were used for modeling monthly runoff for a large catchment area. The model functions in two consecutive steps: (i) estimation of monthly precipitation values with the aid of Meteosat IR data; and (ii) transformation of the monthly rainfall volumes into the corresponding runoff volumes with the aid of a rainfall–runoff model. For the three available spectral channels of Meteosat (IR, visible, and water vapor), it was shown that only IR information is relevant. Following a modified Arkin method (Dugdale *et al.*, 1991), a monthly cloud-cover index (CCI) was estimated, indicating all pixels having a temperature in the IR channel below a specified threshold value. This threshold value varies from region to region. Monthly precipitation values were then obtained from daily CCI values as a non-linear function of CCI (Papadakis *et al.*, 1993). Monthly runoff computed on the basis of rainfall estimated with the help of satellite imagery compared well with the observed runoff.

4.7.3.1 NRCS CURVE NUMBER METHOD

There have been many efforts to improve the performance of existing hydrologic models by the use of RS data for improved parameter estimation. A very well received approach is for the improvement of the Natural Resources Conservation Service (NRCS) runoff curve number (CN) model (US Department of Agriculture, 1972). Various SCS models allow computation of direct runoff volumes, Q, in terms of land use and soil type without requiring hydrologic data for calibration. The volume of direct runoff is given by the equation

$$Q = \frac{(P - I_a)^2}{(P - I_a) + S} \tag{4.25}$$

where P is volume of rainfall, I_a is initial abstraction, and S is a retention parameter. In the SCS model, $I_a = 0.2S$. Thus Q becomes a function of rainfall and S only. In practice, a runoff CN is defined as the transformation of S according to the relationship $\text{CN} = 1000/(S + 10)$. The CN depends on the hydrologic soil group and land use description. Conventionally, the CN are given in tables. Recent information about the NRCS CN method can be obtained from report TR-55 (Mishra and Singh, 2003).

The introduction of RS data allowed a better estimation of the land use and thus a more reliable estimation of the relevant CN. Two approaches have been used for estimating CN using RS data. In one approach, land cover information was obtained using RS data, which were then combined with general soil data to estimate the CN. Ragan and Jackson (1980) used Landsat multispectral data and this approach for runoff estimation. The other approach made use of Landsat average watershed reflectance ratios of spectral bands 0.5–0.6 μm and 0.8–1.1 μm to directly compute the CN without using ancillary soil data (Blanchard, 1974). Remote sensing data can be a valuable tool in runoff prediction especially in areas experiencing rapid land use changes. Chandra *et al.* (1984) used Landsat imagery and aerial photographs to derive land use and vegetal cover data for the Upper Yamuna basin, and then obtained morphometric and relief characteristics of the basin. These were in turn used to derive runoff coefficients for various land uses for use in simulation of runoff, employing a rational formula.

4.7.3.2 DISTRIBUTED HYDROLOGIC MODELS

Inputs for many distributed hydrologic models can be obtained from RS (Kite and Pietroniro, 1996). Some of the hydrologic models are listed below.

- STORM – Storage, Treatment, Overflow Runoff Model
- SWMM – Storm Water Management Model
- DR3M-QUAL – Distributed Routing, Rainfall, Runoff Model – Quality
- CREAMS/GLEAMS – Chemical, Runoff, and Erosion from Agricultural Management Systems/Groundwater Loading Effects of Agricultural Management Systems model
- EPIC – Erosion/ Productivity Impact Calculator
- SWRRB – Simulator for Water Resources in Rural Basins
- PRZM – Pesticide Root Zone Management model
- AGNPS – Agricultural Non Point Source pollution model
- SWAT – Soil and Water Assessment Tool

Although all these models give runoff as output, each of them focuses on a different component(s). For example, SWRRB is more suitable for rural watersheds whereas SWMM is more suitable for urban watersheds. SWAT is more versatile and is applied to different watersheds and river basins in different parts of the globe. Application of SWAT by using inputs from RS DEMs in a GIS framework is demonstrated through a case study in Chapter 6.

4.7.4 Use of recent satellite data

Schmugge *et al.* (2002) discussed various applications of RS in hydrology. Some of the components considered were land surface temperature from thermal IR data, surface soil moisture from passive microwave data, snow cover using both visible and microwave data, water quality using visible and NIR data, and estimating landscape surface roughness using LIDAR. Methods for estimating the hydrometeorological fluxes, evapotranspiration, and snowmelt runoff using these state variables are also described.

4.7.4.1 SOIL MOISTURE AND OCEAN SALINITY SATELLITE

The European Space Agency (ESA) launched a satellite, SMOS, in November 2009, with the mission to observe soil moisture over the Earth's landmasses and salinity over the oceans (http://www.esa.int/esaLP/ESAMBA2VMOC_LPsmos_0.html). An important aspect of this mission is that it demonstrates a new measuring technique, adopting a completely different approach in the field of observing the Earth from space (ESA, 2002). A novel instrument has been developed that is capable of observing both soil moisture and ocean salinity by capturing images of emitted microwave radiation around the frequency of 1.4 GHz (L-band). SMOS carries the first-ever, polar-orbiting, space-borne, two-dimensional interferometric radiometer.

The data acquired from the SMOS mission will lead to better weather and extreme-event forecasting, and contribute to seasonal-climate forecasting. As a secondary objective, SMOS also provides observations over regions of snow and ice, contributing to studies of the cryosphere.

Although soil moisture in the vadose zone is a small percentage of the total global water budget, soil moisture plays an important role in the global water cycle. However, *in-situ* measurements of soil moisture are sparse. SMOS provides the necessary information. Soil moisture retrieval from SMOS sensor data is documented in study report by ESA (2004).

4.7.4.2 GLOBAL PRECIPITATION MEASUREMENT MISSION

The NASA Precipitation Measurement Missions (PMM) program (http://pmm.gsfc.nasa.gov/) is a measurement-based science program that supports scientific research, RS algorithm development, ground validation, and satellite data utilization. The PMM program includes the Tropical Rainfall Measuring Mission (TRMM) satellite, launched in 1997. The PMM program also includes the Global Precipitation Measurement (GPM) Mission, an international partnership initiated by NASA and JAXA (Japan Aerospace Exploration Agency) to provide the next-generation precipitation observations from space, to be launched in 2013.

The GPM mission will use an international constellation of satellites to study global rain, snow, and ice to better understand the climate, weather, and hydrometeorological processes (http://gpm.gsfc.nasa.gov). GPM objectives include (Everett, 2001) providing satellite coverage and sampling strategy to

- reduce uncertainty in short-term rainfall accumulation,
- resolve the main features of the diurnal precipitation cycle,
- improve probability of detection of extreme rain events.

GPM's coverage region is 90° N to 90° S latitude and it is proposed to have a 3-hour revisit interval.

To measure precipitation with improved accuracy and to better understand its role in climate, NASA and NASDA (National Space Development Agency of Japan) planned and carried out the TRMM. The TRMM was the first satellite dedicated to rainfall measurement, and is the only satellite that carries weather radar. The TRMM has provided a wealth of knowledge on severe tropical storms such as hurricanes and short-duration climate shifts such as El Niño. Though the TRMM has exceeded expectations, the mission is still inherently limited in fully understanding the complete role of precipitation in the hydrologic cycle. A critical element driving the scientific objectives of GPM is to understand what scientific problems the TRMM has not been able to address. The TRMM has a limited view of the Earth, ranging from 36° N to 36° S, and samples rain relatively infrequently, passing over the same location about once a day. The TRMM also cannot measure frozen precipitation and is insensitive to light rainfall. Recently, investigators have attempted to merge multiple satellite estimates to minimize the shortcomings and capitalize on the strengths of the various techniques. The TRMM satellite serves as a reliable calibrator of the merged rainfall. These integrated rain products provide a valuable test bed for developing the concept of the GPM constellation of satellites. The GPM satellite is scheduled to be launched in 2013.

4.7.4.3 TROPICAL RAINFALL MEASURING MISSION

The TRMM is designed to provide near real-time precipitation profiles as well as surface rainfall estimates for operational purposes within the tropical and subtropical regions. Two-thirds of global precipitation falls within the tropical regions between 30° N and 30° S (Adler and Negri, 1988; Simpson *et al.*, 1988). In addition, heavily populated developing countries are located in this region. Thus, monitoring of rainfall in these regions is essential. The TRMM has two unique attributes that make it ideal for observing tropical rainfall systems. They are (i) a suite of complementary observing instruments (precipitation radar) and (ii) its orbital characteristics (low-altitude, non-sun-synchronous, 35° tropical inclination). Orbital characteristics of the TRMM enable it to provide sampling in the tropics that is far more frequent and more spatially comprehensive than that obtained from standard polar-orbiting satellites (NAS, 2006). The TRMM is the first satellite Earth observation mission to monitor tropical rainfall, which closely influences the global climate and environmental change. The TRMM is commonly referred to as the "flying rain gauge" and is mainly used to obtain improved measurements of tropical precipitation by means of adding information derived from passive microwave and active microwave sensors together with other sensors operating in the visible and IR portion of the spectrum (NASDA, 2001). The TRMM has three main instruments onboard, namely, the Precipitation Radar (PR), the TRMM Microwave Imager (TMI), and the Visible and Infrared Scanner (VIRS). These are used to obtain tropical and subtropical rainfall measurements, rain profiles, and brightness temperature. The TRMM has the only passive microwave instrument (TMI) in an inclined orbit and the only rain radar (PR) in space. Two other instruments are flown onboard the TRMM, namely, the Cloud and Earth Radiant Energy Sensor (CERES) and the Lightning Imaging Sensor (LIS) to measure the radiation budget and global distribution of lightning, respectively (Kummerow *et al.*, 1998; Levizzani and Amorati, 2002). Data from these instruments can be used either individually or in combination with one another. Estimation of spatio-temporal rainfall distribution using TRMM data is demonstrated in Jeniffer *et al.* (2010). Javanmard *et al.* (2010) compared high-resolution gridded precipitation data (0.25° × 0.25° latitude/longitude) with satellite rainfall estimates from the TRMM over Iran.

4.7.4.4 MEGHA-TROPIQUES

Megha-Tropiques is a French–Indian satellite mission (ISRO, 2009; http://smsc.cnes.fr/MEGHAT/index.htm) designed to study convective systems, focusing on analysis of the water cycle with water vapor distribution and transport, convective systems, life cycle, and energy exchanges in the tropical belt. The tropical areas are where the most intensive energy exchanges occur: radiative exchanges, latent heat exchanges, and transport of constituents and energy through dynamic processes. The science goal is to increase understanding of the energetic and hydrologic processes in the tropics and the way they influence the global circulation of the atmosphere and oceans and climate variability. While the main purposes of developing a tropical data set are to support climate prediction studies and validate climate and weather models over tropical areas, the mission will also provide relevant data for global climate understanding, as tropical processes may also affect the global climate. The basic objectives of the Megha-Tropiques mission are the following:

- Provide simultaneous measurements of several elements of the atmospheric water cycle: water vapor, clouds, condensed water in clouds, precipitation, and evaporation.
- Measure the corresponding radiative budget at the top of the atmosphere.
- Ensure high temporal sampling in order to characterize the life cycle of the convective system and to obtain significant statistics.

The payload on the Megha-Tropiques satellite consists of the following:

- MADRAS – Microwave Analysis and Detection of Rain and Atmospheric Structures – a microwave imager aimed mainly at studying precipitation and cloud properties.
- SAPHIR – Sondeur Atmosphérique du Profil d'Humidité Intertropicale par Radiométrie – a six-channel microwave radiometer for the retrieval of water vapor vertical profiles and horizontal distribution.
- ScaRab – Scanner for Radiation Budget – a radiometer devoted to the measurement of outgoing radiative fluxes at the top of the atmosphere.
- GPS-ROS (Radio Occultation Sounder) – GPS receiver to measure the vertical profile of temperature and humidity at the point of radio occultation.

The Megha-Tropiques mission will provide the following geophysical parameters:

- cloud condensed water content,
- cloud ice content,
- convective–stratiform cloud discrimination,
- rain rate,
- latent heat release,
- integrated water vapor content,
- radiative fluxes at the top of the atmosphere,
- sea surface wind.

4.8 DEMONSTRATION OF IMAGE PROCESSING USING MATLAB

MATLAB (MATrix LABoratory) integrates computation, visualization, and programming in an easy-to-use environment where problems and solutions are expressed in familiar mathematical notation (http://www.mathworks.com/products/matlab/). MATLAB features a family of application-specific solutions called toolboxes. Toolboxes are comprehensive collections of MATLAB functions (M-files) that extend the MATLAB environment to solve particular classes of problems. MATLAB toolboxes are available for signal processing, control systems, neural networks, fuzzy logic, wavelets, simulation, image processing, and many others. The image processing toolbox has extensive functions for many operations for image restoration, enhancement, and information extraction. Some of the basic features of the image processing toolbox are explained and demonstrated with the help of satellite imagery obtained from IRS LISS III data for Uttara Kannada district, Karnataka, India.

IRS LISS III has four spectral bands, which are referred to as bands 2 (green), 3 (red), 4 (NIR), and 5 (SWIR). In this section, these images are referenced by file names "image*.jpg" where * represents the respective band number. For example, band 4 data are stored in file "image4.jpg". In this section, the text in italics refers to MATLAB commands.

4.8.1 Basic operations with MATLAB image processing toolbox

4.8.1.1 READ AND DISPLAY AN IMAGE

Clear the MATLAB workspace of any variables and close the open figure windows. To read an image, use the *imread* command. Let's read a JPEG image named image4.jpg, and store it in an array named I.

I = imread(image4.jpg);

Now call *imshow* to display I.

imshow(I)

Figure 4.16 Image displayed using *imshow*.

The image is displayed as shown in Figure 4.16. This image is IRS LISS III band 4 (NIR) data showing a portion of Uttara Kannada district in Karnataka state, India.

Some features in the image are the Arabian Sea on the left, a river in the top half, and dense vegetation. Small white patches in the image are clouds.

To check the image status in the memory, use the *whos* command to see how I is stored in memory.

whos

MATLAB responds with

Name	Size	Bytes	Class
I	342 × 342	116964	uint8

The grand total is 116964 elements using 116964 bytes; "uint8" represents an unsigned integer with 8 bits.

4.8.1.2 HISTOGRAM EQUALIZATION

As can be seen, image4.jpg is in low contrast, i.e., although pixels can be in the intensity range of 0 to 255, they are distributed in a narrow range. To see the distribution of intensities in image4.jpg in its current state, a histogram can be created by calling the *imhist* function. (Precede the call to *imhist* with the *figure* command so that the histogram does not overwrite the display of the image in the current figure window.)

figure, imhist (I) % Display a histogram of I in a new figure (Figure 4.17).

As can be noticed, the intensity range is rather narrow. It does not cover the potential range of [0, 255], and is missing the high values that would result in good contrast. Most of the DN values are in the range of 0 to 100. A non-linear contrast stretching algorithm, histogram equalization (Sabbins, 1986), can be used to improve the contrast in the image. We can call *histeq* to spread the intensity values over the full range, thereby improving the contrast of I, and store the modified image in the variable I2.

I2 = *histeq*(I);

Display the new histogram equalized image, I2, in a new figure window (Figure 4.18).

figure, imshow(I2)

We can store the newly adjusted image I2 back to disk. If it is to be saved as a PNG file, use *imwrite* and specify a filename that includes the extension "png."

imwrite(I2, 'image4.png')

The contents of the newly written file can be checked using *imfinfo* function to see what was written to disk.

imfinfo('image4.png')

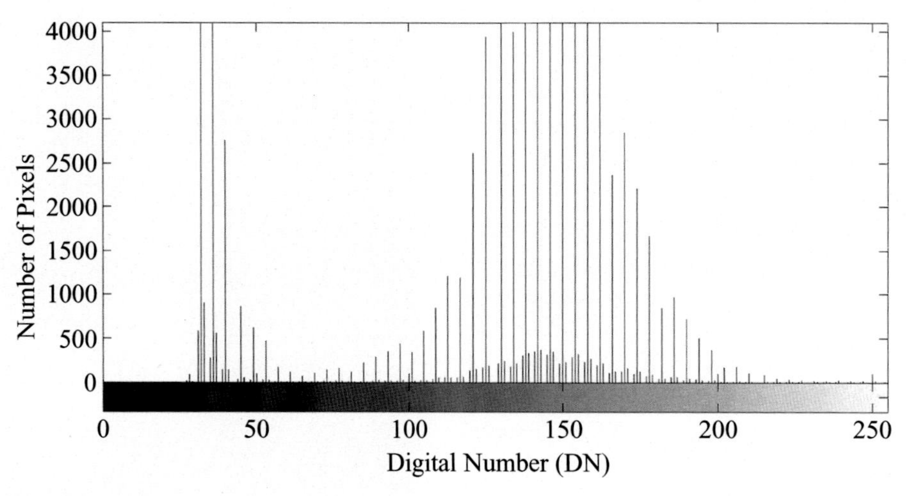

Figure 4.17 Histogram of original image.

Figure 4.18 Histogram equalized image.

4.8.2 Images in MATLAB and the image processing toolbox

The basic data structure in MATLAB is the array of an ordered set of real or complex elements. This object is naturally suited to the representation of images, real-valued ordered sets of color or intensity data. MATLAB stores most images as two-dimensional arrays, in which each element of the matrix corresponds to a single pixel in the displayed image.

For example, an image composed of 200 rows and 300 columns of different colored dots would be stored in MATLAB as a 200×300 matrix. Some images, such as RGB, require a three-dimensional array, where the first plane in the third dimension represents the red pixel intensities, the second plane represents the red and green pixel intensities, and the third plane represents the blue pixel intensities.

This convention facilitates working with images in MATLAB similarly to working with any other type of matrix data, and renders the full power of MATLAB available for image processing applications. For example, a single pixel can be selected from an image matrix using normal matrix subscripting.

```
I(2, 15)
```

This command returns the value of the pixel at row 2, column 15 of the image I.

MATLAB supports the following graphics file formats:

BMP	Microsoft Windows Bitmap
HDF	Hierarchical Data Format
JPEG	Joint Photographic Experts Group
PCX	Paintbrush
PNG	Portable Network Graphics
TIFF	Tagged Image File Format
XWD	X Window Dump

4.8.2.1 CONVERTING IMAGE STORAGE CLASSES

uint8 (8-bit unsigned integer) and *uint16* data can be converted to double precision using the MATLAB function, double. However, converting between storage classes changes the way MATLAB and the toolbox interpret the image data. If it is desired to interpret the resulting array properly as image data, the original data should be rescaled or offset to suit the conversion.

For easier conversion of storage classes, use one of these toolbox functions: *im2double*, *im2uint8*, and *im2uint16*. These functions automatically handle the rescaling and offsetting of the original data. For example, the following command converts a double-precision RGB image with data in the range [0, 1] to a uint8 RGB image with data in the range [0, 255].

RGB2 = *im2uint8*(RGB1);

4.8.2.2 CONVERTING GRAPHICS FILE FORMATS

To change the graphics format of an image, use *imread* to read in the image and then save the image with *imwrite*, specifying the appropriate format. For example, to convert an image from BMP to PNG, read the BMP image using *imread*, convert the storage class if necessary, and then write the image using *imwrite*, with "PNG" specified as your target format.

bitmap = *imread*('image4.bmp');
imwrite(bitmap,'image4.png');

4.8.3 Image arithmetic

Image arithmetic is the implementation of standard arithmetical operations, such as addition, subtraction, multiplication, and division, on images. Image arithmetic has many uses in image processing, both as a preliminary step and in more complex operations. For example, image subtraction can be used to detect differences between two or more images of the same scene or object.

4.8.3.1 ADDING IMAGES

To add two images or add a constant value to an image, use the *imadd* function. *imadd* adds the value of each pixel in one of the input images with the corresponding pixel in the other input image and returns the sum in the corresponding pixel of the output image. Image addition has many uses in image processing. For example, the following code fragment uses addition to superimpose one image on top of another. The images must be of the same size and class.

I = *imread*('image3.jpg');
J = *imread*('image4.jpg');
K = *imadd*(I,J); *imshow*(K)

An added image is shown in Figure 4.19. In this figure LISS III bands 3 and 4 (i.e., red band and NIR band) are added. It may be noted that DN values in K will be rescaled to original image 8-bit system (0–255). Addition can also be used to brighten an image by adding a constant value to each pixel. For example, the following code brightens image4.jpg.

I = *imread*('image4.jpg');
J = *imadd*(I,50);

Figure 4.19 Image after adding two images.

4.8.3.2 SUBTRACTING IMAGES

To subtract one image from another, or subtract a constant value from an image, use the *imsubtract* function. *imsubtract* subtracts each pixel value in one of the input images from the corresponding pixel in the other input image and returns the result in the corresponding pixel in an output image.

X = *imread*('image4.jpg'); J = *imread*('image3.jpg');
K = *imsubtract*(X,J);

4.8.3.3 MULTIPLYING IMAGES

To multiply two images, use the *immultiply* function. *immultiply* does an element-by-element multiplication of each corresponding pixel in a pair of input images and returns the product of these multiplications in the corresponding pixel in an output image. Image multiplication by a constant, referred to as scaling, is a common image processing operation. When used with a scaling factor greater than one, scaling brightens an image; a factor less than one darkens an image. Scaling generally produces a much more natural brightening/darkening effect than simply adding an offset to the pixels, since it preserves the relative contrast of the image better.

For example, the code below scales an image by a constant factor.

I = *imread*('image4.jpg');
J = *immultiply*(I,3.0);
figure, *imshow*(J);   % shown in Figure 4.20

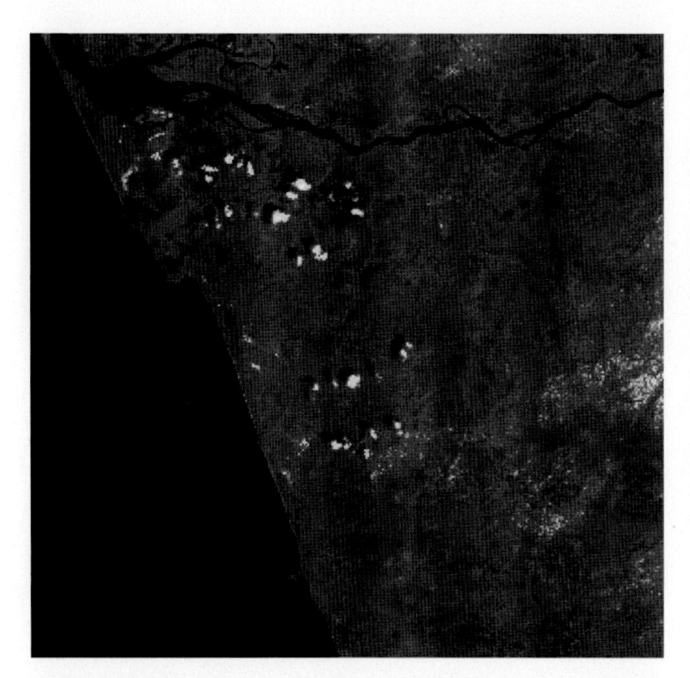

Figure 4.20 Image values multiplied by an integer 3.

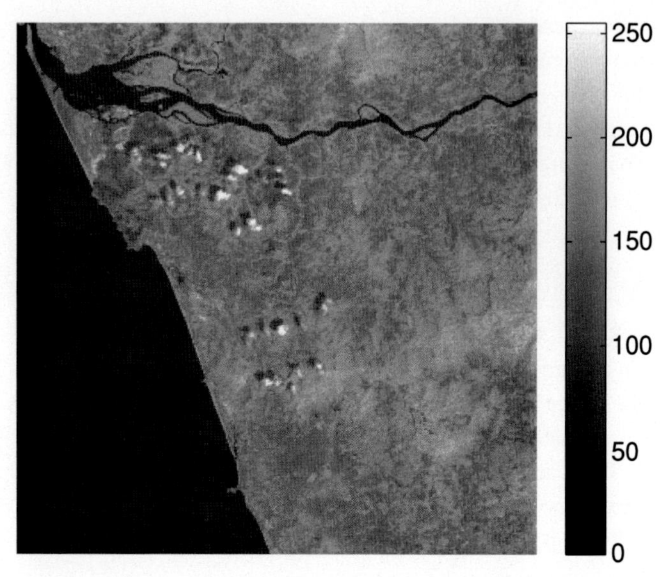

Figure 4.21 Image with color bar.

4.8.3.4 DIVIDING IMAGES

To divide two images, use the *imdivide* function. The *imdivide* function does an element-by-element division of each corresponding pixel in a pair of input images. The *imdivide* function returns the result in the corresponding pixel in an output image. Image division, like image subtraction, can be used to detect changes between two images.

However, instead of giving the absolute change for each pixel, division gives the fractional change or ratio between corresponding pixel values. Image division results in ratio images.

4.8.4 Pixel coordinates

Generally, the most convenient method for expressing locations in an image is to use pixel coordinates. In this coordinate system, the image is treated as a grid of discrete elements, ordered from top to bottom and left to right. For pixel coordinates, the first component r (the row) increases downward, while the second component c (the column) increases to the right. Pixel coordinates are integer values and range between 1 and the length of the row or column.

4.8.5 Special display techniques

In addition to *imshow*, the toolbox includes functions that perform specialized display operations, or exercise more direct control over the display format.

4.8.5.1 ADDING A COLOR BAR

The *colorbar* function can be used to add a color bar or grey tone legend to an axes object. If a *colorbar* is added to an axes object that contains an image object, the *colorbar* indicates the data values referring to different colors or intensities in the image as shown in Figure 4.21.

F = *imread*('image4.jpg');
imshow(F), *colorbar*

4.8.5.2 IMAGE RESIZING

To change the size of an image, use the *imresize* function. *imresize* accepts two primary arguments: (i) the image to be resized and (ii) the magnification factor.

The command below decreases the size of the image by 0.5 times.

F = *imread*('image4.jpg'); J = *imresize*(F,0.5);

Using *imresize*, one can also specify the actual size of the output image. The command below creates an output image of size 100 × 150.

Y = *imresize*(X, [100 150])

4.8.5.3 IMAGE ROTATION

To rotate an image, the *imrotate* function can be used. *imrotate* accepts two primary arguments: (i) the image to be rotated and (ii) the rotation angle. The rotation angle should be specified in degrees. For a positive value, *imrotate* rotates the image counterclockwise; and for a negative value, *imrotate* rotates the image clockwise. For example, these commands rotate an image by 35 degrees counterclockwise (Figure 4.22).

F = *imread*('image4.jpg');
J = *imrotate*(I,35,'bilinear');
figure, imshow(J)

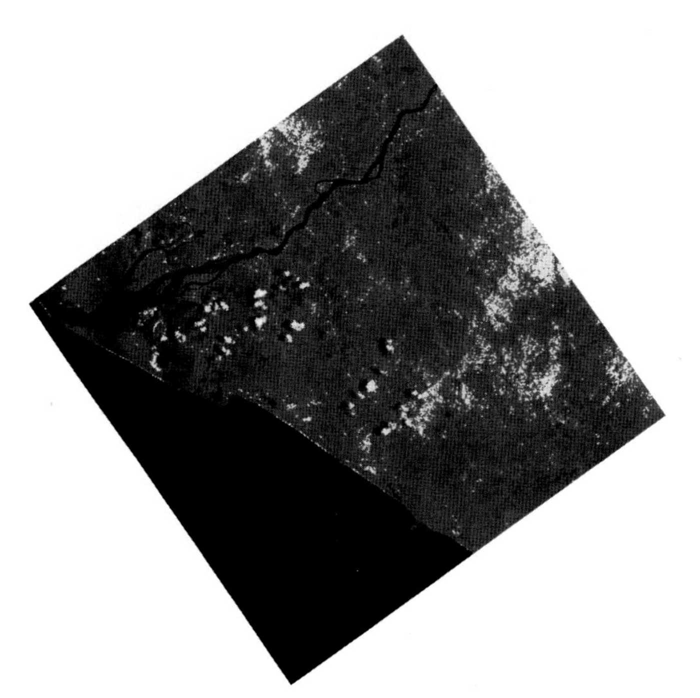

Figure 4.22 Image rotated by 35 degrees.

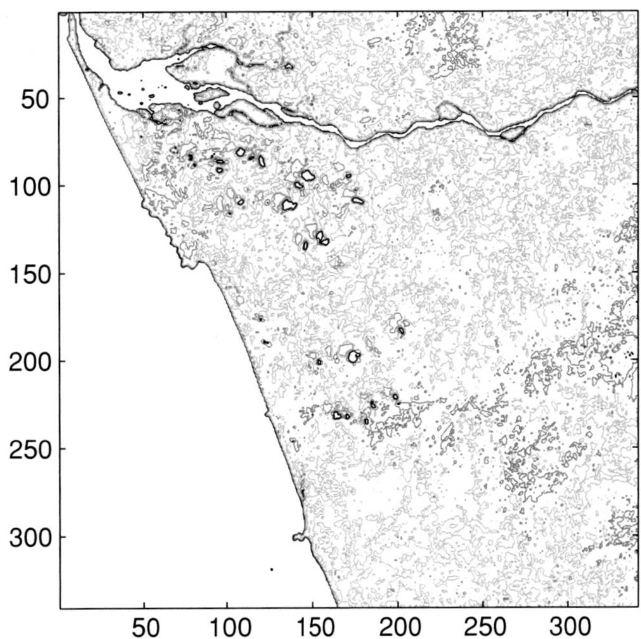

Figure 4.23 Contour plot of an image. See also color plates.

4.8.5.4 IMAGE CROPPING

To extract a rectangular portion of an image, the *imcrop* function can be used. *imcrop* accepts two primary arguments: (i) the image to be cropped and (ii) the coordinates of a rectangle that defines the crop area.

If *imcrop* is called without specifying the crop rectangle, the cursor changes to a cross-hair when it is over the image. Click on one corner of the region to be selected and, while holding down the mouse button, drag across the image. *imcrop* draws a rectangle around the selected area. When the mouse button is released, *imcrop* creates a new image from the selected region.

4.8.6 Analyzing and enhancing images

The image processing toolbox supports a range of standard image processing operations for analyzing and enhancing images. Its functions simplify several categories of tasks, including:

- Obtaining pixel values and statistics, which are numerical summaries of data in an image.
- Analyzing images to extract information about their essential structure.
- Enhancing images to make certain features easier to see or to reduce noise.

4.8.6.1 PIXEL SELECTION

The toolbox includes two functions that provide information about the color data values of image pixels specified. The *pixval* function interactively displays the data values for pixels as the cursor is moved over the image. *pixval* can also display the Euclidean distance between two pixels. The *impixel* function returns the data values for a selected pixel or set of pixels. One can supply the coordinates of the pixels as input arguments, or select pixels using a mouse.

imshow I
vals = *impixel*

4.8.6.2 STATISTICS OF IMAGES

One can compute standard statistics of an image using the *mean2*, *std2*, and *corr2* functions. *mean2* and *std2* compute the mean and standard deviation of the elements of a matrix. *corr2* computes the correlation coefficient between two matrices of the same size.

4.8.6.3 IMAGE CONTOURS

One can use the toolbox function *imcontour* to display a contour plot of the data in an intensity image. This function is similar to the *contour* function, but it automatically sets up the axes so their orientation and aspect ratio match the image. The following example displays a contour plot (contours of same DN values) of image4.jpg as shown in Figure 4.23.

I = *imread*('image4.jpg');
figure, imcontour(I)

4.8.6.4 EDGE DETECTION

One can use the *edge* function to detect edges, which are those places in an image that correspond to object boundaries. To find

Figure 4.24 Edge detection image.

edges, this function looks for places in the image where the intensity changes rapidly, using one of these two criteria: For some of the edge estimators, it can be specified whether the operation should be sensitive to horizontal or vertical edges, or both. *edge* returns a binary image containing 1s where edges are found and 0s elsewhere.

One of the most powerful edge detection methods that *edge* provides is the Canny method. The Canny method differs from the other edge detection methods in that it uses two different thresholds (to detect strong and weak edges), and includes the weak edges in the output only if they are connected to strong edges. This method is therefore less likely than the others to be "fooled" by noise, and more likely to detect true weak edges. The example below illustrates the power of the Canny edge detector. It shows the results of applying the Canny edge detector to the band 4 image (Figure 4.24).

```
F = imread('image4.jpg');
BW1 = edge(F,'sobel');
BW2 = edge(F,'canny');
figure, imshow(BW2)
```

4.8.7 Image enhancement

Image enhancement techniques are used to improve an image, where "improve" is sometimes defined objectively (e.g., increase the signal-to-noise ratio), and sometimes subjectively (e.g., make certain features easier to see by modifying the colors or intensities).

4.8.7.1 INTENSITY ADJUSTMENT

Intensity adjustment is a technique for mapping an image's intensity values to a new range. For example, image4.jpg is a low-contrast image. The histogram of image4.jpg indicates that there are very few values above 80. If the data values are remapped to fill the entire intensity range [0, 255], one can increase the contrast of the image. This kind of adjustment can be achieved with the *imadjust* function in addition to the *histeq* function already explained. The general syntax of *imadjust* is

J = *imadjust* (I, [low_inhigh_in], [low_outhigh_out])

Where, low_in and high_in are the intensities in the input image, which are mapped to low_out, and high_out in the output image. For example, the code below performs the adjustment described above.

```
I = imread('image4.jpg');
J = imadjust(I,[0.0 0.3],[0 1]);
```

The first vector passed to *imadjust*, [0.0 0.3], specifies the low and high intensity values of the image. The second vector, [0 1], specifies the scale over which you want to map them. Thus, the example maps the intensity value 0.0 in the input image to 0 in the output image, and 0.3 to 1. Note that one must specify the intensities as values between 0 and 1 regardless of the class of I. If I is in *uint8*, the values supplied are multiplied by 255 to determine the actual values to be used.

To use *imadjust*, one must typically perform two steps:

1. View the histogram of the image to determine the intensity value limits.
2. Specify these limits as a fraction between 0.0 and 1.0 to pass them to *imadjust* in the [low_in high_in] vector.

The MATLAB image processing toolbox has many more capabilities and only a small portion of them are explained here.

4.9 CONCLUSIONS AND FUTURE SCOPE

The basic concepts of RS, the various sensors available, and different image processing techniques are explained in this chapter. Remote sensing applications in modeling different components of the hydrologic cycle are discussed in detail, indicating the strong potential of using RS for water resources planning and management.

Active microwave RS from satellites offers the potential of (almost) all-weather application due to the penetration capabilities through clouds and shadows when compared to optical sensors.

Advances in microwave RS will also provide vital information required for soil moisture modeling in the vadose zone. However, the necessary algorithms are not universally applicable.

The current usage of RS data in hydrologic modeling is relatively low. One reason is that most hydrologic models in operational use are not designed to use spatially distributed data, which is a prerequisite to the sensible use of RS data. Research is needed into the development of generalized algorithms and into the design of hydrologic models more suited to the routine use of RS data. Another reason for the low usage of RS data in hydrologic modeling is the lack of appropriate education and training. Operational agencies and consultants predominantly use traditional techniques. Potential users should be properly trained and appraised about the advantages of using RS.

Satellite RS can also be an appropriate tool to help alleviate some of the hydrometric data collection and management problems facing many developing countries. Progress in water resources research depends on the availability of adequate data for model development and validation. Remote sensing plays a vital role in this process and can successfully address some of the previously intractable problems.

EXERCISES

4.1 Discuss the potential for using the digital image enhancement procedures explained in this chapter to model various components of the hydrologic cycle.

4.2 Consider a portion of an image given in an array of 8×10 with the following DN values:
   (a) Rows 1–2: 30
   (b) Rows 3–5: 35
   (c) Rows 6–8: 30
   Apply the Laplacian filter and check whether it has any effect in edge enhancement.

4.3 For the data given in Exercise 4.2, apply the N89E directional filter and compare the contrast ratio of enhancement with that obtained from Exercise 4.2.

4.4 For the data given in Exercise 4.2, apply the N45W directional filter and comment on edge enhancement.

4.5 Consider an image array of 6×15 as follows:
   (a) Columns 1–3: 50
   (b) Columns 4–5: 60
   (c) Columns 6–10: 50
   (d) Columns 11–13: 40
   (e) Columns 14–15: 50
   Apply the Laplacian filter. Also apply any two weighting factors and compare edge enhancement capabilities.

4.6 For the data given for the example problem in Section 4.4.5.2 for directional filters, apply the Laplacian filter and compare edge enhancement capabilities.

4.7 Explain the procedure to assess the performance of a classification algorithm using the confusion matrix and the scope for deriving various statistical measures using the matrix.

4.8 Obtain Landsat TM images (bands 1–7) for any region from the website http://glovis.usgs.gov/. Using the image processing toolbox of MATLAB, perform the following tasks.
   (a) Show the histograms of all the images.
   (b) Show the scatter plot of band 3 versus band 4 and comment.
   (c) Contrast stretch bands 3 and 4 using *histeq* function and comment.
   (d) For a standard FCC, get the pixel values at row = 15 and column = 45 for each of the six bands (bands 1–5 and 7) and comment on the likely feature at that pixel.
   (e) Produce an NDVI (normalized difference vegetation index) image and also show the color bar.
   (f) Produce a band 5 – band 4 image.
   (g) From a band 4 image, approximately estimate the area occupied by water bodies (use spatial resolution of Landsat TM data).

4.9 Challenging problems to be solved using the image processing toolbox of MATLAB for the Landsat images obtained from http://glovis.usgs.gov/.
   (a) Produce a standard FCC.
   (b) Produce PC images using data from six bands (excluding band 6) and comment about image compression.
   (c) Produce an FCC of first three PC images.
   (d) Derive ISH images.
   (e) Density slice an NDVI image and show vegetation in different tones of green color.

5 Geographic information systems for hydrologic modeling

Geographic information systems (GIS) are powerful technology capable of integrating information from many sources and bringing it to a common platform to carry out query analysis for specific purposes. Data from different sources can include land use/land cover maps prepared from satellite data, soil maps, political boundaries, water supply layout, and elevation data procured manually or by satellites in raster form, etc. This chapter provides a background to basic data handling in GIS and different types of digital elevation models (DEMs). Delineation of sub-watershed boundaries is explained briefly together with a distributed hydrologic model. Three case studies are presented along with the D8 algorithm, which is commonly followed for watershed delineation using raster-based DEM. Use of different data products for flood analysis using GIS is discussed with examples. Web-based GIS is introduced along with its applications in hydrology.

5.1 INTRODUCTION

GIS technology is capable of stacking, analyzing, and retrieving large amounts of data having geographical information within it. By geographical information, we refer to the x, y, z coordinates of land surfaces represented in a user-defined coordinate system. GIS does not create new information but it manipulates existing information to answer the required queries. GIS caters to the needs of various branches of engineering and sciences, such as cartography, RS, water resources, transportation, environmental, navigation, etc.

Topographic information plays a major role in distributed hydrologic models as they treat the hydrologic processes and their interactions at relatively small scales compared to the lumped ones. The elevation information, when integrated inside a GIS platform, allows one to estimate the slope and *aspect* (compass direction that a slope faces), which are of utmost importance in flow modeling and watershed delineation studies. The need to map the surface of the Earth for various applications led to the science of surveying. Earlier techniques employed to map surface elevation used equipment such as theodolites, prismatic compasses, and levelling instruments. These were later replaced by their digital versions, thereby eliminating the errors associated with manual recording of data. As more sophisticated equipment came up, the older instruments began to be replaced. At present, total station and differential global positioning systems (GPS) are commonly in use to give an accurate three-dimensional representation of the terrain. GIS and GPS are two terms that a beginner is likely to misinterpret. To make the distinction between GIS and GPS clear, consider the case of a surveying experiment. In plane-table surveying, the surveyor makes an on-site sketch of the problem under consideration on a map, i.e., data are being recorded there. In the office, the map is completed by adding appropriate legends and made visibly pleasing. In this step, the data recorded previously are displayed for ease of interpretation. In short, GPS is the technology that records data in the field and GIS is a platform that extracts useful information from the recorded data.

Manual surveying techniques often suffer due to difficulty in setting up the tripod when dealing with rugged topography. Moreover, human-induced errors also distort the data on-site. These problems were overcome by stereo-photogrammetric techniques and radar interferometery. This chapter introduces DEMs using elevation data and their use in hydrologic modeling using GIS. Hydrologic models can be developed to predict runoff from precipitation and snowmelt on GIS. It is important to learn about data representation in GIS for understanding hydrologic modeling using GIS. This chapter starts with data representation, followed by a discussion as to how these data can be effectively used for query analysis. Case studies dealing with DEMs are also presented. Finally, its use for flood studies is discussed in Section 5.9.

5.2 REPRESENTATION OF SPATIAL OBJECTS IN GIS

Information within the GIS platform can consist of spatial data, non-spatial data, or ancillary data. Spatial data, as the name suggests, give importance to the spatial relationships between data points in a three-dimensional space. For example, if the control

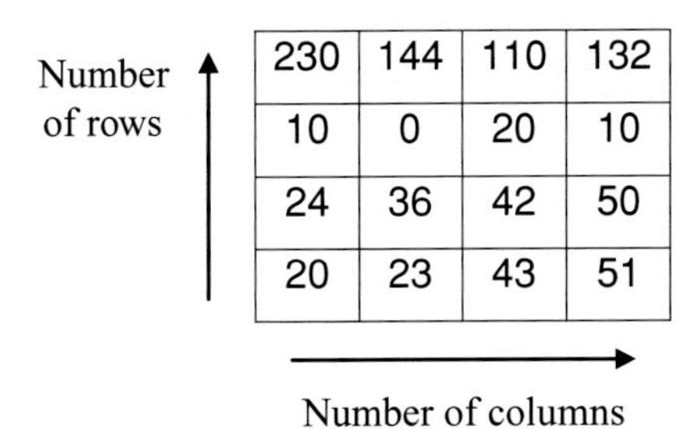

230	144	110	132
10	0	20	10
24	36	42	50
20	23	43	51

Figure 5.1 Raster data representation.

volume (CV) considered for a particular hydrologic modeling is a pentagon, spatial data would contain details about the:

- location, which is specified by the center of gravity of the pentagon,
- shape, which is the pentagon, and
- size, explained by the length of the five sides of the pentagon.

Proper understanding of data formats used in GIS plays a crucial role in interpretation of the end products. This section is devoted to the different formats of data representation followed in GIS.

The spatial data within GIS can be represented in raster, vector, or TIN (triangulated irregular network) format. These formats are used to store and retrieve geographical information.

5.2.1 Raster format

Raster format refers to representation of information in the form of square cells called pixels. As these cells are organized in an array form, they can be identified with row and column numbers. Each of the cells will have a single value. Digital images in any format (jpeg, tiff, gif, etc.) loaded into GIS are the best examples of raster data. For example, a satellite image of India having a spatial resolution of 0.4 × 0.4 meters would mean that the digital numbers (DN) within each pixel are an aggregate of reflectance values from an area of 0.4 × 0.4 meters on the ground. Row and column numbers provide location coordinates for each pixel in a digital image. In Figure 5.1, a location coordinate of (3, 2) would point at the pixel in the third row, second column, which indicates number 0. In this case, the raster is part of a satellite image, a pixel value of 0 can usually be interpreted as the existence of a water body in the area on the ground represented by this pixel.

5.2.2 Vector format

In a vector format, features on the Earth are represented as points, lines, areas (polygons), or TINs. Points are represented with the help of location coordinates (x, y); lines as a sequence of points; and areas or polygons as a set of interconnected closed lines, as shown in Figure 5.2. For example, within a control volume, wells can be represented as points, streets as lines, lakes as polygons,

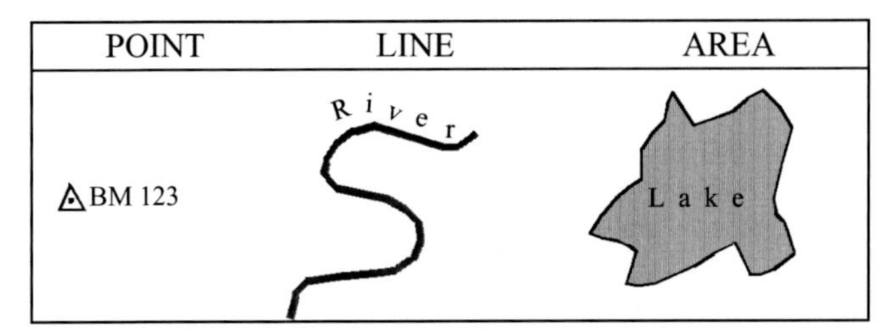

Figure 5.2 Vector data representation of different features.

and elevation of the entire terrain as a TIN. Data are stored in two files generally, one containing the location information and the other containing information on the attributes.

5.2.3 Triangulated irregular network format

TIN refers to an irregularly arranged network of non-overlapping triangles that are used to represent a surface. Instead of analyzing a cell (as in raster) or a feature (as in vector), a triangle representing a particular area is analyzed. TIN offers many advantages over raster in presenting a surface. Efficient representation of ground by a raster depends on the pixel size, whereas in TIN the size of triangles can be altered to represent a particular underlying surface. In other words, the size and shape of each triangle can be varied to represent different levels of terrain of a mountainous region. TIN has the ability to describe a surface at different levels of resolution. Raster-based representation becomes defunct in capturing the different complexities of relief in terrain such as highlands or mountains. In Figure 5.3, a TIN representation of elevation information is shown. Elevation information is obtained from the Shuttle Radar Topography Mission (SRTM). In the zoomed view, note the subtle variations in topography depicted with the help of varying triangles. TIN is popular mainly because of its simplicity in representation and efficient storage capacity.

5.2.4 Comparison of raster and vector formats

As mentioned earlier, information representation in GIS can be in raster, vector, or TIN formats. The suitability of each depends on the application and, more importantly, on the accuracy demanded. The three methods have their own advantages and disadvantages.

In the raster format, terrain is viewed with the help of grids overlying the area considered. These grids may represent x, y coordinates or latitude and longitude or, simply put, the location of each cell within an array of cells in two dimensions. As a consequence, the accuracy of the raster format depends on the cell resolution. Contrary to this, the vector format deals with a more accurate representation of the features within the terrain. This is achieved by a process known as digitization, which involves manual tracing of the required feature boundaries within a digital image with respect to a specified scale (e.g., 1:10,000). Raster

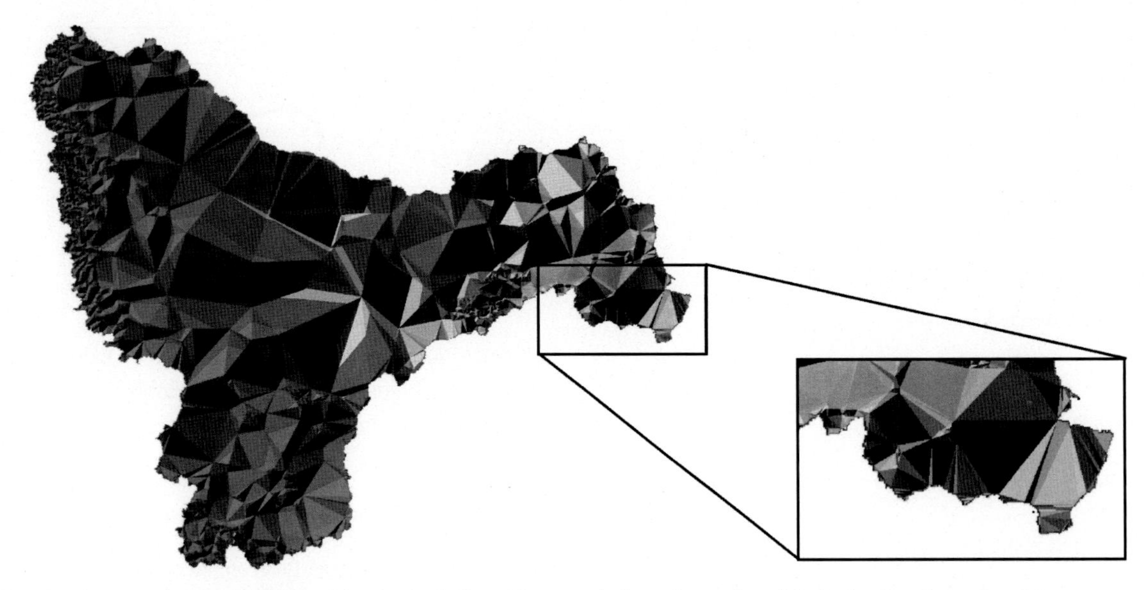

Figure 5.3 TIN representation of the Krishna River basin, India, with zoomed view of a portion of the basin. See also color plates.

formats are efficient when comparing information represented by equal cell sizes. On the other hand, vector formats are efficient even when the data being compared are of different scales. An analogy to this is capturing a digital photograph of a person (raster format) or manually sketching using point, line, and polygon features (vector format). Lastly, raster files are generally very large but vector files are mostly much smaller.

5.3 GIS FOR PROXIMITY ANALYSIS

In many modeling exercises, query analysis is the desired end result required by the user. How to find the most optimal solution for any problem or the shortest route to reach a particular destination are examples of queries that are normally encountered. *Proximity analysis* tools are employed for query analysis. Proximity, as the word implies, defines the degree of closeness between chosen data in the spatial domain both quantitatively and qualitatively. This is largely used to answer queries and bring out the optimum solutions; for example, finding out the shortest path to reach a particular destination. GIS comprises several tools for proximity analysis, some of which are discussed in this section.

5.3.1 Vector distance tools: buffer

Buffer is largely known as an area-expanding technique. This tool is normally present in many GIS packages, such as ERDAS, ENVI, and ArcGIS. The buffer tool creates a zone of user-defined width on either side of the feature class being investigated (point, line, or polygon). This is a very useful tool with multi-faceted applications. A general example would be to find out the degree of encroachment by roadside vendors on either side of a crowded pavement. An example in hydrology is to find the settlements in a floodplain within 100 meters of a river. Buffer zones are normally generated adjacent to stream networks to protect water quality, fish habitat, and other resources. These are termed buffer strips (riparian wetlands) and they are extensively used for the protection of water quality by designing forest riparian buffer strips, for land preservation initiatives, watershed development planning, etc. Figure 5.4 shows the stream networks of the Krishna basin, India, after buffering to 100 meters, with a zoomed view of the streams. Notice in Figure 5.4(b) how an envelope is created after buffering. This may not be visible in Figure 5.4(a).

5.3.2 Raster-based distance tools

Tools operate on raster format mainly to show the distance of cells from a set of features or to allocate each cell to the closest feature. This tool comes with a number of additional functions such as Euclidean distance (for allocating polygons around the nearest point feature class) and Euclidean direction (for depicting the direction of the closest point feature class). Available in a number of software packages, such as IDRISI, ArcGIS, etc., this tool calculates the Euclidean distance of each cell to the nearest among target cells and depicts the same in a separate image. Figure 5.5 shows a Euclidean distance image for randomly generated location coordinates. Figure 5.6 shows the Euclidean direction image of the same set of data points.

In addition to the raster-based distance tools, there are various other functions in the same toolset, such as *cost distance*, *cost allocation*, *cost path*, *cost back link*, etc., that deal with minimizing cost while analyzing a cost surface. Several other tools that specialize in finding out the least-cost path are *path distance*, *path distance allocation*, *path distance back link*, etc.

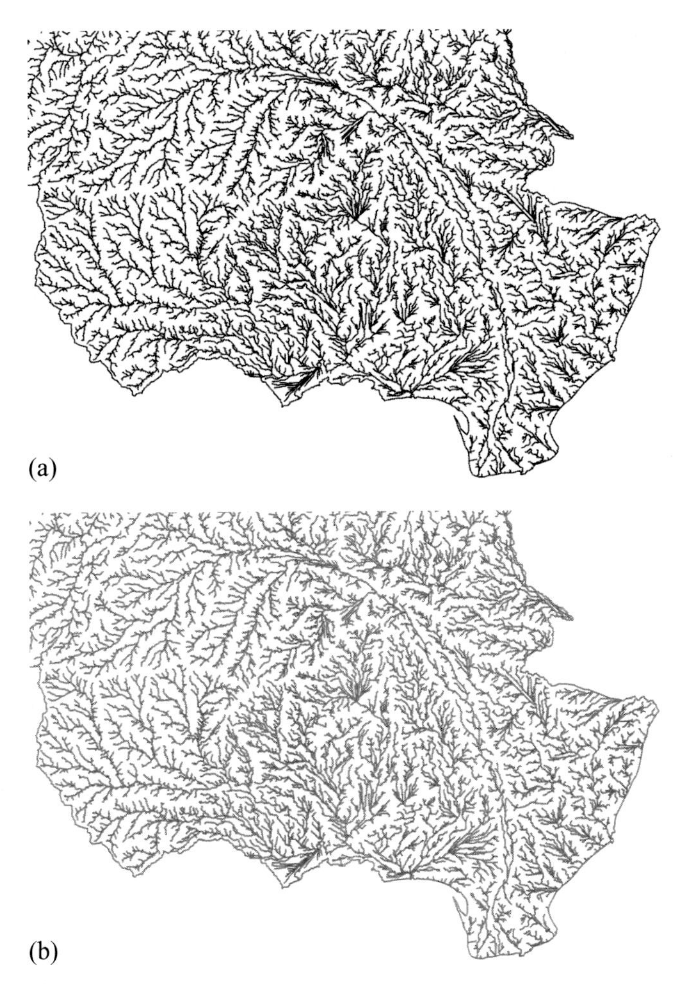

Figure 5.4 Line feature class showing stream networks of eastern Krishna basin: (a) before buffering, (b) after buffering to 100 m.

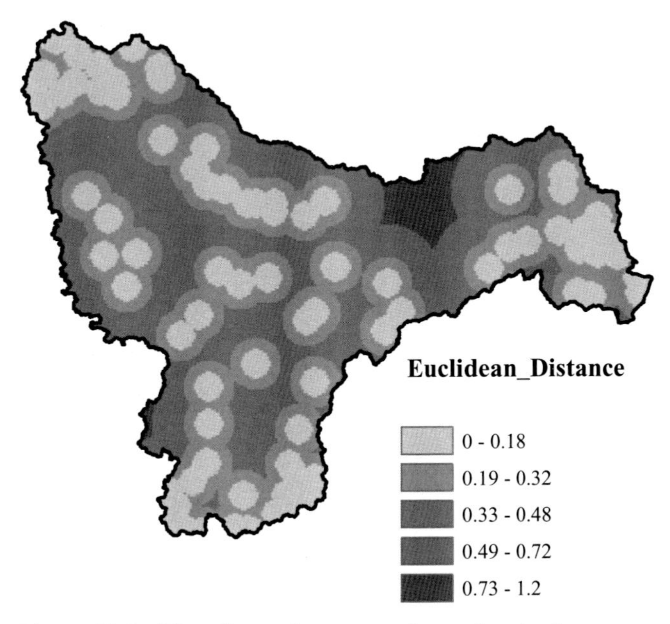

Figure 5.5 Euclidean distance image created around randomly generated point feature class. See also color plates.

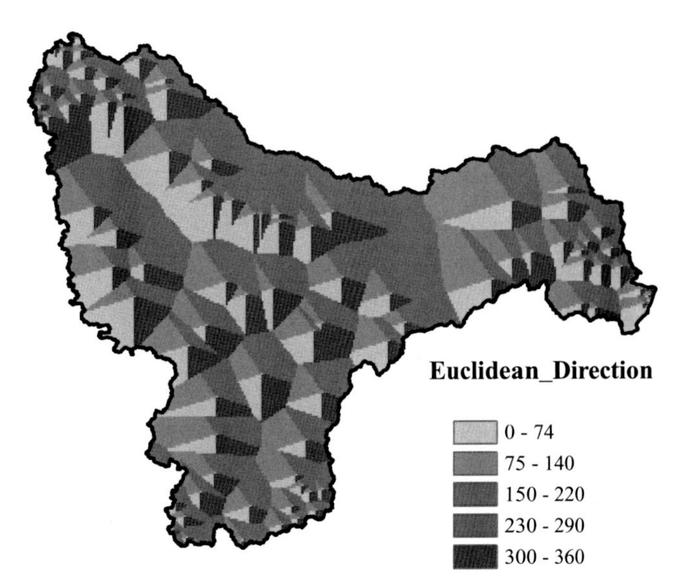

Figure 5.6 Euclidean direction image created for the set of data points used in Figure 5.5. See also color plates.

If we have a network such as the feature class of a stream network and rain-gauge stations in any basin, then the network analysis toolset is the best to use to extract information about the network. It helps one to find the shortest path among the rain-gauge stations.

5.4 DIGITAL ELEVATION MODELS

A DEM is defined as "any digital representation of the continuous variation of *relief* over space" (Burrough, 1986). Here *relief* refers to the height of the Earth's surface with respect to the datum considered. Another way to understand DEMs is as arrays of cells, each representing elevation values with respect to a particular projection system. Readers are likely to get confused between digital surface model (DSM), digital terrain model (DTM), and DEM. Though they may all mean elevation in raster format, a DEM in reality is a superset of both DTM and DSM. A DEM gives the bare Earth elevations free of vegetation, buildings, and other non-ground objects with reference to a datum such as mean sea level (MSL). A DSM includes the tops of buildings, power lines, trees, and all objects as seen in a synoptic view. A DTM is a relief map and indicates the nature of the land with reference to elevation such as rugged terrain, hilly terrain, etc. In runoff models, topography plays an important role as it defines the surface water movement across a watershed. Catchment characteristics that influence the generation of runoff are width, shape, channel or drainage network, slope gradient, slope aspect, slope curvature, upslope length, specific catchment area, and upslope contributing area. Sections 5.4.2 and 5.5 discuss the different types of DEMs and their application in providing information regarding *relief*, flow *direction*, and *slope*.

5.4.1 Representation of digital elevation data

In a digital environment, information about height can be represented either as contour lines, or in a grid format, or as a TIN. Each of these formats, such as grid-based DEM, TIN, and contour, are discussed in detail in this section.

5.4.2 Types of digital elevation models

5.4.2.1 GRID-BASED DIGITAL ELEVATION MODELS

A grid-based DEM (GDEM) represents the elevation of the terrain at discrete points in a regular, rectangular grid. GDEMs that are readily available and simple to use have widespread application in distributed hydrologic models. They yield readily usable data sets that represent the elevation of a surface as a function of geographic location at regularly spaced horizontal (square) grids. Since the GDEM data are stored in the form of a simple matrix, values can be accessed easily without having to resort to a graphical index and interpolation procedures.

In a hydrologic model, the type of terrain attributes required and their role vary according to the process being modeled. The terrain attributes can be divided into primary and secondary attributes. The primary attributes are extracted directly from GDEM, while the secondary attributes are estimated from the primary attributes. The most common primary terrain attributes extracted from DEM include morphological features such as drainage network and catchment parameters such as area, width, shape, and slope. Secondary attributes could be compound topographic index (CTI), which is a function of slope and upstream contributing area per unit width orthogonal to flow direction, elongation ratio, which is a watershed parameter developed to reflect basin shape, etc. Resolution of a DEM plays an important role in any modeling process. The resolution of elevation data represents the horizontal accuracy of a GDEM. The accuracy of a GDEM or its derived products depends on:

- Source of the elevation data: such as the techniques for measuring elevation, the nature of the locations, and the density of samples.
- Structure of the elevation data: such as grid, contour, or TIN.
- Horizontal resolution and vertical precision at which the elevation data are represented.
- Topographic complexity of the landscape being represented.
- Algorithms used to calculate the different terrain attributes.

GDEMs are useful in modeling runoff, erosion, and evaporation; for estimating the volume of proposed reservoirs; and for determining the probability of landslides. Further, GDEMs can be used in analyzing changes with relation to the topographic properties; for instance, analyzing the relation between crop patterns and terrain exposure. Many studies have analyzed the catchment characteristics using GDEMs. These include: (i) automatic delineation of catchment areas (O'Callaghan and Mark, 1984; Martz and De Jong, 1988); (ii) development of terrain characteristics (Moore *et al.*, 1991) and drainage networks (Fairchild and Leymarie, 1991); (iii) detection of channel heads (Montgomery and Dietrich, 1988); (iv) estimating soil moisture (Beven and Kirkby, 1979; O'Loughlin, 1986; English *et al.*, 2004); (v) determination of flow accumulation (Peuker and Douglas, 1975), flow direction, and routing (Tarboton, 1997, 2002); and (vi) automated extraction of parameters for hydrologic or hydraulic modeling (Doan, 2000; Ackerman, 2002).

Ground depressions play a major role in the collection and storage of incoming precipitation, thereby modifying the runoff response and sediment yield of a watershed. In the early stages of hydrologic modeling, depression storages were empirically evaluated or included as a percentage of the storage associated with the upper soil zones (Horton, 1939; Boughton, 1966) or indirectly estimated from the total volume of storage. With the availability of GDEMs of finer resolution, the geometric properties (terrain attributes such as depth, surface area, and volume) of the depressions can be more accurately determined and used in estimating the depression storage.

Catchment-scale soil erosion modeling is useful for efficient soil conservation planning for decreasing erosion-related problems (Jetten *et al.*, 2003) and sedimentation of reservoirs. The important terrain attributes influencing erosion used in the erosion models are slope angle, slope length, aspect, and wetness index (WI) derived from GDEM. In the past, empirical relations were used for computation of slope angle and slope length in many erosion models based on universal soil loss equations (Mitasova *et al.*, 2007). The empirical relations are now replaced by terrain attributes derived from high-resolution GDEM.

5.4.2.2 TRIANGULATED IRREGULAR NETWORK

In TIN, the elevations are represented at the vertices of irregularly shaped triangles, which may be few in number when the surface is flat but may be numerous for a surface with steep slope. A TIN is created by running an algorithm over a raster to capture the nodes required for TIN. Even though several methods exist, the Delaunay triangulation method is the most preferred one for generating TIN. A TIN for the Krishna basin in India created using USGS DEM data (http://www.usgs.gov) is shown in Figure 5.7. It can be observed from this figure how the topographical variations are depicted with the use of large triangles where change in slope is small, and small triangles of different shapes and sizes where there are large fluctuations in slope.

For hydrologic modeling, under certain circumstances some researchers (Braun and Sambridge, 1997; Turcotte *et al.*, 2001)

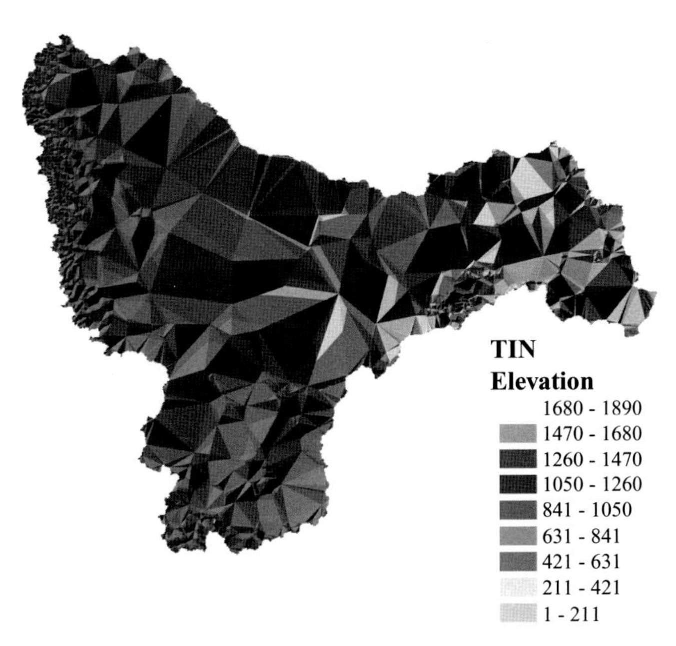

Figure 5.7 TIN for the Krishna basin created from USGS DEM data. See also color plates.

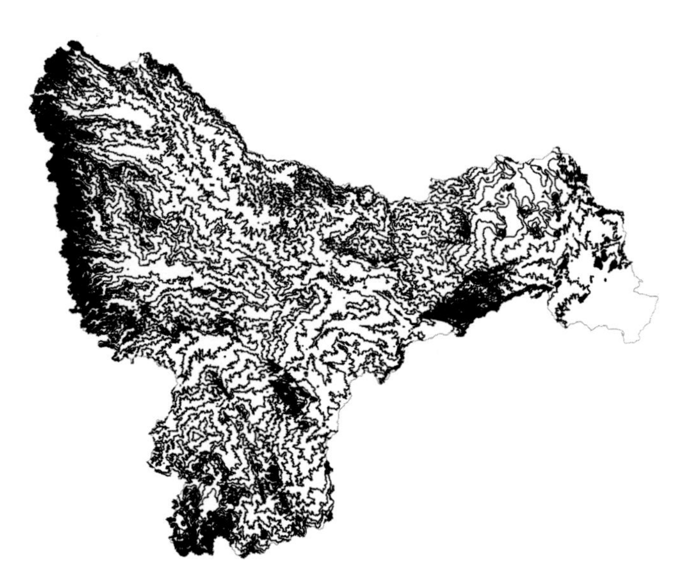

Figure 5.8 Contour map for the Krishna basin created from USGS DEM data.

have observed that artifacts of the underlying regularity of rasters may manifest themselves in the model output (Slingsby, 2003). For example, a DEM-based modeling in watershed delineation within GIS computes flow directions using the D8 algorithm wherein direction of flow is considered to vary along 45° increments. Using TIN instead of a DEM raster allows flow paths to be computed along the steepest lines of descent of TIN facets (Jones *et al.*, 1990).

5.4.2.3 CONTOUR

Contours are another way of representing elevation in the GIS platform. Contours represent points having equal heights/elevations with respect to a particular datum such as MSL. They can be created by either digitizing benchmarks from maps or directly from the DEM using the *contours* options available in GIS-related packages such as ArcGIS, ERDAS, etc. Figure 5.8 shows a contour map for the Krishna basin, India. On comparing with Figure 5.7, it may be noticed how dense contours lie along the western and eastern regions of the basin, pointing to large slope fluctuations.

5.5 APPLICATIONS OF DIGITAL ELEVATION MODELING

Watershed or drainage basin or catchment area is that part of a terrain responsible for draining surface water to a particular outlet. The size and shape of watersheds will be different for different basins. For example, the watershed for a river can be larger than that for a lake. The location of watershed boundaries or divides is essential for hydrologic modeling studies. Topographic maps help in sketching the watershed boundaries, as these give accurate representation of relief using contour lines.

Contour lines play an important role in delineation. The first step is to assume the direction of flow lines, which is necessary for identifying the outlet of the watershed. The outlet is the location where the water from the given catchment drains out. The next step is to identify the flow lines (to the outlet), and the adjacent streams and tributaries that contribute to the flow at the outlet. For this, the locations of high elevation points around the contributing water bodies are to be identified. Visualization is the key factor in determining the slope direction of the terrain. The watershed boundary is the line that has land sloping in opposite directions on either side.

Initially, a rough sketch needs to be developed for the watershed boundary. Once the sketch has been prepared, checking can be carried out by selecting certain check points/test points, as is normally done in surveying. The flow of water can be traced from these test points on the terrain. If the point selected lies outside the watershed boundary, its flow direction would lead to some other stream perpendicularly crossing all the contour lines. On the other hand, if the test point selected lies inside the watershed boundary, tracing the flow would directly lead to the watershed outlet matching with the rough draft. If the checking does not agree with the rough draft of the watershed prepared, changes have to be made to avoid the check/test point that proved the draft wrong. DEMs are applied to obtain many specific attributes desired by the user as described in the following sub-sections.

5.5.1 Drainage pattern extraction from digital elevation model

DEMs can be used to extract drainage patterns of a basin required for flow routing in most distributed hydrologic models. The algorithm used must be capable of identifying the slope variation and possible direction of flow of water. There are several functions in GIS that enable analysis of the terrain in different perspectives. For example, the function *Sinks* helps to identify the locations where water is likely to accumulate. It denotes a comparative depression in elevation with respect to the adjoining cells. Similarly, the flow direction and flow accumulation rasters can be created to assist in identifying the drainage patterns. A detailed overview of the D8 algorithm for drainage pattern extraction is provided in the next sub-section. An illustration of drainage pattern extraction is presented with the help of three case studies in Section 5.5.3.

5.5.2 D8 algorithm

Flow direction is a very important from the point of view of hydrology. Most GIS implementations use the D8 algorithm (Jenson and Domingue, 1988) to determine flow path. This algorithm is briefly discussed here. Flow directions are estimated from elevation differences between the given grid cell and its eight neighboring cells, hence the name D8 for the algorithm.

A depression in the DEM downloaded from the USGS website has to be removed before using it further. A DEM that is free of sinks is termed a depressionless DEM. The steps for delineating a watershed from a depressionless DEM are listed below.

(a) FLOW DIRECTION GRID

In watershed analysis using raster-based DEM, water from each cell is assumed to flow or drain into one of its eight neighboring cells, which are towards left, right, up, down, and the four diagonal directions. The general flow direction code, or the eight-direction pour point model, followed for each direction from the center cell is shown in Figure 5.9. From the figure it can be inferred that, if the direction of steepest drop is to the top of the current processing cell, its flow direction would be coded as 64 and so on. The number against each of the directions shown in Figure 5.9 represents the direction in which water travels to enter the nearest cell. The numbers have been set by convention from the 2^x series where $x = 0, 1, 2$, etc.

Using the eight-direction pour point model, the flow direction grid can be obtained. The flow direction tool, when applied on a raster-based DEM, will generate an output raster with values ranging from 1 to 255. If a cell is lower than its eight adjacent neighbors, that cell is given the value of its lowest neighbor. On the contrary, if a cell has the same change in elevation values in multiple directions and is part of a sink, the value assigned to that cell will be the sum of those directions. Consider a small example as shown in Figure 5.10, where a small sample of a DEM data in raster format is given. Assume the cell size to be 1 unit in dimension.

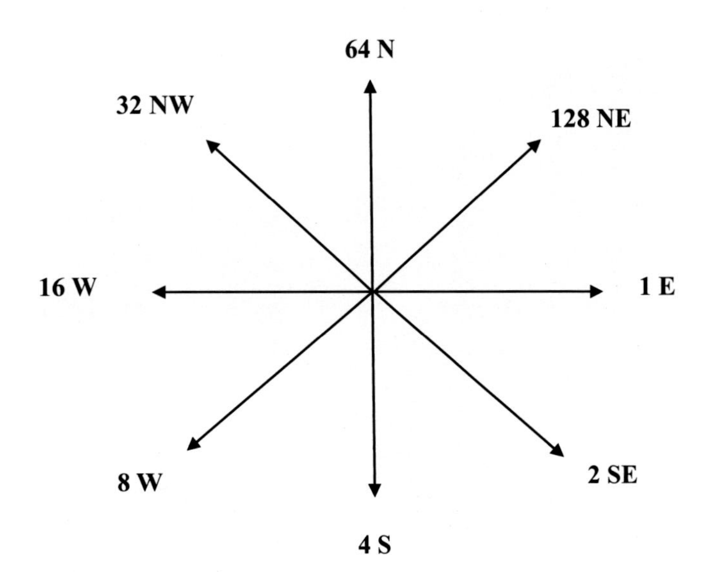

Figure 5.9 Eight directions for the pour point model.

78	72	69	71	58
74	**67**	**56**	49	46
69	**53**	**44**	37	38
64	**58**	**55**	22	31
68	61	47	21	16

Figure 5.10 Small sample of DEM data.

Focus on the cell with elevation 67. The possible flow directions would be towards the three adjacent cells (numbered 56, 53, and 44) of lower elevation values. Water will flow in the direction in which the greatest elevation decrease per unit distance is obtained. Slope can be calculated along the diagonal by taking the difference between destination cell value and original cell value, and dividing by $\sqrt{2}$ (i.e, the distance between cell centers assuming each cell is 1 unit long on each side). The slope calculations are shown in Figure 5.11. In the example shown in Figure 5.11(b), the diagonal slope is the greatest and therefore water would flow towards the bottom-right cell. As a result, the center cell (with elevation 67) is allotted a flow direction value of 2. The same procedure is repeated for each of the cells in a DEM raster. The new grid obtained is called the flow direction grid, as shown in Figure 5.12(b).

(b) FLOW NETWORK

Once the flow direction grid has been obtained, the flow network can be created by extending the lines of steepest descent within each cell, as shown in Figure 5.13.

78	72	69	71	58
74	**67**	**56**	49	46
69	**53**	**44**	37	38
64	**58**	**55**	22	31
68	61	47	21	16

$$\text{Slope} = \frac{67 - 56}{1} = 11.00$$

78	72	69	71	58
74	**67**	**56**	49	46
69	**53**	**44**	37	38
64	**58**	**55**	22	31
68	61	47	21	16

$$\text{Slope} = \frac{67 - 44}{\sqrt{2}} = 16.26$$

78	72	69	71	58
74	**67**	**56**	49	46
69	**53**	**44**	37	38
64	**58**	**55**	22	31
68	61	47	21	16

$$\text{Slope} = \frac{67 - 53}{1} = 14.00$$

Figure 5.11 Slope direction in (a) 0° from 67, (b) 45° from 67, (c) 90° from 67.

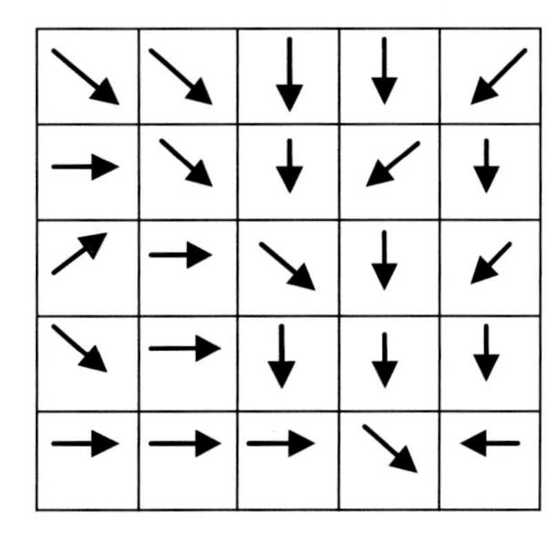

2	2	4	4	8
1	2	4	8	4
128	1	2	4	8
2	1	4	4	4
1	1	1	2	16

Figure 5.12 Structure of a flow direction grid (b) when the direction of steepest descent or flow path is as shown in (a).

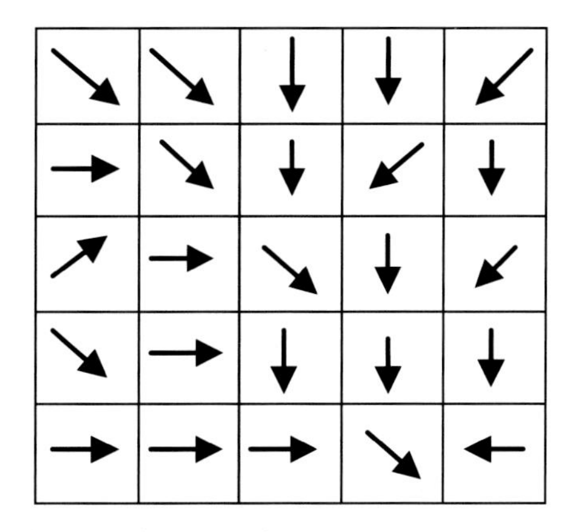

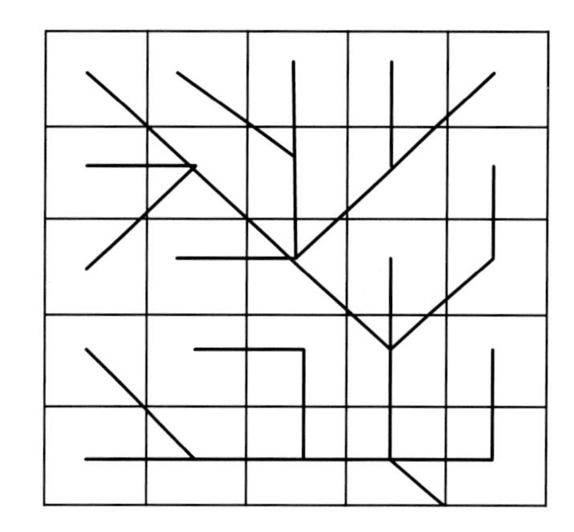

Figure 5.13 Flow directions and flow network.

(c) FLOW ACCUMULATION GRID

In the flow accumulation grid, the number in each cell denotes the number of all cells that flow to that particular cell. Here, the weight for each cell is assumed to be 1. Figure 5.14 clearly illustrates the flow accumulation grid.

(d) DELINEATED STREAM

To delineate a stream from a flow accumulation grid, it is necessary to specify a threshold on the number of cells contributing. Figure 5.15 highlights the flow accumulation for a threshold of five cells.

(e) STREAM LINKS

On specifying the threshold number, the stream links associated with this threshold are obtained with the help of the flow network grid obtained in step (b), as highlighted in Figure 5.16.

All the above steps to extract sub-watersheds from a raster-based DEM are shown in the form of a flowchart in Figure 5.17.

0	0	0	0	0
0	3	2	2	0
0	0	11	0	1
0	0	1	15	0
0	2	5	24	1

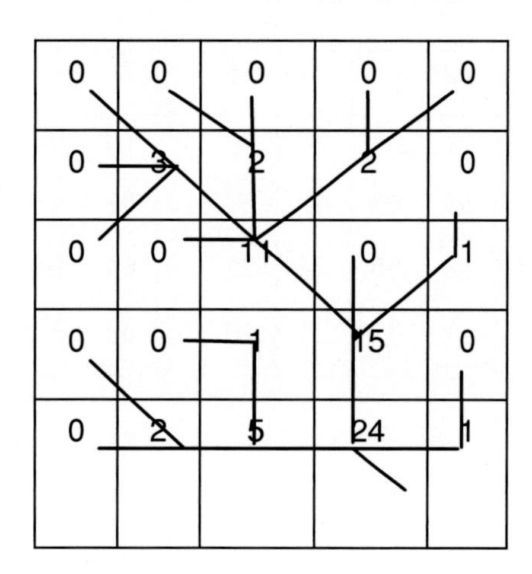

Figure 5.14 Flow accumulation grid.

0	0	0	0	0
0	3	2	2	0
0	0	**11**	0	1
0	0	1	**15**	0
0	2	**5**	**24**	1

Figure 5.15 Flow accumulation for a five-cell threshold.

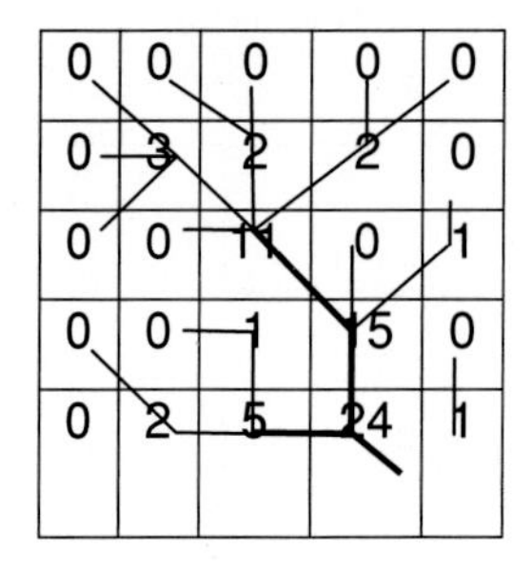

Figure 5.16 Stream network for a five-cell threshold (heavy lines).

5.5.3 Case studies

Case studies for watershed delineation using SRTM data for three basins in India, namely Malaprabha, Krishna, and Cauvery, are discussed in this section. Before analyzing the steps followed, a brief discussion about the basins considered is presented.

The Malaprabha basin is the catchment of the Malaprabha River, upstream of the Malaprabha Reservoir, having an area of 2093.46 km^2 situated between latitudes 15° 30′ N to 15° 56′ N and longitudes 74° 12″ E to 75° 8′ E. It lies in the extreme western part of the Krishna River basin in Karnataka state, India. The Krishna basin is India's fourth largest river basin and covers 258,948 km^2 in southern India. The basin is mostly flat, except for the Western Ghats and some hills towards the center and northeast regions of the basin. Cauvery basin, with catchment area of 72,000 km^2, is known for its extensive irrigation system and providing the domestic water supply to Bangalore and other cities in Karnataka state.

(a) DIGITAL ELEVATION MODELS

The SRTM DEMs, from the USGS website, are downloaded in raster format for the two basins considered and are presented in Figure 5.18. As can be noticed in Figure 5.18, while the DEM of the Malaprabha basin displays high elevation areas along the central basin region, the Krishna basin DEM is found to be undulating and flanked by a regular line of ranges of the Western Ghats towards the center and northeast regions. The Cauvery basin exhibits high elevation areas along the northwestern region. Detailed steps involved in extracting flow direction and flow direction images are illustrated only for the Krishna and Cauvery basins as the Malaprabha basin is part of the Krishna basin.

(b) FLOW DIRECTION

As the SRTM DEM is in a raster format, to determine flow along the connected grid cells, the *flow direction* image of the basins is generated within the GIS platform using the D8 algorithm explained in Section 5.5.2 and is shown in Figure 5.19.

From Figure 5.19, it can be observed that, for the Krishna basin, the *flow direction* of the majority of cells near the north is towards the bottom (color red indicates direction 4, which means *flow direction* is towards south). For the Cauvery basin, such a prominent *flow direction* path is not visible.

(c) FLOW ACCUMULATION

From the flow direction image, a *flow accumulation* image is created. Flow accumulation images show the area required to create sufficient runoff so as to incise the underlying bedrock and create a first-order stream channel. In ArcGIS, a flow accumulation raster is created by trial and error of threshold values. The number of cells contributing to the stream networks is limited using the

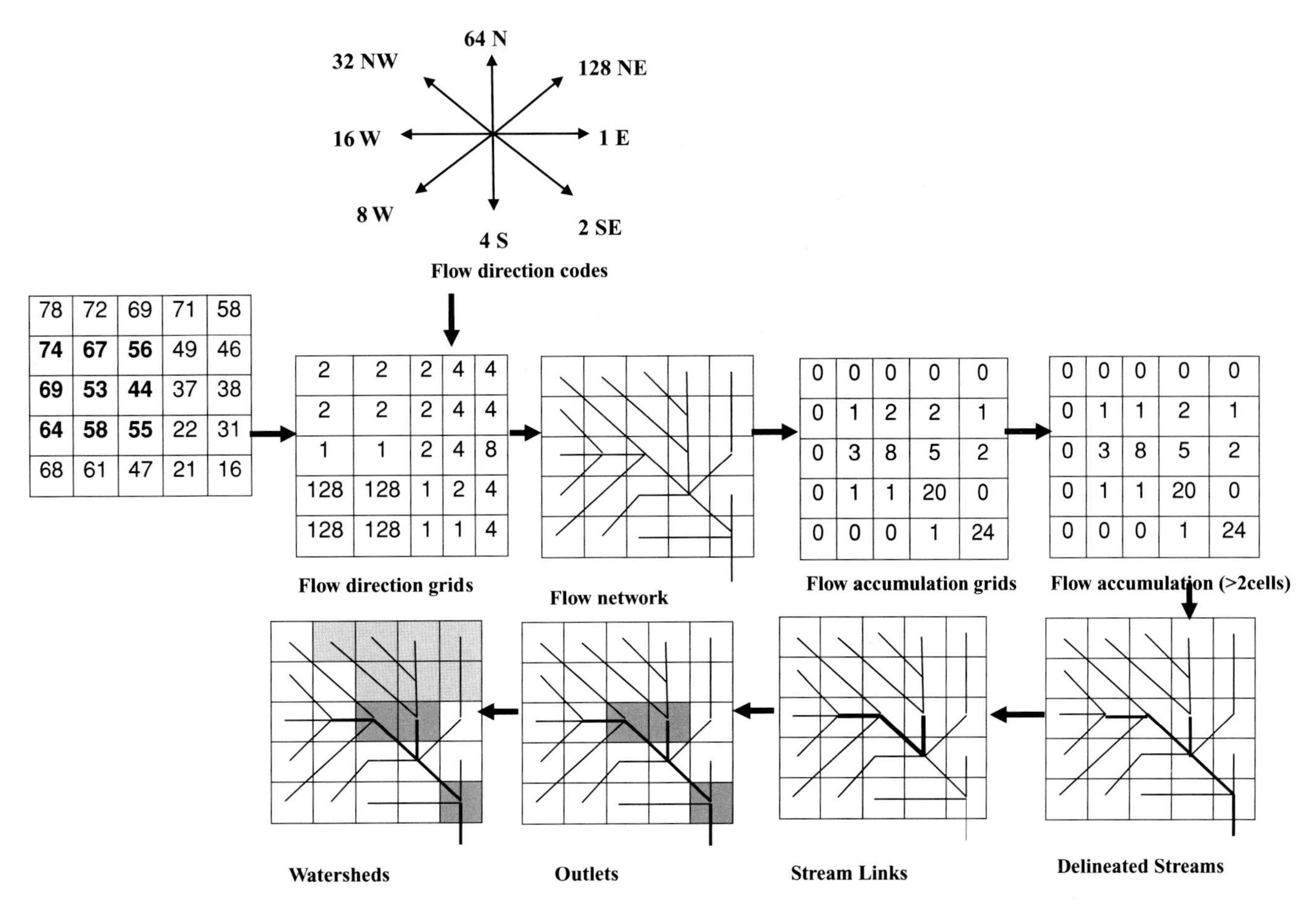

Figure 5.17 Steps in extracting sub-watersheds from a DEM using the D8 algorithm.

conditional tool and the normal practice is to set the threshold value at the first break-point value of classification to estimate the optimum threshold value. The end product of flow accumulation images for the Krishna and Cauvery basins is shown in Figure 5.20. It can be seen that while the lines are clearly visible in the Krishna basin, they are just faintly visible in the Cauvery basin.

(d) STREAM NETWORK

A vector data set showing a drainage network is derived based on combined information from the flow accumulation and flow direction data sets. The stream network vectors obtained using the ArcView soil and water assessment tool (AVSWAT) for the Krishna basin and using ArcGIS for the Cauvery basin are shown in Figure 5.21. It can be observed that the Cauvery basin shows a denser stream network than the Malaprabha basin. The difference might be due to the larger Cauvery basin area and also due to setting of a lower threshold value.

(e) POUR POINTS

In the next step, pour point locations are created. The placement of pour point locations is the most important step as far as watershed delineation is concerned. If the locations of hydrometric gauging stations are not available, pour points need to be created manually. The final product is a vector data set, created with probable outlets of drainage sub-basins in the drainage network. Figure 5.22 shows the location of sub-basin outlets or pour point locations for the Cauvery and Krishna basins.

(f) WATERSHED DELINEATION

The watershed delineation is carried out for the three basins following the methodology explained in Section 5.5.2 and a data set with delineated drainage sub-basins is derived from the drainage network in combination with the data set of outlets of drainage basins as shown in Figure 5.23.

5.6 OTHER SOURCES OF DIGITAL ELEVATION DATA

5.6.1 LIDAR

Light detection and ranging (LIDAR) sensors operate on the same principle as laser equipment. Pulses from a laser are scattered across the terrain from aboard an aircraft. The arrival times of

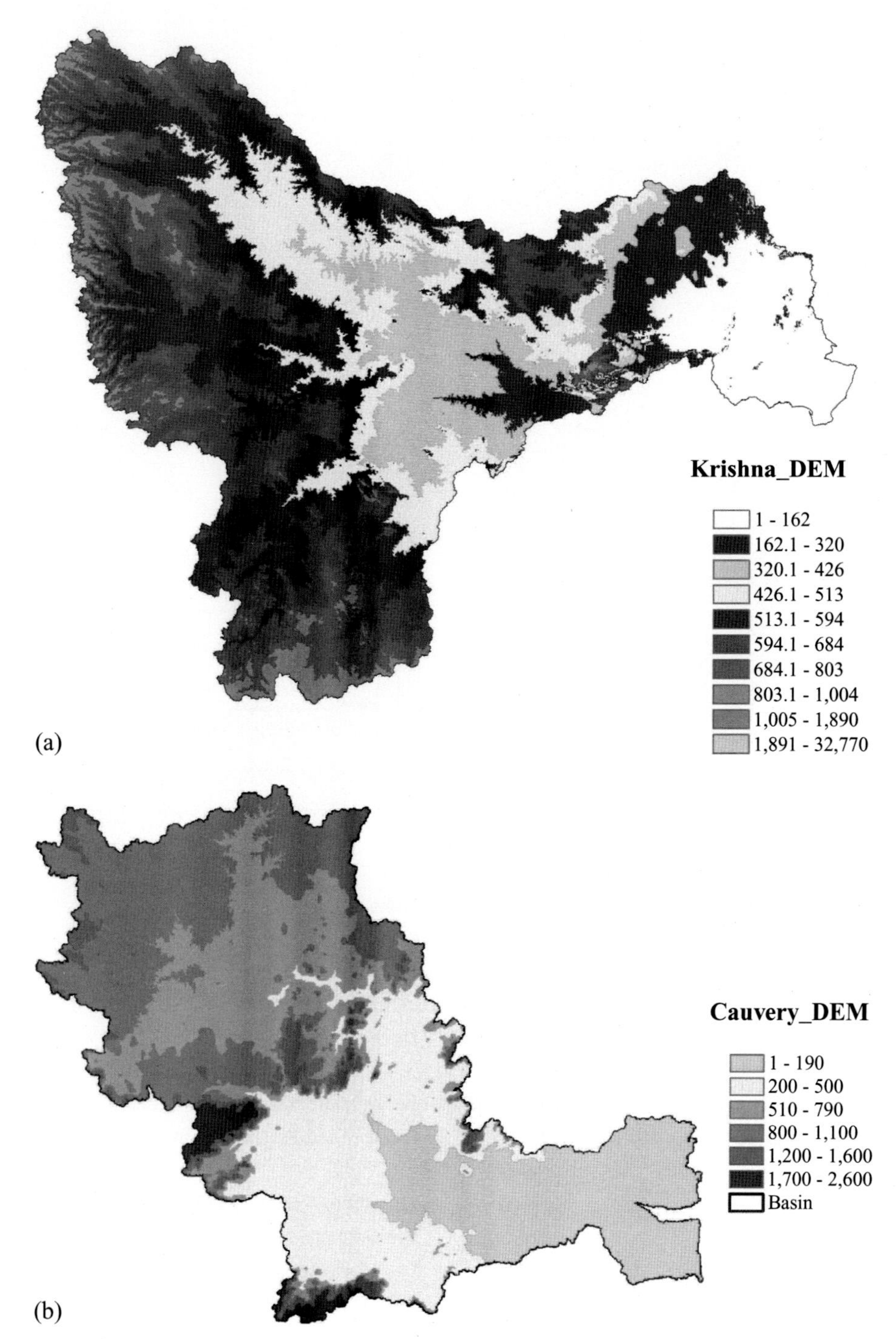

Figure 5.18 SRTM data for (a) Krishna basin and (b) Cauvery basin. See also color plates.

returning pulses are used to determine the location coordinates. The advantages of LIDAR are manifold. It enables DEM generation of a large area within a short period of time with minimum human dependence. The data, once generated, can be processed using Bentley software. Once the highly dense point cloud from LIDAR is obtained, before gridding the data, filtering of vegetation points and outliers is essential. MicroStation Bentley has in-built algorithms to cater to this requirement. This provides an advantage over GPS, in which information under a canopy is not available accurately. LIDAR resolution is the crucial factor in determining a topographic surface accurately. High resolution regarding data refers to more data or point cloud per given area. The disadvantage of procuring high-resolution LIDAR data is the expense involved in data collection.

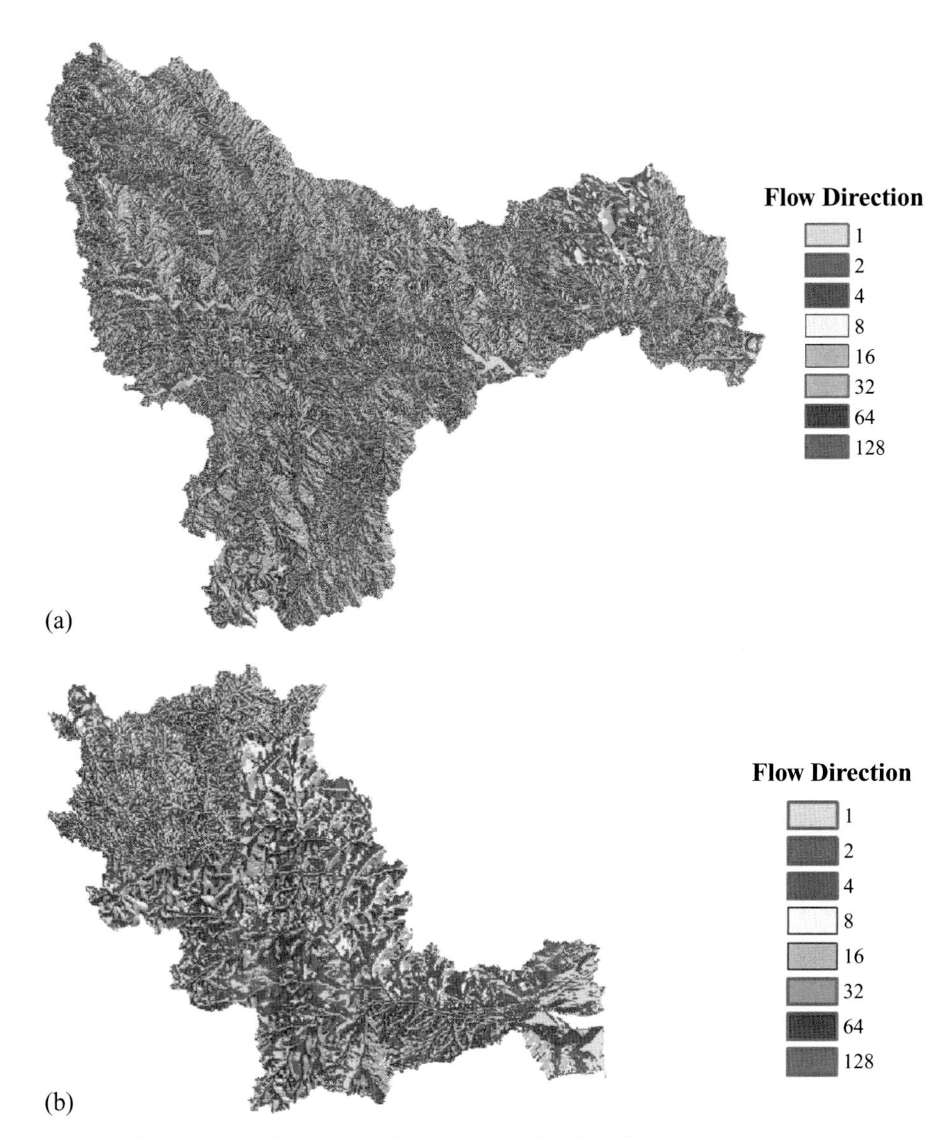

Figure 5.19 Flow direction image for (a) Krishna basin and (b) Cauvery basin. See also color plates.

5.6.2 ASTER

The Advanced Spaceborne Thermal Emission and Reflection Radiometer (ASTER) is an imaging instrument flying on Terra, a satellite launched in December 1999 as part of NASA's Earth Observing System (EOS). ASTER is a cooperative effort between NASA, Japan's Ministry of Economy, Trade and Industry (METI) and Japan's Earth Remote Sensing Data Analysis Center (ERSDAC). ASTER is being used to obtain detailed maps of land surface temperature, reflectance, and elevation (http://asterweb.jpl.nasa.gov/). ASTER's three subsystems are: the visible and near infrared (VNIR), the shortwave infrared (SWIR), and the thermal infrared (TIR). The ASTER global digital elevation model (GDEM) provides digital elevation data at approximately 30 m spatial resolution for 1 × 1 degree tiles in GeoTIFF format. The ASTER GDEM is freely distributed by METI (Japan) and NASA (USA) through the Earth Remote Sensing Data Analysis Center (ERSDAC) and the NASA Land Processes Distributed Active Archive Center (LP DAAC) (https://lpdaac.usgs.gov/lpdaac/products/aster_products_table). Algorithms to use these products were created by the ASTER Science Team, and are implemented at the LP DAAC. (http://eospso.gsfc.nasa.gov/eos_homepage/for_scientists/atbd/viewInstrument.php?instrument=24).

5.6.3 Radar interferometry

Interferometery is the technique used to survey large areas giving moderately accurate values of elevation. Its only disadvantage is that observation under a canopy is not possible. The principle of data acquisition in interferometric methods is

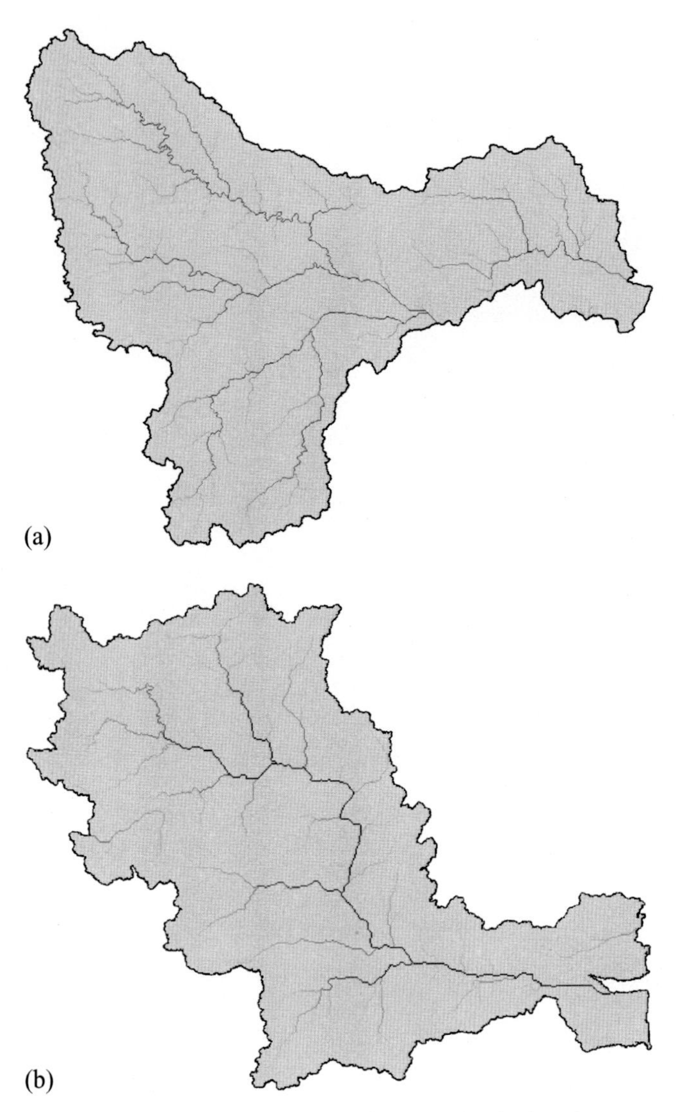

Figure 5.20 Flow accumulation raster for (a) Krishna basin and (b) Cauvery basin.

similar to stereo-photographic techniques. When the same area is viewed from different orbits from a satellite at different times, the differences in phase values from the scattered signals may be used to derive terrain information. Only images from the satellite having the same orbital track must be used in this procedure. The term interferometery means combining. A combination of two waves undergoes constructive interference when their frequencies are the same. In the case of different frequencies, they combine to interfere destructively. When electromagnetic radiation from a satellite or radar strikes a terrain, both its amplitude and phase undergo a change. Phase change may be due to terrain properties such as material, dielectric property, etc. Radar interferometry makes use of the phase changes of the received radiation to measure terrain height. Two or more radar images can be used effectively to generate a DEM. This technique is largely used for hazard monitoring, such as movement of crustal plates in earthquake-prone areas, land subsidence, glacial movement, flood monitoring, etc., as they give a high accuracy of down to a centimeter in elevation.

5.6.4 Use of Shuttle Radar Topography Mission data

The SRTM, launched in the year 2000, consisted of a radar system (two radar antennas) aboard the space shuttle Endeavour. Topographic/elevation data were created using an instrument known as interferometric synthetic aperture radar (IFSAR). The SRTM-generated elevation data are available at resolution of 1 arc second (approximately 30 meters) for the USA only, and at resolution of 3 arc seconds (approximately 90 meters) for the remaining part of the globe. These data can be accessed freely from the website http://seamless.usgs.gov/Run.htm. The data can be accessed using different methods described in the website and can be imported to a GIS platform. The SRTM grid files normally come in the World Geodetic System 1984 (WGS84) projection system. Each point of an SRTM grid file is normally assigned an elevation value (in meters). These grid files, once imported in a GIS platform, can be used for a variety of applications involving visualization of three-dimensional terrain. The use of a DEM for watershed delineation has been discussed in Section 5.5.2. Another function used with the SRTM DEM in hydrology is the *aspect* function. It highlights the drainages and relief patterns over a terrain. Overlapping an *aspect* layer over a DEM with reduced transparency will clearly display the drainage networks, as shown in Figure 5.24.

5.7 COMBINING DIGITAL IMAGES AND MAPS

This section deals with the methodologies available in the GIS platform to combine the information content from both digital images and maps. Before the methods are discussed, it is appropriate to have a basic understanding about digital images and maps in general.

In most countries, topographic maps are prepared by government departments, for example, in India, topographic maps are provided by the Survey of India department, one of the oldest central agencies vested with the responsibility of surveying and mapping throughout the length and breadth of India. In China, the State Bureau of Surveying and Cartography provides topographic maps at 1:25,000 and 1:50,000 scales. The National Survey and Cadastre of Denmark is credited with producing topographic data for Greenland, Denmark, and the Faroe Islands, etc. Other types of maps include planimetric maps that show horizontal positioning of features, special purpose maps that are utility specific, such as vegetation, transportation, and so on. In addition there are physical maps showing state boundaries, cadastral maps, etc.

Digital images in a GIS are those that are presented as pixels. A digital image can be represented in certain standard formats

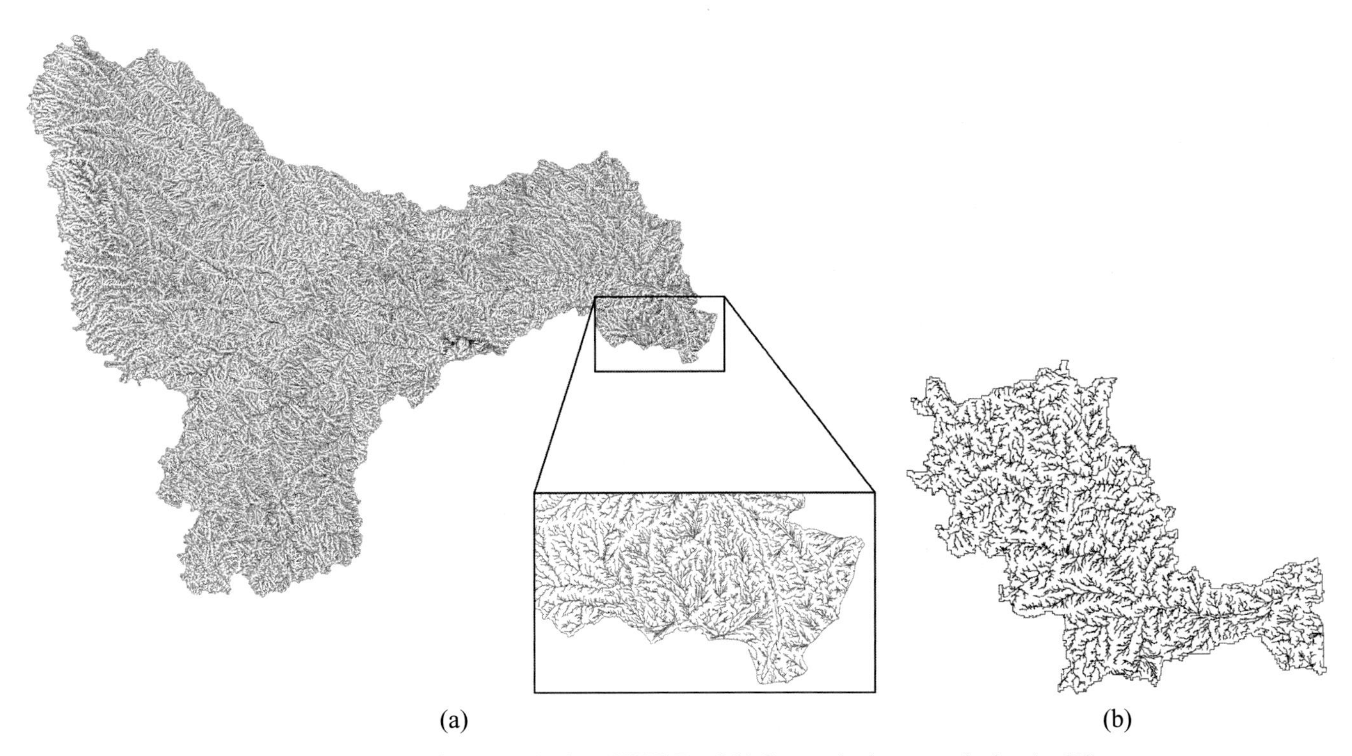

Figure 5.21 Stream network of (a) Krishna basin extracted using AVSWAT and (b) Cauvery basin extracted using ArcGIS.

such as jpg, tiff, gif, png, etc. Some formats may involve lossless compression (e.g., jpeg), whereas some may support compression with loss of information, which minimizes the amount of data to be handled by a computer during display. ArcGIS, for example has a number of tool sets that operate on raster format for query analysis, with many file formats that can be exported to and imported from other platforms as well.

Data in several forms can be represented individually in the GIS platform in the form of maps or images. Each data format has its own merits and demerits. The aim is to extract useful information from the maps into images for better understandability.

Consider a toposheet that provides contour information and *relief* (i.e., vertical and horizontal representation of land surface) information. If there is some means of extracting the existing elevation information from toposheets and adding it to the satellite imagery of the same area, a three-dimensional model of the terrain can be generated. This can prove to be extremely useful in many flood-related studies. As another example, if the soil maps could be integrated with the DEM of the same terrain, it would be very useful in rainfall–runoff studies.

A point shape file, such as of benchmarks from a toposheet, may represent the point features selected in the map or images in a binary format. A line shape file could be the stream network system from a physical map, and a polygon shape file could be the watershed/catchment area of a prominent reservoir. Once created, the shape files can be saved in a layer format and added to an existing database. Thus, data are stored in layers of a compact form that can be edited. For example, over a soil map of any river basin, one can overlay shape files of land cover, wetlands, floodplains, etc., as shown in Figure 5.25. This facilitates a thorough study and visualization of overland flow. Such data integration methodologies (Ehlers *et al.*, 1991) are important when dealing with natural hazards such as landslides, floods, etc.

5.8 INTEGRATION OF SPATIAL, NON-SPATIAL, AND ANCILLARY DATA INTO A DISTRIBUTED HYDROLOGIC MODEL

Spatial data representation in the GIS platform has already been discussed in the previous sections. This section provides a brief explanation of non-spatial and ancillary data and their integration into a distributed hydrologic model.

Non-spatial data, as the name suggests, refer to data that are independent of location-based information. A catchment, digitized as a polygonal shape file, can be described by spatial data. This is because the data are dependent on the overall shape of the catchment, the location (x, y, z) of its boundaries, its orientation, etc., whereas, its non-spatial/attribute/characteristic information can include details such as name of the catchment, number of tributaries joining it, the total area, length, etc.

Ancillary data, on the other hand, cover those data that tend to complement the existing information. For example, if our

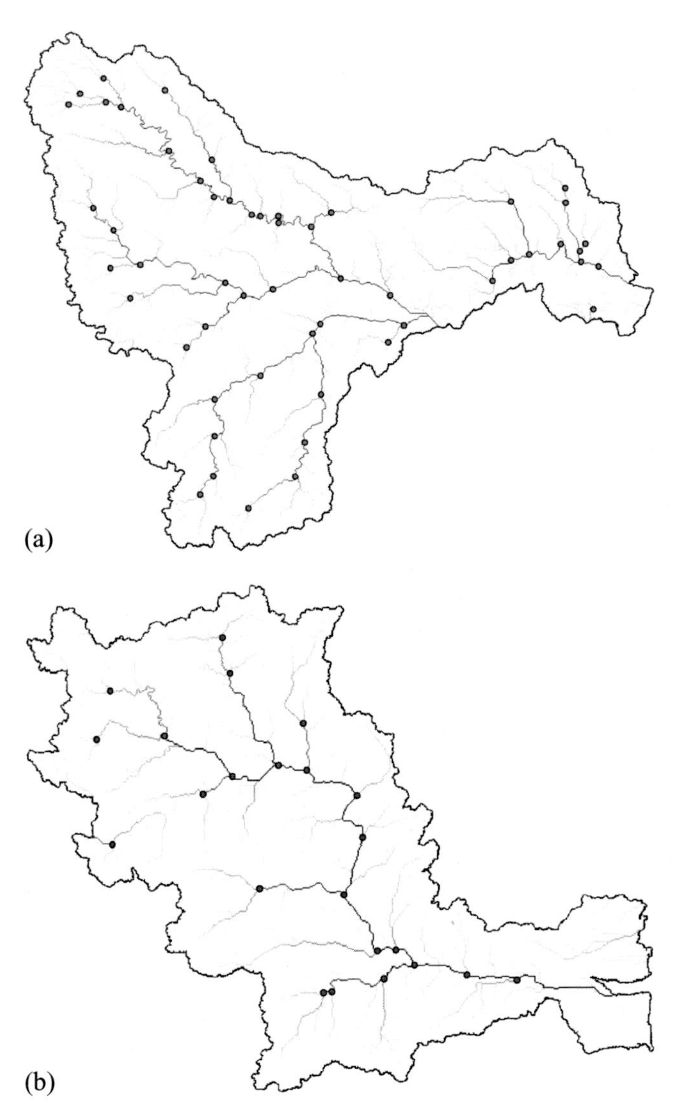

Figure 5.22 Drainage sub-basin outlets that are fixed in the catchment of (a) the Krishna basin and (b) the Cauvery basin by using ArcGIS.

primary concern is with rainfall–runoff modeling for a catchment, then a soil map of the catchment, elevation information (such as total station survey data) or SRTM images can serve as suitable ancillary data.

For any basin, if the contour lines and valley/ridge lines are known, a hydrologically useful DEM can be generated by interpolating the elevation data from contours. Procedures are available to automatically delineate a continuous channel network out of both a DEM and channel data digitized from a contour map. From this channel network and DEM, many hydrologic attributes can be calculated, such as *aspect*, Horton–Strahler stream order, etc. From each grid block, runoff can be computed using a hydrologic model. Routing of the resulting runoff and inflow from upstream channels can be carried out using a buffering technique.

5.9 GIS AND REMOTE SENSING FOR FLOOD ZONE MAPPING

For planning flood management measures, the most important requirement is the availability of reliable, accurate, and timely information. Through the years, different methods have been attempted to obtain this information. Previously, flood studies were dependent on ground surveys, which were cumbersome and time consuming. The problem was aggravated further in areas where manual surveying was not possible. Then came the era of aerial photographs, which suffered from limitations due to cloud cover. At present, technological advances have brought forth a series of techniques that can complement the existing methods of collection of data. Satellite images provide useful and timely information for flood studies. Images acquired in different spectral bands during, before, and after a flood event can provide valuable information about flood occurrence, intensity, and progress of flood inundation. However, unless there is a database management system for storing and integrating the voluminous data from different sources, the full potential of satellite images cannot be utilized. A GIS provides such a platform. Topics related to applications of GIS to flood management studies are briefly addressed next.

5.9.1 Flood inundation mapping

For planning flood relief operations, continuous availability of information on flood-affected areas at frequent intervals of time is required. This information is provided by satellite RS. By studying satellite images taken during a flood, the extent of the area inundated during the progress or recession of the flood can be visualized (Showalter, 2001). Land use/land cover information can be obtained in more detail from the standard false color composite (FCC) when compared to the true color composite (TCC). Figure 5.26 shows two FCCs prepared from the Resourcesat (AWiFS) satellite showing the original course (southwest trending in left panel) of the Kosi River, India, and that developed after the breach (southwards in right panel) due to flood in 2008 (Bhatt *et al.*, 2010). The amount of inundation during the flood period can be very clearly noticed from Figure 5.26. Merging of panchromatic images with multispectral images (explained in Chapter 4, Section 4.4.6.1) will considerably enhance the flood zone mapping and damage assessment. In Figure 5.27 (top left panel) a panchromatic image from CARTOSAT is merged with multispectral images of IRS LISS IV to prepare a FCC (Bhatt *et al.*, 2010). The amount of inundation and flood damage assessment can be estimated by overlaying the two FCC images, before and during the flood event, in GIS (Figure 5.27). In addition, if a political or town/city boundary map is overlaid on these images in a GIS, regional flood damage assessment can be made.

5.9.2 Near real-time monitoring of floods

For monitoring the impact of natural disasters and to plan relief operations, the use of near real-time data is a prerequisite.

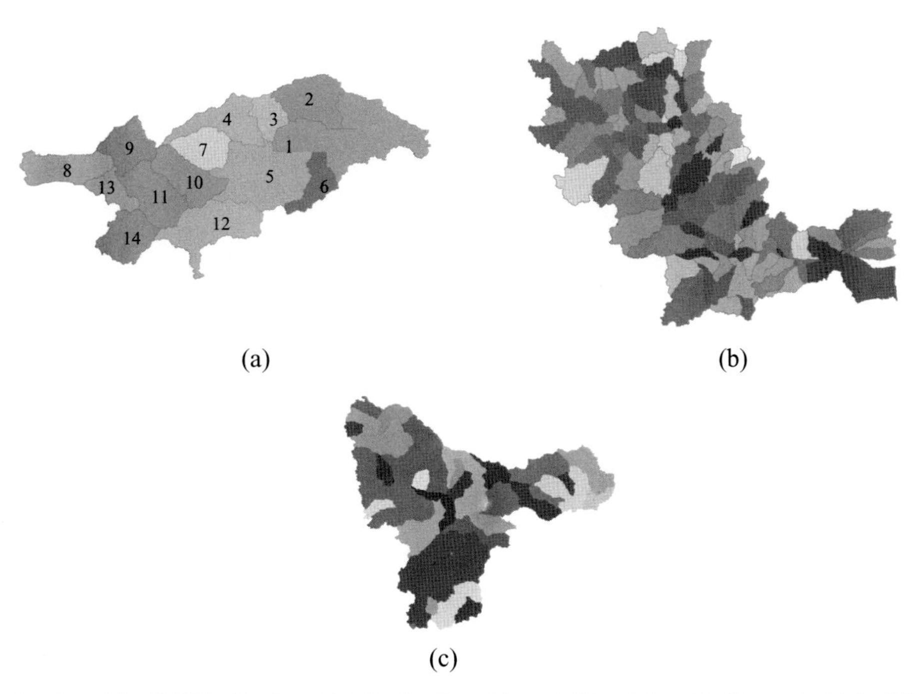

Figure 5.23 (a) Sub-basins formed by AVSWAT for the outlets fixed in the catchment of the Malaprabha Reservoir. (b) ArcGIS-generated sub-watersheds for the Cauvery basin. (c) ArcGIS-generated sub-watersheds for the Krishna basin overlain with stream networks. See also color plates.

Figure 5.24 Aspect layer over DEM from SRTM data.

Near real-time data refer to information that is delivered soon after collection. In India, the National Remote Sensing Centre (NRSC), Hyderabad, and ISRO have been operationally using satellite RS for flood disaster management for more than two decades (Bhanumurthy *et al.*, 2010). As per the National Flood Commission report, around 40 million hectares of land in India are prone to floods (Bhanumurthy *et al.*, 2010). Whenever a flood event occurs in India, government-based agencies provide maps showing the flood-affected areas and flood damage statistics in near real time. Using different sources of information, the

Figure 5.25 Integration of layers in a GIS environment. See also color plates.

rainfall activity and subsequent flood condition is monitored on a daily basis. This section introduces the information accessible from the India Meteorological Department (IMD) as well as the TRMM. The IMD provides climate- and rainfall-related information, numerical weather prediction models, etc., which are valuable when dealing with disaster management. Kalpana-1, the first meteorological satellite launched by ISRO has a VHRR scanning radiometer providing three-band images in the visible, thermal IR, and water vapor IR. Near real-time meteorological data from Kalpana-1 images are used to study the cloud cover over the country. Thick cloud cover indicates the possibility of heavy rainfall and flooding under certain situations.

Information about rainfall is also available from the TRMM. The products available include not only the raw brightness values captured at different frequencies, but also the rainfall rate (both convective and stratiform) in mm/hr. These data products are given in a gridded format, which can be downloaded free of charge. In addition, information about merged rainfall products is also available. Hydrologic models may be tuned to use these existing high-quality data products to arrive at better prediction, facilitating near real-time monitoring of floods.

The procedure usually followed to observe floods using satellite imagery involves capturing information at different intervals of time and detecting the changes in inundated areas. To provide warning at an appropriate time and to make relief operations successful, data need to be acquired during the rise, peak, and fall of a flood wave. Presently there exists a huge database of optical satellite data from the IRS series, NOAA, Terra (a multi-national NASA research satellite), and microwave data from RADARSAT (Canadian satellite)/ERS SAR/Envisat (launched by ESA). The microwave images captured at different polarizations, incidence angles, resolutions, and swaths offer flexibility to map floods from different perspectives. A list of sensors available along with their use is provided in Table 5.1 (Bhanumurthy *et al.*, 2010).

To make use of the full capacity of the sensors available, it is essential to understand the properties of these images and the underlying logic to interpret any useful information. Interpretation of water bodies using optical images is based on the fact that water absorbs or transmits most of the electromagnetic energy (except in the blue and green bands), which results in less reflection, leading to a dark color on many FCCs. When it comes to flood water, turbulent movement of the water leads to transport of different concentrations of sediments, which reflect energy in different bands. This leads to patches of light shades along the path of an advancing flood. This is in fact used to interpret flood movement. But the only hindrance is the presence of cloud cover, which makes it difficult to observe the area beneath. This problem was overcome with the availability of microwave data. Its all-weather capability greatly complements the existing optical images. Radarsat provides microwave images. Integration of RS imagery (optical and microwave images) over a cadastral map of villages over the region in a GIS framework will facilitate damage assessment and planning for mitigation masures. Figure 5.28 shows the temporal progression of flood inundation during August 20–29, 2008, in the Kosi River, India (Bhatt *et al.*, 2010). This map was prepared by integrating Radarset SAR images, IRS P6 AWiFS images, and cadastral maps in a GIS framework.

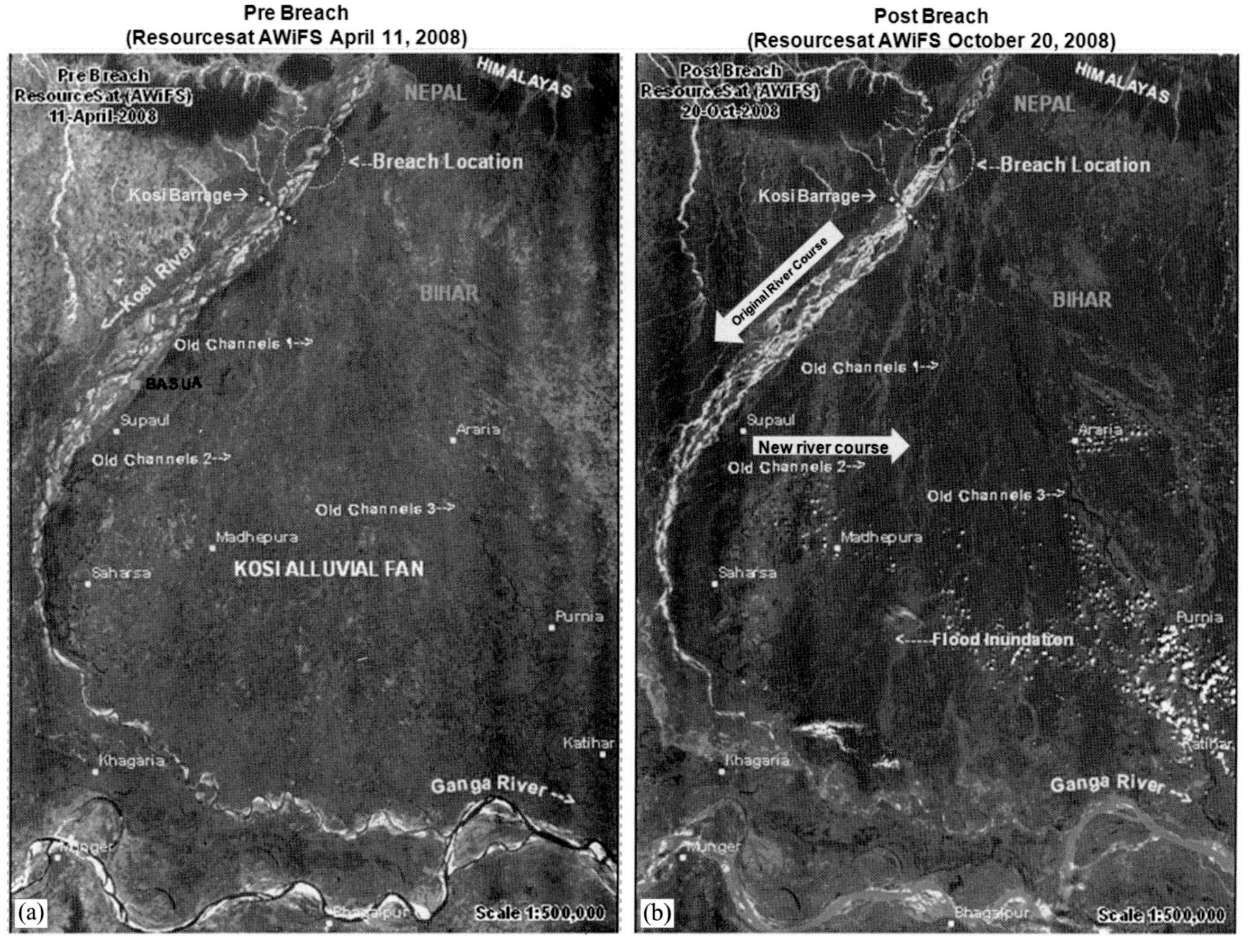

Figure 5.26 Satellite images (Resourcesat AWiFS) showing the original course (southwest trending in left panel) of the Kosi River, India, and that developed after the breach (southwards in right panel) due to flood in 2008 (Bhatt *et al.*, 2010). See also color plates.

Extracting the statistics of flood damage from an optical or microwave image involves a series of interdependent procedures. In many cases, data from both these complementary sources are clubbed together to get a better perspective of the flood map. A flowchart is shown in Figure 5.29 depicting the series of steps required to convert optical and microwave images into flood maps and to assess the damage statistics (Bhanumurthy *et al.*, 2010).

When using optical satellite data, after georeferencing (adding location coordinates with respect to a particular projection), the satellite image is classified by either supervised or unsupervised methods to differentiate three layers, namely, water, cloud, and cloud shadow. Outliers need to be removed by using a mask before integrating. When dealing with SAR images, the first step would be speckle filtering to remove the inherent salt-and-pepper noise (grainy appearance) of these images. The next step would be geolocation and classification into water layers. After applying a mask, these data from SAR are integrated with the pre-flood river bank along with the corrected data from the optical satellite image. From this, a flood inundation layer is created from which a flood map is created and damage statistics are estimated (Oliver and Quegan, 2004).

Satellite data can be used to extract the river course before the flood season. Once imported into a GIS environment, the existing databases, such as land cover maps, basin utility maps, etc., can be overlaid on top of one another to create a flood map. For a quick analysis to be possible, other information such as administrative boundaries, road, rail network, and crop area are kept ready in the GIS environment. For example, Figure 5.30(a) shows the breach on the Kosi River, India, which occurred in August 2008 leading to large-scale loss of human life and property. On August 18, 2008, the Kosi River burst through its eastern embankment about 13 km upstream in Nepal, thereby running through a new course 15–20 km wide. At its peak, the intensity of flow went up to

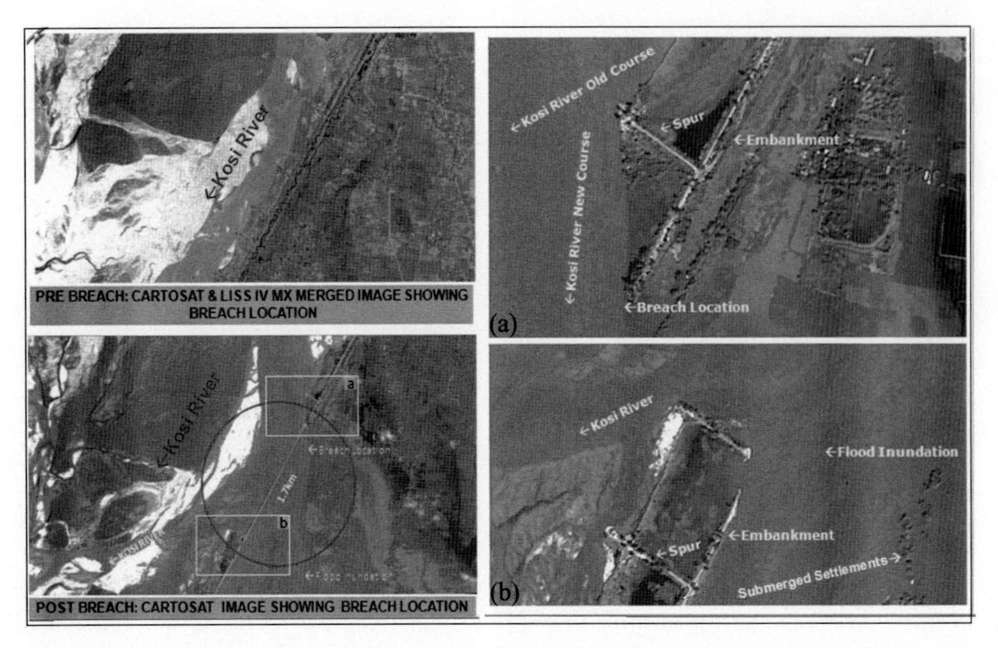

Figure 5.27 IRS CARTOSAT and LISS IV images showing before and after a flood event in the Kosi River, India (Bhatt *et al.*, 2010). See also color plates.

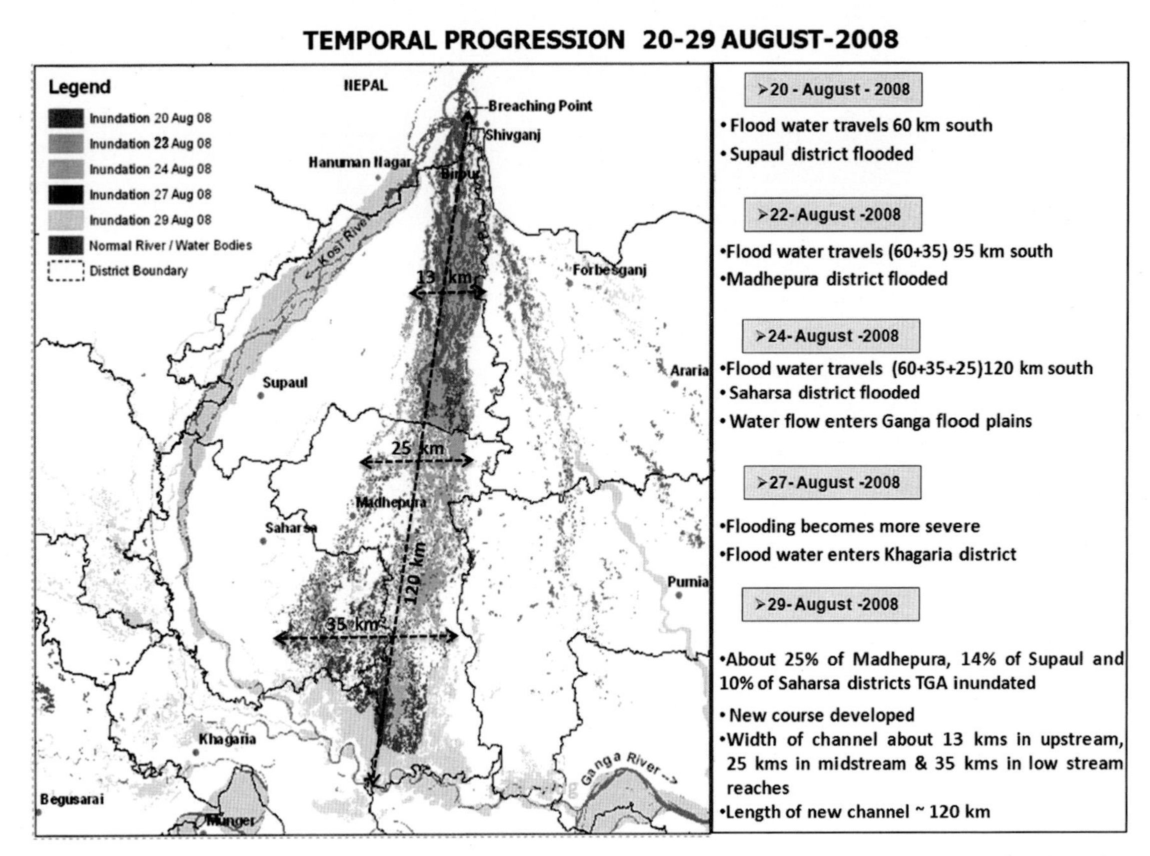

Figure 5.28 Temporal progression of Kosi River flood inundation in India during August 20–29, 2008 (Bhatt *et al.*, 2010). See also color plates.

Table 5.1 *List of satellite sensors with their use for flood monitoring*

Sl no.	Satellite	Sensor/mode	Spatial resolution (m)	Spectral resolution (μm)	Swath (km)	Use
1	IRS-P6	AWiFS	56	B2: 0.52–0.59 B3: 0.62–0.68 B4: 0.77–0.86 B5: 1.55–1.70	740	Regional-level flood mapping
2	IRS-P6	LISS-III	23.5	B2: 0.52–0.59 B3: 0.62–0.68 B4: 0.77–0.86 B5: 1.55–1.70	141	District-level flood mapping
3	IRS-P6	LISS-IV	5.8 at nadir	B2: 0.52–0.59 B3: 0.62–0.68 B4: 0.77–0.86	23.9	Detailed-level mapping
4	IRS-1D	WiFS	188	B3: 0.62–0.68 B4: 0.77–0.86	810	Regional-level flood mapping
5	IRS-1D	LISS-III	23.5	B2: 0.52–0.59 B3: 0.62–0.68 B4: 0.77–0.86 B5: 1.55–1.70	141	Detailed-level mapping
6	Aqua/Terra	MODIS	250	36 in visible, NIR,and thermal	2330	Regional-level mapping
7	IRS-P4	OCM	360	Eight narrow bands in visible and NIR	1420	Regional-level mapping
8	Cartosat-1	PAN	2.5	0.5–0.85	30	Detailed-level mapping
9	Cartosat-2	PAN	1	0.45–0.85	9.6	Detailed-level mapping
10	Radarsat-1	SAR/ ScanSAR Wide	100	C-band (5.3 cm; HH polarization)	500	Regional-level mapping
11	Radarsat-1	SAR/ScanSAR Narrow	50	C-band (5.3 cm)	300	District-level mapping
12	Radarsat-1	Standard	25	C-band	100	District-level mapping
13	Radarsat-1	Fine beam	8	C-band (5.3 cm)	50	Detailed-level mapping
14	ERS	SAR	25	C-band; VV polarization	100	District-level mapping

(Bhanumurthy *et al.*, 2010)

166,000 cubic feet per second (cusec) compared with the regular flow of 25,744 cusec. This created major flooding in Nepal and Bihar state in India. A total of 3.3 million people were affected in Bihar state alone (Needs Assessment Report, 2010). Snapshots of this flooding captured by the RADARSAT and IRS P6 satellites are shown in Figures 5.30(a) to (f) (Bhatt *et al.*, 2010).

From the snapshots shown in Figure 5.30, it may be seen that a significant amount of information, such as topography of area inundated, can be inferred from real-time satellite data, which helps in disaster management. The post-flood satellite data can be analyzed to delineate the active river channel and river bank lines. These may be used further to determine the extent of erosion and deposition. Consequently, erosion cum deposition maps can be created in a GIS environment. Once the flood maps are created, the damage statistics can be sent to the concerned departments for planning relief and rehabilitation measures on the ground.

5.9.3 Modeling using LIDAR data

With finer and better vertical accuracy of a DEM, improved analysis can be undertaken using it. Presently a DEM of the entire globe can be freely accessed from SRTM (approximately 90 m spatial resolution) and ASTER (30 m spatial resolution). Some studies may, however, demand a DEM at a much finer spatial scale. This can be made possible by the use of LIDAR data. Though generating LIDARdata is a costly exercise, once procured, the exact terrain height information can be put to full use for applications involving better vertical accuracy. For example, after extracting the sub-watersheds using a DEM, the distribution

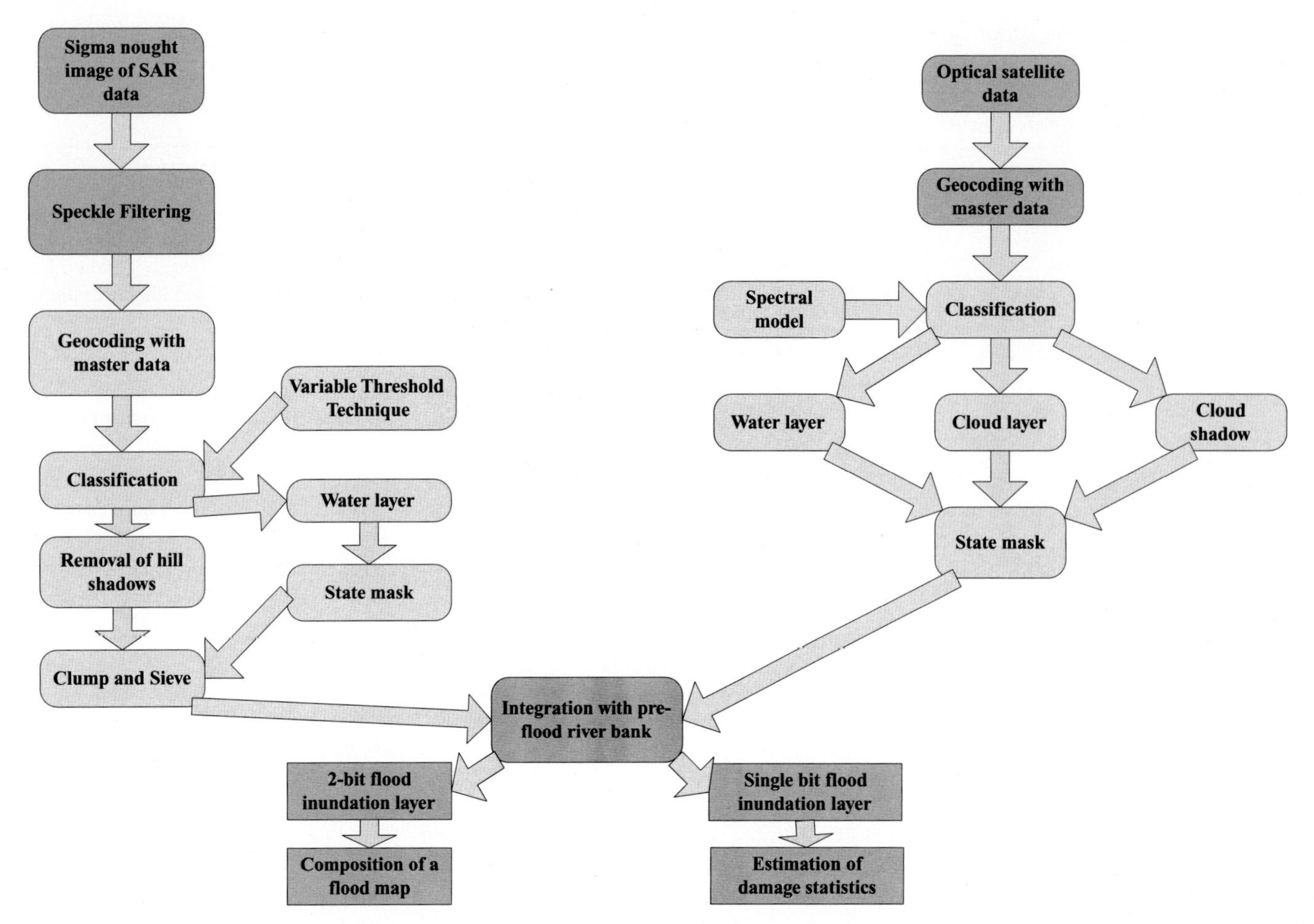

Figure 5.29 Flowchart for creation of flood map by integration of both optical and microwave images (Modified from: Bhanumurthy *et al.*, 2010).

of peak flow rates along the drainage network at different nodes can be estimated. These peak flow rates are compared with the existing capacity of storm-water drains to determine whether the area is likely to be flooded or not.

5.10 WEB-BASED GIS

For monitoring the spread of flood-affected areas, comprehensive data collection is essential for timely decision-making. Web-based GIS (Web GIS) is gaining popularity for its capability to store huge amounts of heterogeneous geolocated data sets and process the data for user-specific purposes. These data sets are made available in a visually informative and interactive format. This section discusses Web GIS, which can essentially be considered as an integration of a GIS with the Internet. Before explaining Web GIS, some basic concepts of information exchange via the Internet are discussed in the following section.

Web browsing is all about transferring information to or accessing information from a remote computer in any part of the world. Simply put, the client/user accesses information stored in a server through a network, i.e., the Internet, by following certain rules/protocols. The commonly used protocols are discussed briefly.

5.10.1 Protocols

Protocols are a fixed set of rules that are to be followed when there is a necessity for information exchange between two computer systems. It is necessary that both the source and destination computers abide by these protocols to communicate with each other. Protocols are sometimes grouped into lower-level, upper-level, and application protocols. Some of the common protocols that need to be satisfied are discussed in the following sub-sections.

5.10.1.1 INTERNET PROTOCOL

Internet protocol (IP) is the main protocol involved in sending packets of information via the Internet. IP follows certain rules to send and receive packets of information at the address level. Every computer that is connected to the internet has a unique IP address that distinguishes it from all the other computers on

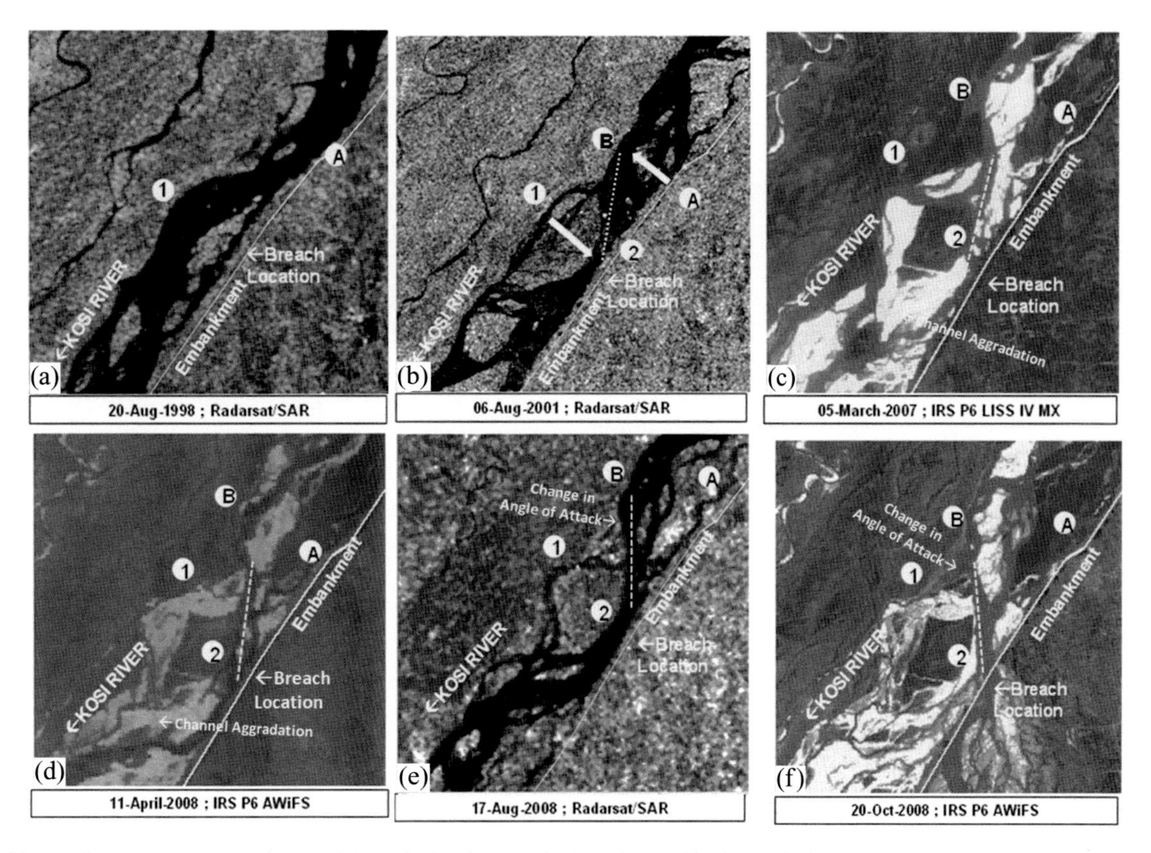

Figure 5.30 Satellite images (RADARSAT and IRS P6) showing the changes observed in the Kosi River course, India, near the breach location during the period 1998–2008. (Dashed lines in the images show changes in angle of attack, whereas numbers/letters encircled in yellow indicate the shift in river course) (Bhatt *et al.*, 2010). See also color plates.

the Internet. When an email or text is sent via the Internet, the information from the host computer is converted into small information modules known as packets. These packets of information contain both the sender's and receiver's IP address. Once this information is relayed within the Internet, it encounters several *gateway computers*, which decode the address and direct it to the adjacent gateway, and so on. This continues until one of the *gateway computers* relates the IP address as belonging to a computer within its nearest neighborhood. It is very similar to a person seeking a route to the office on the first day at work in a new city. In this case, a number of strangers might help the person to find the destination, in much the same way as the *gateway computers* do. The IP addresses can be either static or dynamic in nature. A static IP address, as the name suggests, is fixed. The Internet Service Provider (ISP) may provide static addresses. Dynamic IP address may change at any time. The host computer takes up any "free" address as soon as it logs on to the Internet. This mechanism is normally followed when the users of computers do not need to be identified with the same address. Dynamic addresses are mandatory only if the user is running a web server, email server, FTP, or DNS server on the user's website. For normal browsing of the Internet, sending and receiving emails, uploading and downloading files, etc., a static address is sufficient.

Generally, users find it difficult to remember a series of numbers when referring to an IP address. They may more easily recollect their email ID than their IP address. To overcome this problem, the Domain Name System (DNS) allows IP addresses to be translated into words.

The next question is, when a single email is converted into several packets of information, in which order do they reach the destination IP? Individual packets of information can take any route to reach the receiver address, as they are not interconnected like the bogies of a train. What is the mechanism by which these packets are arranged and delivered in the same order at the receiving end? This mechanism is through the Transmission Control Protocol (TCP). This set of rules ensures that the packets are rearranged in the same order as the order in which they were sent. They actually keep track of the sequence of the information packets in a message.

5.10.1.2 COMMON WEB PROTOCOLS

As discussed earlier, protocol facilitates smooth information exchange and web protocol does the same for transferring data via the Internet. The commonly used web protocols are listed in this sub-section (Bonnici, 2006).

HTTP: HyperText Transfer Protocol. Usually when a web page is browsed, the web browser uses HTTP to transfer data. It is a common standard that enables any browser to connect to any server.

FTP: File Transfer Protocol, as the name indicates, is used to transfer files across the Internet. The difference between HTTP and FTP is that, while the former is used for displaying websites, the latter is more concerned with transferring files from a location on a computer to a location on a different computer. This is a comparatively older protocol, but is still in use.

URL: Uniform Resource Locator. The URL points to a unique address for a file that can be obtained via the Internet. The URL normally contains the name of the protocol used, a domain name, a path name, and a description that specifies the file location. The protocols are denoted by strings followed by the three characters.

For example, consider the website: http://www.esri.com/software/arcgis/index.html. *http* is the protocol used to retrieve the required information; the string www.esri.com that follows http refers to the domain name of the host computer; *software/arcgis* refers to the path for the desired information, which is located in the host computer; *index.html* is the file name. It may be noted that file extensions of both ".htm" and ".html" are acceptable. URLs can be static as well as dynamic in nature. As the name indicates, a static URL is one in which the information content is fixed and doesn't change. A dynamic URL points to web pages whose content varies depending on input and data updates. Dynamic URLs can be spotted by looking for characters such as ?, +, &, etc. in the address bar (an example of a fictitious dynamic URL would look something like: http://www.astronomy.com/product.php?categoryid=1&productid=12). The only disadvantage that they suffer from is that different URLs may be available with the same information content.

5.10.2 Web GIS

Web GIS is the technology that allows information presentation to cater to the needs of a large number of users via the Internet. The biggest advantage of Web GIS is that users are provided with data without the necessity for any commercial GIS software. The data are directly viewed by means of a web browser. Information within the server cannot be deleted, or added. Such rights would be restricted to the administrator managing the GIS server alone. Static maps are readily available via the Internet, but Web GIS is more than that. It is a sophisticated architecture working in the background to provide geo-related information free of cost to a whole set of users with no location restrictions. The Web allows wide accessibility and visual interaction with the data, which enables clients to view the updates. The only disadvantage it has is the problem of speed. Streaming huge amounts of graphical data over the Internet can sometimes turn out to be an extremely slow affair, especially during network congestions. On the other hand, worries about the high cost of maintaining a GIS system and its upgrading are fading with the arrival of Web GIS. To understand how a typical Web GIS system works, it is necessary to examine its architecture.

In the simplified architecture of Web GIS, a client seeks access to spatial information through a web browser. This request is sent to a middleware that interprets the requirements of the client. The request is further sent to the Web GIS software, which is also present on the server. The Web GIS software conducts a query analysis on its database to extract the specific attribute data required by the client. Once this is completed, the appropriate spatial entity is sent back to the middleware for reinterpretation, after which it is streamed back to the client. Because of data streaming, a file (audio or video) can be played back without completing its downloading.

It is also equally important to analyze the data format in which these large chunks of information are being transferred through the Internet. The huge amounts of graphical data in the server need to be converted into either raster or vector format before making them accessible via the Web. Both the formats (raster and vector) have their own advantages and disadvantages. Since the volume of raster data is much higher than that of the vector data, usually conversion to the latter is preferred. When vector data are involved, highlighting a single object is possible and data processing requires less time. Esri's ArcIMS is an example of a web-based product (www.esri.com) that is capable of providing data to clients in both raster and vector forms. The raster data are provided in common formats such as jpeg, gif, tif, etc. To receive the raster data, the client does not have to use any additional software. Through a web browser, the data are sent from the server to the client. For vector data the client needs to have a Java plug-in installed on the system. Streaming of vector data is made possible by ArcXML.

After becoming familiar with the architecture of Web GIS and the mode of data transfer, it is necessary to understand the "middleware." Common Gateway Interface (CGI) is the name given to the program that links the web browser with the server. This is initiated when information that is to be accessed by a client is transmitted through HTTP.

To sum up, the basic components of a Web GIS system include: the type of Web GIS server, network of client computers, programming language linking client and server (for example Practical Extraction and Report Language, PERL), and the large amounts of spatial and non-spatial data in raster or vector format or both.

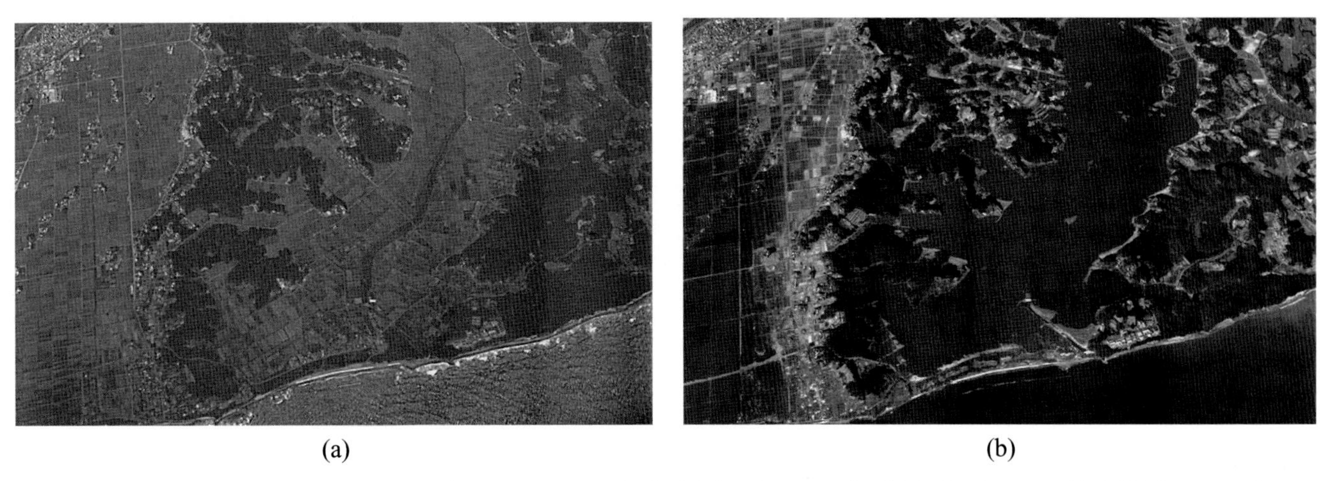

(a) (b)

Figure 5.31 Google Earth images showing Sendai Airport, Japan (a) in 2003 before the tsunami and (b) after the tsunami in 2011. See also color plates.

5.10.3 Use of Google Earth

One of the freely available programs that help us to understand Web GIS better is Google Earth. It is the most popular virtual browser freely accessible through the Internet that allows two- and three-dimensional visualization of geography. The free version of Google Earth can be downloaded from the website http://earth.google.com/. This is extremely useful as a teaching tool and for students to learn more about the geographical aspects of the terrain. The satellite images used by Google Earth are captured at different times. Integration to a common format enables draping of these images over the terrain. The main advantage of Google Earth is that it allows placement of *placemarks* atop the virtual globe. Navigation in Google Earth can be an enriching experience for students. Though it is a very useful tool, it lacks the capacity of a GIS system. This is mainly because query analysis for application-specific purposes cannot be performed in Google Earth. The available versions of Google Earth can be used to create data in KML (Keyhole Markup Language) format. The Professional version of Google Earth enables importing of Esri shapefiles and MapInfo tab files. It also offers tools for the creation of new data from Google's servers.

Google teamed up with a satellite imagery company called GeoEye to publish before and after images of the tsunami that struck Japan on March 11, 2011. The Google Earth images in Figure 5.31 show the extent of damage caused by the calamity.

5.10.4 Real-time decision support systems (RT-DSS) for hydrologic modeling using Web GIS

How quickly can decisions be made to tackle a natural disaster? What are the information databases on which the decisions are to be based? These are some of the questions that this section addresses. Real-time GIS implies that as soon as any data are created, they are fed directly to a GIS server for monitoring and decision-making, such as during a calamity. Some of the application areas in which a real-time decision support system via Web GIS can be an effective tool are flood relief operations, water supply management, optimizing reservoir operations, etc. Though there are numerous areas in which a real-time decision support system has proved useful, its specific use for watershed delineation in hydrology is discussed in detail here.

Hydrologic analysis based on real-time data assists the users to study the hydrologic impacts on life and the environment. Analysis in hydrology is always centered around spatially related data within a watershed or basin. For any hydrologic study of a small basin, a large number of factors, such as soil type, land cover type, soil moisture, groundwater table, etc., need to be considered. Previously, such analysis used manual survey data and existing maps. This was not only cumbersome but also time consuming. The development of GIS has provided a platform to integrate data from different sources to study the current status of a catchment. Hydrologic modeling and model implementation efforts based on digital maps using GIS have become a generalized approach for hydrologic analysis (Turcotte *et al.*, 2001).

The process to access real-time data sets starts with the user specifying the area of interest through a webpage. HyperText Markup Language (html), Java applet, Javascript, C, MapServer, Common Gateway Interface (CGI), etc. are programming languages that help in the creation of a web-based interface for the user to query. Once the specification part is completed by the user, the system chosen tries to link up with the server via the network, which follows certain protocols. The system acts as a medium to make the server understand what the client requires. By the use of an appropriate language, large amounts of data from the server can be transferred back to the client.

For example, to set up a watershed delineation system using Web GIS, large sets of data are required. These data may consist of DEMs, stream network information, land use maps, soil maps, etc. in vector or raster format or both. Secondly, an appropriate programming language has to be selected suitable to the application requirement. The most important aspect in watershed delineation is the selection of pour point locations. Since the information transfer is via the Web, users normally have a tendency to zoom in/out/pan to select their pour point locations if this information is not available a priori.

The MapServer Web GIS tool can be selected as the CGI engine for developing the Web GIS map–user interface. MapServer, a CGI, is an OpenSource development environment for building spatially enabled internet applications (Choi *et al.*, 2001). Also important is the availability of ancillary data so that the user can select pour points to the nearest possible accuracy. Finally, the server should take care of the delineation process using flow direction grids and pour point locations. For these, inbuilt algorithms can be used. The watershed image generated can be streamed back to the users within a very short time. Watershed delineation processing is often one of the more time-consuming jobs of hydrogeographical analyses using GIS. The time required for processing has been a main reason why the application has not been available for internet use (Pandey *et al.*, 2000). The total time, from the beginning of a user query to the time when the finished watershed map is provided, is a practical measure of the overall performance of the web-based watershed delineation system used.

An example of StreamStats, which is a USGS web-based GIS application for water resources planning and management, is discussed here briefly (http://streamstats.usgs.gov). StreamStats was created by the USGS with Esri and is built using Esri's ArcHydro tools. A map-based interface is made available for site selection. A user can zoom in by various methods within the interface to select the locations of his/her interest. Upon selection of a station within the database, the user is provided with the previously published information about that particular station. When an ungauged site is selected within the USA in StreamStats, it determines the drainage-basin characteristics and streamflow statistics for that site by solving regression equations (Guthrie *et al.*, 2009). The regression equations that are used to determine the flow statistics for ungauged sites were developed using computed streamflow statistics and measured basin characteristics for a selected group of stream gaging stations. Different states in the USA have been segregated into hydrologic regions based on the climate and physical characteristics, and separate regression equations have been developed for each region.

Users can also identify stream reaches towards either the upstream or downstream from the user-selected sites. The query builder within StreamStats expects input from the user in the form of the function to be used, state to be targeted, output format, input coordinates of the area of interest, etc. Given these inputs, a URL is built and is used to send the request to the StreamStats server. The output is presented to the user in the form of a map frame. The function ordered by the user is executed and the output returned in an appropriate format. The area-averaged estimates of streamflow statistics for the user-selected site are also displayed.

StreamStats is compatible with Microsoft Internet Explorer version 5 and higher versions. This application provides direct access to comprehensive data sets. The functionalities of StreamStats can be used as web services so that websites or desktop GIS can access the same information. Web services have been developed for drainage-area delineation, regression estimates for ungauged sites, stream gauging station information, and stream-reach address determinations (http://water.usgs.gov/osw/streamstats/).

StreamStats is used not only in watershed delineation but also in the design of water supplies, for wastewater discharge permitting, and for riparian wetland protection.

However, StreamStats suffers from certain drawbacks: streamflow estimates by StreamStats assume natural flow conditions at the site. If human intervention, such as reservoir regulation, affects the flow at the selected site, the output needs to be adjusted by the user to take into account such changes. When one or more of the basin characteristics are employed to solve the regression equations pertaining to an ungauged basin, extrapolation is involved.

In addition to Streamstats, there are numerous projects undertaken by different countries that provide disaster management using web-based GIS. For example, the Transnational Internet Map Information System on Flooding (TIMIS) is jointly being developed by seven project partners in Luxembourg, France, and Germany to improve flood forecasts for the Mosel basin and to develop flood warning systems for small river catchment areas. TIMIS takes up about one hundred rivers and streams as input and compiles flood hazard maps.

Another project launched in 2010 by the ISRO and the Central Water Commission is the India–WRIS (Water Resources Information System), which conglomerates all the information related to water resources of the nation into a single window. The layers and attributes used are watershed atlas, administrative layers, water resources projects, thematic layers, and environmental data.

5.11 SUMMARY

Various techniques in GIS involving topological, network, and cartographic modeling, map overlay, and geostatistics are useful tools for hydrologic modeling. DEMs can serve as a valuable asset for delineating drainage basins and extracting topographic information such as slope, drainage divides, and drainage networks. Satellite RS data provide information on land use/land cover, soil, and flood zone mapping. Merely accumulating data from different sources is not really useful. Methods to synthesize models and extract useful information are being used to improve decision-making. The available data from RS platforms, when integrated into the database in the GIS platform, can be used for effective management of water resources, flood damage assessment, and mitigation.

This chapter introduced DEMs in general. The varied potential of DEMs within the GIS platform is shown with the help of case studies of three basins, namely, the Malaprabha, Krishna, and Cauvery basins in India. The use of DEM to visualize the terrain and study its various topographical factors is stressed. It is shown that elevation data, properly handled in GIS, can serve as a great asset for hydrologic modeling. The mechanism of information exchange through the Internet is discussed. The Web GIS architecture is explained, and also how users can access spatial data within a GIS server through the click of a button on the Web. Real-time decision support systems to delineate watersheds have also been explained with an example.

EXERCISES

5.1 Download the HYDRO1K Asia data from the USGS website from the following link for any region of interest.

http://eros.usgs.gov/#/Find_Data/Products_and_Data_Available/Elevation_Products

Download the products with the following information: drainage basin, elevation, flow direction, slope, and stream data.

(a) From the elevation data, create a TIN in GIS.
(b) From the shapefile of stream data, create a region of 100 m around the stream data. (*Hint*: Use proximity tools.)
(c) From the point feature class created in (b), use raster distance methods to create the Euclidean distance image.
(d) Load the flow direction and drainage basin data in GIS and interpret it.

5.2 Digital elevation data (in m) with respect to mean sea level for a small catchment area are given in the table below. Assume that the boundary for the area shown is at a constant higher elevation of about 80 m. Perform the following tasks.

(a) Estimate the flow direction for each grid element and show the flow direction for each element.
(b) Using the D8 DEM algorithm, specify the flow direction as a number for each grid element.
(c) Identify the number of upstream grid cells contributing to each grid cell and indicate that number in each grid element.
(d) Identify the flow paths showing tributaries and main stream and also identify the mouth of the basin (catchment).
(e) Identify the sub-catchment boundaries.

65	45	50	45	60	70
50	45	43	42	50	55
50	48	41	40	60	65
55	50	40	35	45	50
50	40	35	30	50	55
45	42	28	20	40	50

6 Case studies and future perspectives

This chapter deals with synthesizing the models and methodologies presented in earlier chapters. Two case studies are presented to demonstrate statistical downscaling, quantification of uncertainties, and uses of RS and GIS in impact assessment. Perspectives on future research in hydrology in the context of climate change are provided in the last section.

6.1 CASE STUDY: MALAPRABHA RESERVOIR CATCHMENT, INDIA

Most of the topics explained in earlier chapters, such as RS, GIS, DEM, hydrologic models, climate change, and downscaling, are integrated and demonstrated through a case study to obtain streamflow projections for the Intergovernmental Panel on Climate Change (IPCC) Special Report on Emissions Scenarios (SRES) using a global climate model (GCM) output and a support vector machine (SVM) based a downscaling approach.

Modeling the streamflow response of a river basin is necessary as changes in streamflow affect almost every aspect of human well-being. Recently, with increase in awareness of climate change, assessment of climate change impact on streamflows is gaining importance. As explained in Chapter 5, GCMs are widely used to simulate climate conditions on Earth, several decades into the future, for various constructed climate-change scenarios. The climate variables simulated by GCM are at coarse spatial resolution and hence referred to as large-scale atmospheric variables (LSAV). The GCMs are generally run at coarse spatial resolution, and therefore they are not capable of simulating climate conditions and hydrologic processes at a relatively finer spatial scale (e.g., watershed, small river basin). Against this backdrop, a few methods have been developed in the past for translating GCM-simulated LSAV to future streamflows at relatively finer watershed scale.

Better understanding of climate and its changes, and hydrologic processes and their relationships within a catchment would lead to evolution of better GCMs, downscaling methodologies, and hydrologic models. Nevertheless, as the knowledge on any of those processes can never be considered complete, there will always be a factor of uncertainty associated with projected streamflows. Depending on the choice of emissions scenarios and downscaling methods, future streamflow projections may vary.

This study investigates the differences in future streamflow projections for the Malaprabha Reservoir catchment, India, resulting from two different methods for generating them from third-generation Canadian coupled GCM (CGCM3) data sets for four emissions scenarios from the IPCC Fourth Assessment Report (AR4): A1B, A2, B1, and COMMIT. The first method involves downscaling CGCM3 data to hydrometeorological variables at catchment scale and using them as input to a hydrologic/empirical model to obtain projections of streamflows from the catchment. The second method involves directly downscaling CGCM3 data to streamflows in the catchment. An SVM technique has been chosen for downscaling.

6.1.1 Streamflow projection methods

Broadly, there are three methods available for projecting future streamflows. In method 1, either future projections of runoff from GCMs are routed through a study area to obtain future projections of streamflows, or the future projections of precipitation and temperature from GCMs are input to hydrologic models developed for the study area to yield future projections of streamflows (Arora and Boer, 2001; Xu *et al.*, 2005; Nohara *et al.*, 2006; Wetherald, 2009). Models based on method 2 downscale LSAV to surface climate variables (hydrometeorological variables) at river-basin scale and then use them as inputs to a hydrologic model developed for the river basin to derive corresponding projections of streamflows. In method 3, future projections of LSAV are directly downscaled to streamflow in a river basin (Ghosh and Mujumdar, 2008; Tisseuil *et al.*, 2010).

The advantage of method 1 is that it uses GCM output (LSAV) that is readily available and easy to use. The shortcoming of method 1 is the simplistic representation of the hydrologic cycle within a GCM (Xu *et al.*, 2005), and consequent error in estimating streamflows. Projections obtained using models of method 2 have better representation of the hydrologic cycle and hydrologic models can be tailored to fit the rainfall–runoff relations for a river basin. Method 3 is simple, easy, and faster when compared to method 2, but models based on this method may not be able to capture the relationships between climate features influencing hydrometeorology of a river basin and the underlying mechanisms governing streamflow generation from the basin. So in this study, method 2 is employed to obtain streamflow projections. Further, the choice of predictor variables (input to model) can significantly affect the projections of streamflows.

The objective of this study is to compare future streamflows projected over the Malaprabha Reservoir catchment in India by method 2 using LSAV simulated by CGCM3 for four AR4 emissions scenarios (A1B, A2, B1, and COMMIT). The scenarios that are recommended in SRES AR4 are referred to as emissions scenarios in this section. Inferences drawn on the differences in streamflow projections for the choice of emissions scenario are assessed.

The soil and water assessment tool (SWAT; Neitsch *et al.*, 2001) hydrologic model is used to capture the relationship. The SVM technique has been chosen for downscaling. It has the advantage of implementing the structural risk minimization principle, which provides a well-defined quantitative measure of the capacity (which is the ability to learn any training set without error) of a learned function to generalize over unknown test data. Further, the global optimum solution can be obtained with SVM.

6.1.2 SWAT and AVSWAT

SWAT is a river basin scale model developed by Dr. Jeff Arnold for the United States Department of Agriculture (USDA) Agricultural Research Service (Neitsch *et al.*, 2001). It can predict the impact of land management practices on water, sediment, and agricultural chemical yields in large complex watersheds with varying soils, land use, and management conditions over long periods of time. SWAT is a physically based, distributed, continuous-time model that operates on a daily time scale. Physical processes associated with water movement, sediment movement, crop growth, nutrient cycling, etc. are directly modeled by SWAT.

For modeling purposes, a watershed is partitioned into a number of sub-basins, which are then further subdivided into hydrologic response units (HRUs). The use of sub-basins in a simulation model is particularly beneficial when different areas of the watershed are dominated by land uses and soils dissimilar enough in properties to impact hydrology. Simulation of the hydrology of a watershed is separated into two major parts in SWAT. The first part deals with the land phase of the hydrologic cycle, which considers the amount of water, sediment, nutrient, and pesticide loadings in each sub-basin. The second part deals with the routing phase that considers the movement of water and sediments through the channel network to the outlet.

AVSWAT-2000 (version 1.0) (Di Luzio *et al.*, 2002) is an ArcView extension and a graphical user interface (GUI) of the SWAT model. The two systems, ArcView and SWAT, are dealt with as two independent master components in the integration system. The conceptual design of the integrated system includes an add-on external user interface and a shared internal database to couple the two systems (Figure 6.1).

AVSWAT is organized in a sequence of several linked tools grouped into eight modules (Di Luzio *et al.*, 2002):

1. watershed delineation,
2. definition of HRU,
3. definition of the weather stations,
4. AVSWAT databases,
5. input parameterization, editing, and scenario management,
6. model execution,
7. read and map-chart results, and
8. calibration tool.

The basic map inputs required for the AVSWAT include digital elevation maps, soil maps, land use/land cover maps, hydrography (stream lines), and time series of weather variables with their locations.

6.1.3 Study region and data used

The study region is the catchment of the Malaprabha River, upstream of the Malaprabha Dam. It has an area of 2093.46 km^2 situated between latitude 15° 30′ N to 15 56′ N and longitude 74° 12′ E to 75° 8′ E. It lies in the extreme western part of the Krishna River basin in India, and is one of the main tributaries of the Krishna. The source of water for the study area is precipitation primarily from the southwest monsoon extending over the period June–September. The location map of the study region is shown in Figure 6.2.

Daily records of rainfall, maximum, minimum, and mean temperatures, wind velocity, and relative humidity for the period January 1978 to December 2000 of 11 gauging stations in the catchment of the Malaprabha Reservoir were collected. A thematic layer showing the locations of the hydrometeorological gauging stations was prepared using their spatial coordinates. Later, hydrometeorological data were assigned to each sub-basin based on its proximity to the gauging station.

The meteorological data used for this study comprised the recorded data at various stations in the study area, the reanalysis data on LSAV extracted from the database prepared by National Centers for Environmental Prediction (NCEP; Kalnay *et al.*, 1996), and CGCM3. The details of the data used are furnished in Table 6.1. In Table 6.1, Ta represents air temperature, Zg is geo-potential height, Hus is specific humidity, Ua is zonal wind velocity, Va is meridional wind velocity, prw is precipitable water, ps is surface pressure, LH is latent heat flux, SH is sensible heat flux, SWR is shortwave radiation flux, LWR is longwave radiation flux. Numbers by the side of these abbreviations indicate different

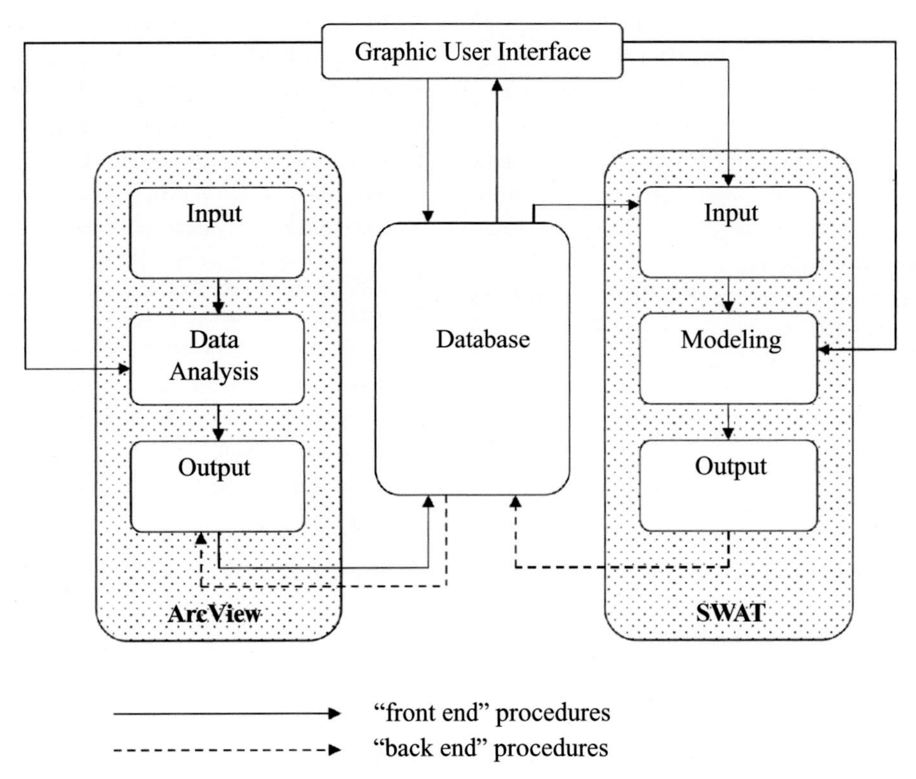

Figure 6.1 Architecture of the interface system coupling ArcView and SWAT.

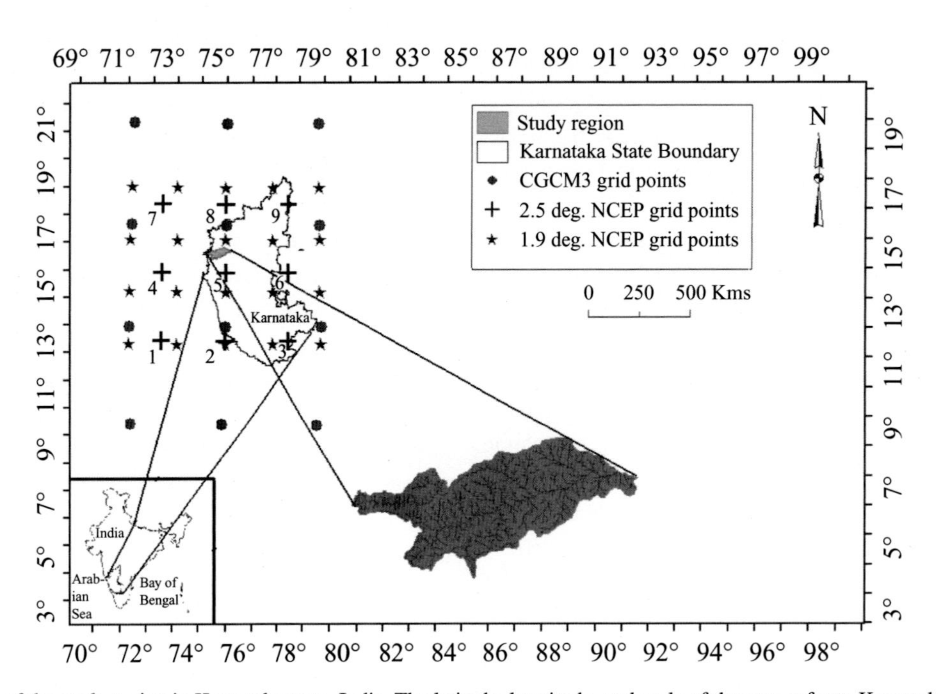

Figure 6.2 Location of the study region in Karnataka state, India. The latitude, longitude, and scale of the map refer to Karnataka state. The data extracted at CGCM3 and 1.9° NCEP grid points are re-gridded to the nine 2.5° NCEP grid points. Among the nine grid points, 1, 4, and 7 are on the Arabian Sea, and the remaining points are on land.

Table 6.1 *Probable predictors selected for downscaling predictands*

Set no.	Predictand	Probable predictor(s) selected from NCEP and CGCM3 monthly data sets for downscaling predictand
1	Precipitation	Ta 925, Ta 700, Ta 500, Ta 200, Zg 925, Zg 500, Zg 200, Hus 925, Hus 850, Ua 925, Ua 200, Va 925, Va 200, prw, ps
2	Maximum temperature	Ta 925, Ua 925, Va 925, LH, SH, SWR, LWR
3	Minimum temperature	Ta 925, Ua 925, Va 925, LH, SH, SWR, LWR
4	Wind speed	Ua 925, Va 925
5	Relative humidity	Ta 925, Hus 925, LH
6	Cloud cover	prw

pressure levels in hectopascals (hPa). For example, Ta 925 represents air temperature at 925 hPa pressure level. The CGCM3 data on LSAV and the NCEP data on atmospheric fluxes are re-gridded to a common 2.5° NCEP grid using the Grid Analysis and Display System (GrADS; Doty and Kinter, 1993).

The mean monthly maximum temperatures in the catchment vary from 25 °C to 34 °C and the average of the mean monthly maximum temperatures is 28 °C. The mean monthly minimum temperatures range from 17 °C to 21 °C. The day temperatures rarely fall below 25 °C. The hottest months are April and May, with mean maximum temperature of 34 °C. December and January are the coldest months with mean minimum temperature of 17 °C. On an annual basis, the diurnal difference between the maximum and the minimum temperatures is 8 to13 °C.

The wind speeds are high during the monsoon season (June to September) and low during November, December, and January. The mean monthly wind speed is 9.6 km/hr during the peak monsoon (July), while in non-monsoon months the mean monthly wind speed varies from 3 to 6 km/hr.

The SRTM DEM data modified for the study region were procured from the International Water Management Institute (IWMI), Hyderabad, India. Land use/land cover map and soil map of the study area were procured from the Karnataka State Remote Sensing Application Center (KSRSAC), Bangalore, India, prepared based on panchromatic and LISS III merged, IRS satellite images. Topomaps of the study region were procured from Survey of India at the finest available scale of 1:50,000.

The runoff curve numbers considered for selected land use, land cover, and soils in the Malaprabha sub-basin were adapted from Mishra and Singh (2003). The available water content, also referred to as plant-available water or available water capacity, is calculated by subtracting the fraction of water present at the permanent wilting point from that present at field capacity.

The basin attributes are obtained using the given basin layer. The SWAT–ArcView interface calculates area, resolution, and geographic coordinate boundaries for the basin and for each sub-basin. The length of the longest stream and the proportion of each sub-basin within the basin are also estimated.

6.1.4 Generation of streamflows with the SWAT model

A brief description of the methodology for generating streamflows (Figure 6.3) using models based on method 2, described in Section 6.1.2, is presented in this section.

Generation of future streamflows by a model considering SWAT involves four stages. The four stages involved in generation of streamflows using the SWAT model are as follows:

Stage 1: SWAT model is calibrated and validated for the study region with inputs from RS, GIS, DEM, and observed hydrometeorological data.
Stage 2: SVM models are developed to downscale monthly sequences of LSAV to monthly sequences of hydrometeorological variables at the river-basin scale (Figure 6.4).
Stage 3: Sequences obtained in stage 1 are disaggregated from monthly to daily scale using KNN technique.
Stage 4: Daily sequences are input to SWAT hydrologic model developed for the study area to generate streamflows.

6.1.4.1 STAGE 1: CALIBRATION AND VALIDATION OF THE SWAT MODEL

The main source of hydrologic input to a catchment is rainfall. Therefore, assessment of its spatial and temporal variation in the study region is necessary before developing a hydrologic simulation model. Average rainfall over the catchment of the Malaprabha Reservoir is estimated at monthly and annual time scales by the SWAT model using the records of the selected rain gauges in the study region. To check the general validity of the estimated rainfall from the SWAT model, the representative rainfall provided by SWAT and that obtained by adopting the Thiessen polygon method, which is a common method for computing average rainfall over an area, are compared. GIS is used for estimation of the areas of the Thiessen polygons. The average rainfall at the monthly time scale for the study region, computed using the SWAT model and the Thiessen polygon method, is found to be correlated fairly well.

Temporal variation of average rainfall in the catchment of the Malaprabha Reservoir and its relationship with the streamflows recorded at the reservoir site are studied. Results show that peak flows observed at the reservoir correspond to heavy rainfall in the catchment.

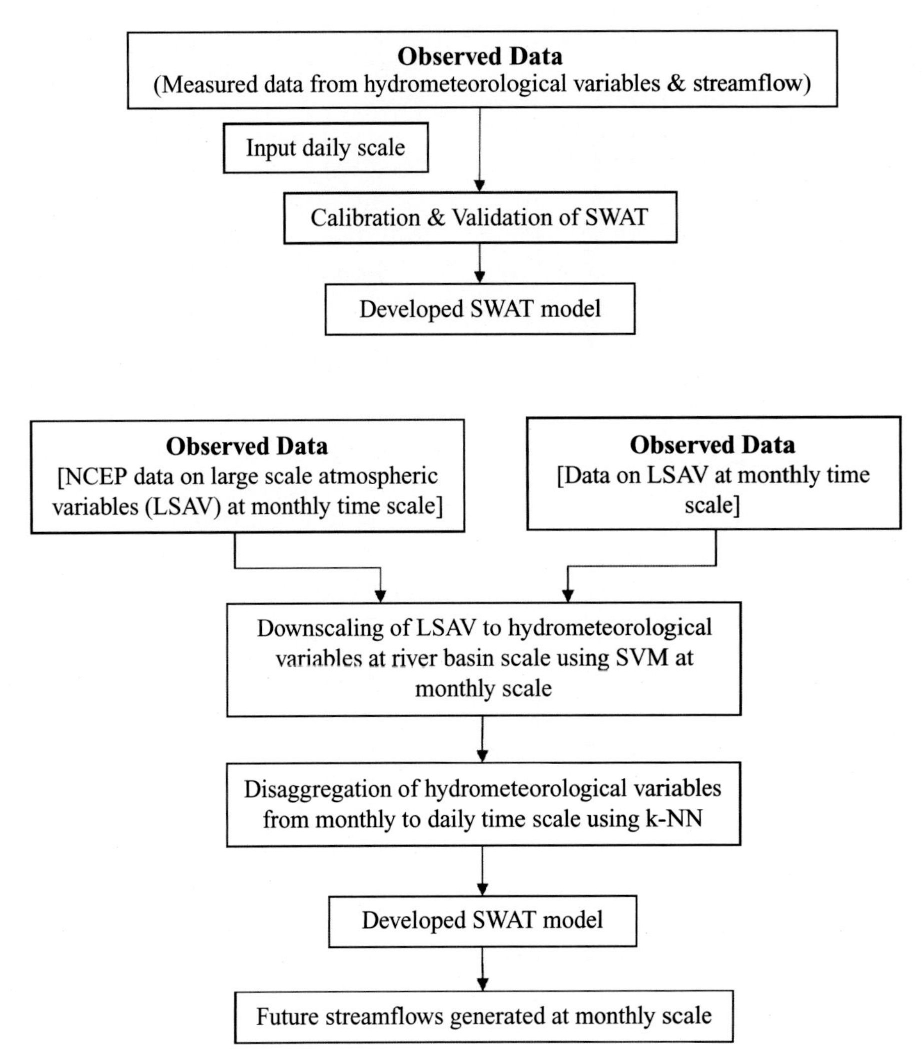

Figure 6.3 Methodology for generating future streamflows by the SWAT model.

The runoff from the catchment is simulated using the SWAT model with an ArcView interface. Firstly, ArcView map themes and database files, which provide necessary information about the watershed, are prepared. The ArcView map themes required for the interface include those of DEM data, land cover, land use, and soil. The database files necessary for the interface include (Di Luzio *et al.*, 2002):

1. Location tables of sub-basin outlet, watershed inlet, gauging stations of precipitation, temperature, solar radiation, wind speed, and relative humidity.
2. Look-up tables of land use and soil.
3. Data tables for precipitation, temperature, solar radiation, wind speed, relative humidity, point discharge (annual, monthly and daily loadings), reservoir inflow (monthly and daily if available), and potential evapotranspiration (if available).

The DEM data (shown in Figure 6.5) are pre-processed using an ArcView interface of the SWAT model, to obtain the stream network in the catchment of the Malaprabha Reservoir. For this purpose, the minimum watershed area (critical source area) is specified as 210 hectares. Subsequently, the stream network is reviewed and drainage basin outlets are fixed through the screen interactive option of the SWAT model. The SWAT model is run forming 14 drainage sub-basins in the Malaprabha Reservoir catchment (Figure 6.6), and the physiographic characteristics of the sub-basins are noted. A vector data set with the longest streams in the delineated sub-basins is shown in Figure 6.7.

The land use/land cover and soil maps of the Malaprabha Reservoir catchment are overlaid on each other to identify hydrologic response units (HRUs). The information about the type of land use/land cover and soil in each HRU, and the number of HRUs in each drainage sub-basin are documented.

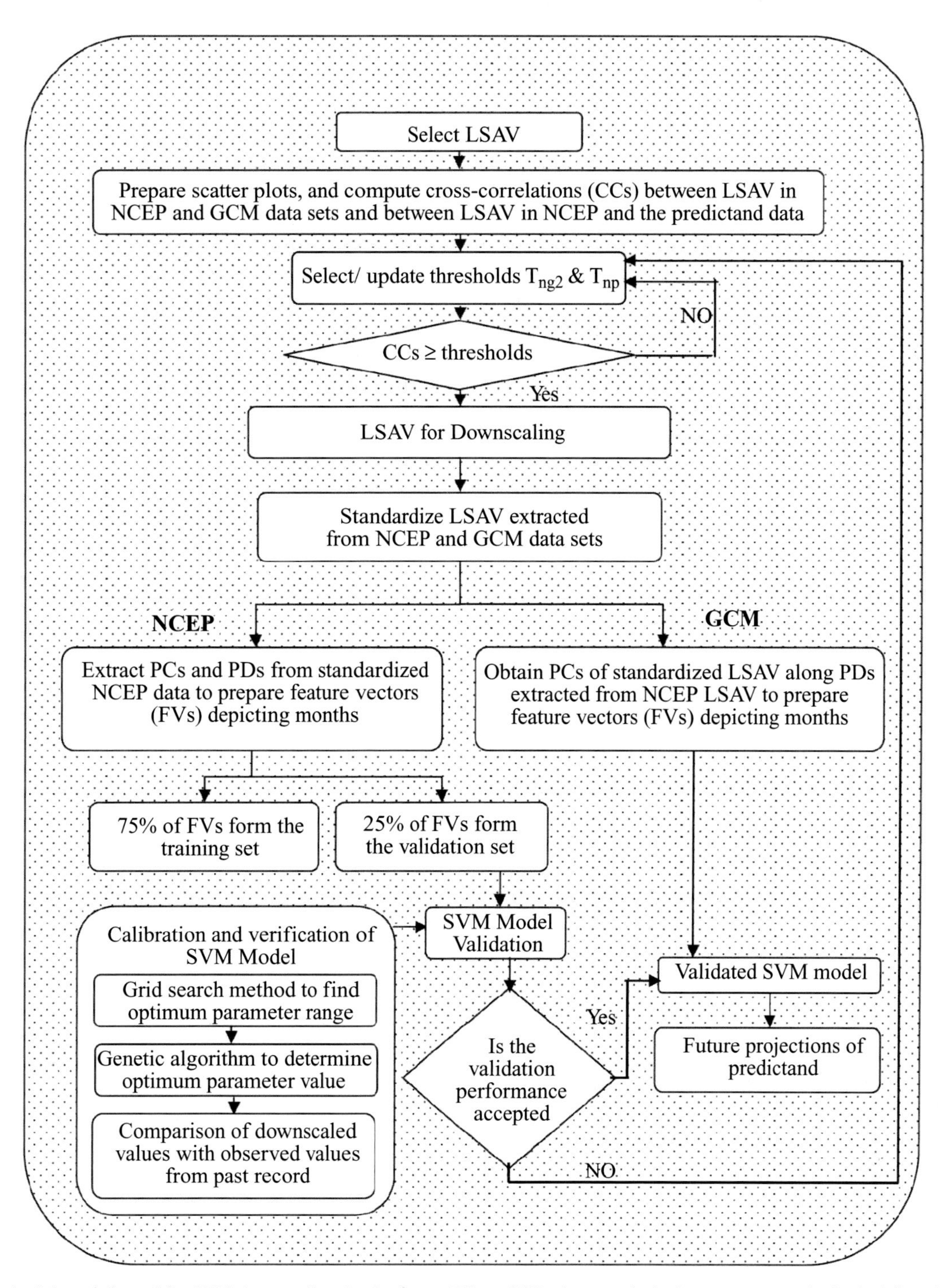

Figure 6.4 Methodology followed for SVM downscaling. In the figure, PCs and PDs denote principal components and principal directions, respectively. T_{ng} is the threshold between predictors in NCEP and GCM data sets. T_{np} denotes the threshold between predictors in NCEP data and the predictand.

The prepared data tables of weather variables are fed into the SWAT model and it is run. The development of the SWAT model involves calibration and validation phases. Traditionally, the first 70% of the available record is selected for training and the remaining 30% is used for validation. In the current study, the data for the period from January 1978 to December 1993 are considered for model calibration, and those for the period from January 1994 to December 2000 are considered for model validation.

The SWAT model provides the amount of water in the land phase of the hydrologic cycle, sediment, nutrient, and pesticide

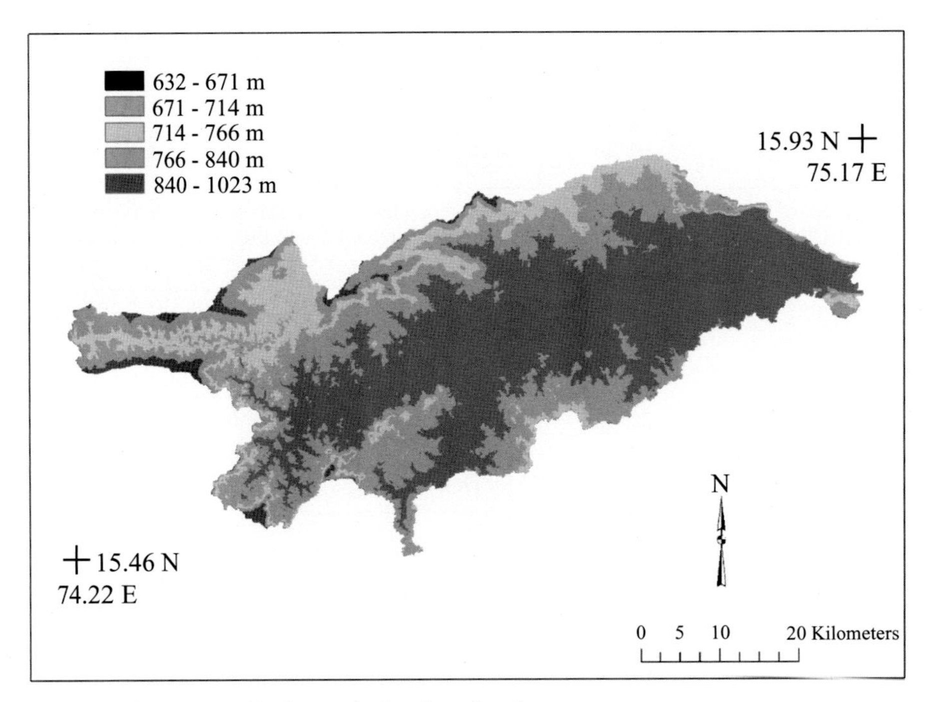

Figure 6.5 DEM of the catchment of the Malaprabha Reservoir. See also color plates.

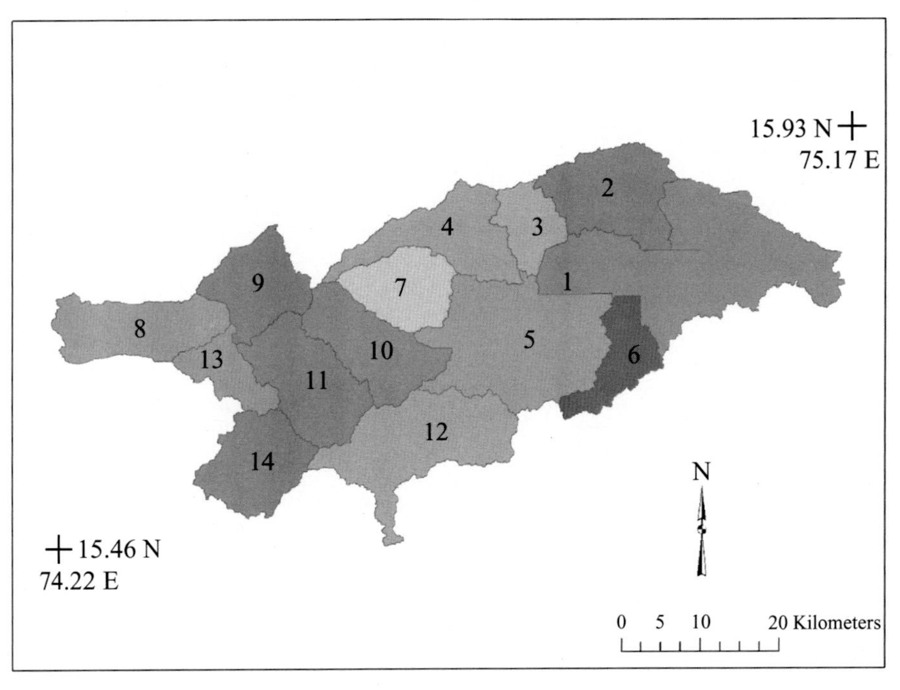

Figure 6.6 Sub-basins formed by AVSWAT for the outlets fixed in the catchment of the Malaprabha Reservoir. See also color plates.

loadings in each sub-basin, and sediment routed through the channel network to the outlet as outputs. However, only the runoff generated by SWAT is considered for validation. In the calibration phase, the runoff simulated by the SWAT model at monthly time scale is compared with that observed at the Malaprabha Reservoir. In general, the model over-predicts or under-predicts the runoff. In a few cases, the model may not simulate intermittent peak flows, possibly due to loss of information because of non-uniformity in spatial distribution of the available rain gauges in the study region.

The SWAT model estimates the water yield from a HRU for a time step, using Equation 6.1. The water leaving a HRU contributes to streamflow in the reach:

$$\text{WYLD} = \text{SURQ} + \text{LATQ} + \text{GWQ} - \text{TLOSS} - \text{Pond abstractions} \quad (6.1)$$

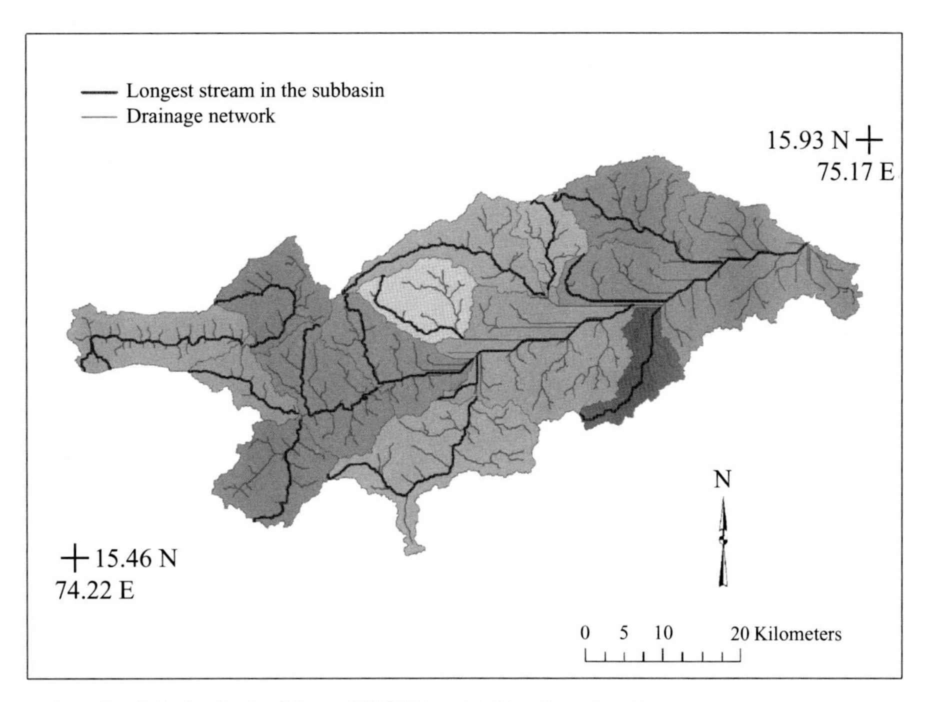

Figure 6.7 Longest stream in each sub-basin obtained from AVSWAT model. See also color plates.

where SURQ, LATQ, and GWQ represent contributions to streamflow in the reach from surface runoff, lateral flow, and groundwater, respectively, during the time step. TLOSS refers to the amount of water lost from tributary channels during transmission. The groundwater is primarily contributed by shallow aquifers. The area-weighted values of rainfall, SURQ, GWQ, and WYLD for the Malaprabha catchment, estimated from the corresponding values of the variables at the 14 drainage subbasins in the catchment, are some of the outputs from this model.

In the present case study, the SWAT model is found to overpredict the runoff. To overcome the problem, possible options include decreasing the curve number (CN), increasing available water capacity (AWC) of soil, the plant uptake compensation factor (EPCO), and the soil evaporation compensation factor (ESCO) (Neitsch *et al.*, 2001).

Soil moisture is depleted from various depths in the soil layer(s) to meet evaporative demand. The parameter ESCO allows users to specify the contribution from different depths of soil in meeting the soil evaporative demand. The default setting in SWAT causes 50% of the soil evaporative demand to be met from the top 10 mm of soil and 95% of the same to be met from the top 100 mm. When ESCO approaches 0, the SWAT model allows more water to be extracted from the lower layers of soil to meet the evaporative demand. On the other hand, as ESCO approaches 1, the model allows less variation from the default setting, indicating a situation in which evaporative demand is met primarily from the top layer of soil.

Further, plants take up water from their root zone to meet transpiration requirements. The parameter EPCO allows users to specify the vertical distribution of plant water uptake within the root zone. The default setting in SWAT allows plants to uptake 50% of water demand from the upper 6% of the root zone. When EPCO approaches 1.0, the SWAT model allows more of the water uptake demand to be met by lower layers in the soil. On the other hand, as EPCO approaches 0, the model allows less variation from the default setting, indicating that plant water uptake occurs primarily within upper root zone.

With a view to examining the sensitivity of the result from the SWAT model to variation of parameters, sensitivity analysis is performed using the latin hypercube one-factor-at-a-time (LH-OAT; Morris, 1991) method to identify sensitive parameters of SWAT. The sensitive parameters are those that cause large changes in the streamflow generated by the model when perturbed. In the LH-OAT sensitivity analysis, the OAT method is repeated for each point sampled in the LH sampling. The details of the method can be found in Griensven (2005). The records for the period 1978–93 were used for auto-calibration of the sensitive parameters by Monte Carlo analysis. In this regard, the ranges of the parameters were divided into initial input range and behavioral range. The initial input range was based on Monte Carlo analysis, and theoretical limits of the parameters. A thousand parameter sets, containing a fairly broad range of parameter values, were initially generated from the initial input range using independent uniform distributions. The streamflows generated using each of the parameter sets were compared with the observed streamflows

Table 6.2 *Ranges of different sensitive parameters considered for calibration of SWAT model and optimal values selected*

Sensitive parameter	Initial range	Behavioral range	Parameter value selected during calibration
Curve number (CN)	−20% to +20% of CN (upper limit 100)	−5% to +5% of CN	75[a]
AWC	−0.04 to +0.04 of AWC	+1 to +0.04	0.04[a]
EPCO	0.1 to 1	0.5 to 0.8	0.75
ESCO	0.1 to 1	0.1 to 0.5	0.4

[a] represents the weighted CN and AWC for the different combinations of soil and land use.

using eight performance measures. The behavioral range of each of the parameters was determined using cutoff for the performance measures. The initial range (Neitsch *et al.*, 2000) and behavioral range of the parameters are listed in Table 6.2. The values of EPCO and ESCO are varied from 0.1 to 1.0 with an increment of 0.1, whereas the value of AWC is varied from −0.04 to +0.04 with an increment of 0.01. From the behavioral parameter range, ten thousand parameter sets were generated from independent uniform distributions. For each combination of the chosen parameters, the runoff simulated by the SWAT model is compared with that observed at the Malaprabha Dam for the calibration period (January 1978 to December 1993), in terms of eight model performance indicators to arrive at optimal set of parameters. Finally, the simulated and observed streamflows were plotted graphically for the short-listed parameter sets to arrive at an optimal parameter set by visual interpretation. Results pertaining to the parameters, EPCO = 0.75, ESCO = 0.4, and AWC = 0.04, were selected from the calibration phase (Table 6.2).

Streamflows simulated by the SWAT model for the validation period after accounting for the combined effects of retention storage and evapotranspiration are shown in Figure 6.8. It can be noticed from the figure that the model performs fairly well with an R^2 value of 0.95 during the validation period (January 1994 to December 2000).

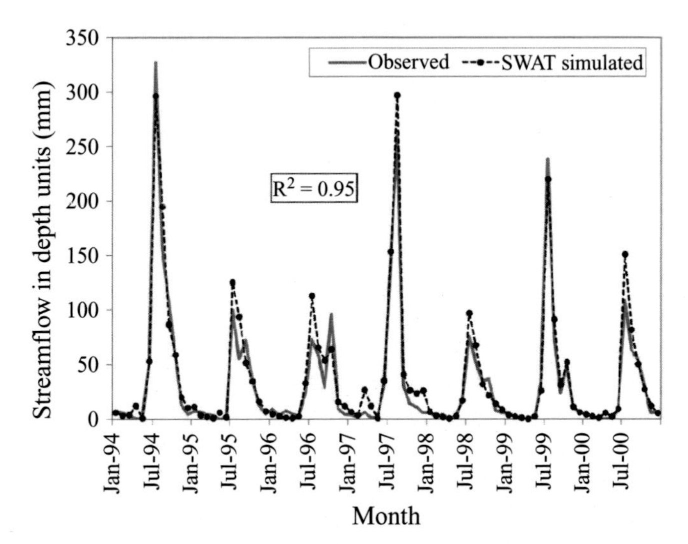

Figure 6.8 Observed and simulated monthly streamflows at the Malaprabha Dam site for the validation period.

6.1.4.2 STAGE 2: DOWNSCALING

In this stage, six SVM models were developed for spatial downscaling of LSAV to six hydrometeorological variables (precipitation, maximum and minimum temperature, relative humidity, wind speed, and cloud cover) for the Malaprabha catchment. To develop a downscaling model for each hydrometeorological variable (predictand), a separate set of probable predictors was selected from the LSAV. The probable predictors selected for each predictand are given in Table 6.1. From the probable predictors, the potential predictors for downscaling each of the six predictands (variables) were selected by specifying two threshold values (T_{ng} and T_{np}). The T_{ng} is for cross-correlation between predictors in NCEP and GCM data sets, whereas T_{np} denotes the same between predictors in the NCEP data set and predictand. From the NCEP data on potential predictors, orthogonal PCs that preserve more than 98% of the variance in the data were extracted, and corresponding principal directions were noted. The PCs of GCM predictor data were extracted along the principal directions of NCEP predictor data. The PCs account for most of the variance in the predictor data and also remove redundancy, if any, among the input data. The use of PCs in the analysis makes the model more stable and reduces the computational load. A feature vector for each month was formed using the PCs extracted for the month. These feature vectors formed using PCs extracted from NCEP data were used for calibration and validation of the SVM models. Of the feature vectors, 75% were used as the training set and the rest were used as the validation set. The relationship between feature vectors in the training set (prepared from potential predictors for a predictand) and historical data of the predictand at catchment scale was captured using SVM. The optimal

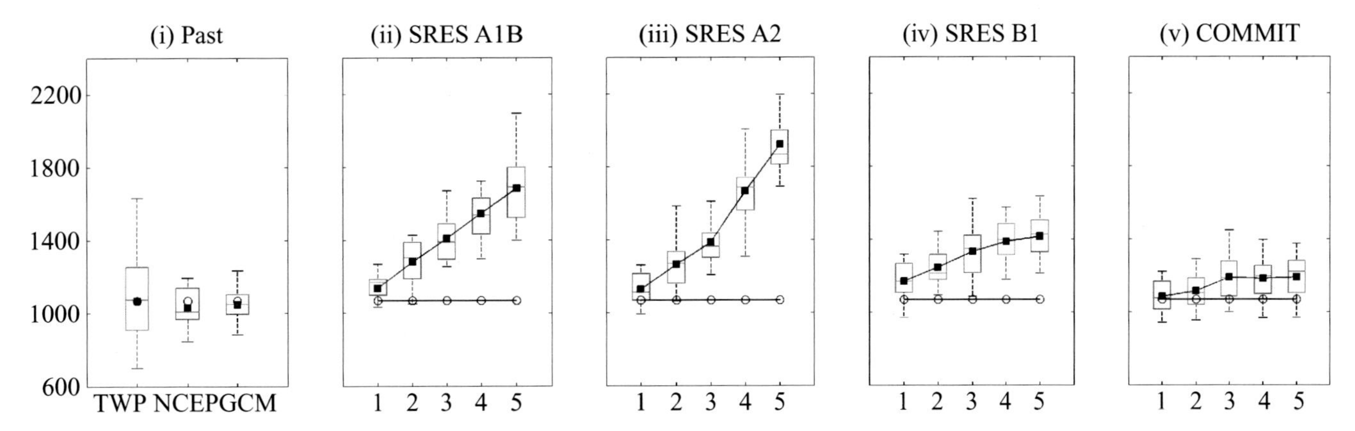

Figure 6.9 Typical results from the SVM-based downscaling model for precipitation.

ranges of SVM parameters, namely kernel width (σ) and penalty term (C), were obtained using the grid search procedure. The range of σ and C having the least normalized mean square error (NMSE) was selected as the optimal parameter range. The optimal value of each parameter was obtained from its optimum range using a genetic algorithm. The developed SVM model was validated by providing feature vectors of the validation set as input and comparing output (downscaled values of predictand) with the observed (historical) values of the predictand at catchment scale for the validation period, using NMSE.

The sequences of each of the downscaled hydrometeorological variables obtained for the past (1978–2000) and future periods (2001–2100) were compared with those observed in the study region using mean values. Typical results obtained for precipitation are shown in the form of box plots in Figure 6.9. Results for the past are shown in Figure 6.9(i), whereas those for 2001–20, 2021–40, 2041–60, 2061–80 and 2081–2100, for the four emissions scenarios A1B, A2, B1, and COMMIT, are shown in (ii), (iii), (iv), and (v), respectively. Small circles in Figure 6.9 denote the observed mean monthly value of precipitation during 1978–2000, and the dark squares represent the downscaled mean monthly value of precipitation. In Figure 6.9(i), NCEP and GCM denote precipitation obtained using the SVM downscaling model for input from predictor variables in NCEP and GCM data respectively, for the past period (1978–2000). In Figure 6.9(ii), (iii), (iv), and (v) the solid line that joins the circles indicates the historical mean value of monthly precipitation, while the line connecting the solid squares within the box plots depicts the mean trend of precipitation projected by the GCM. Similar analysis is performed on all five other variables downscaled using SVM.

Results pertaining to calibration and validation of SVM models developed for downscaling precipitation and maximum and minimum temperature are presented in Anandhi *et al.* (2008, 2009) and those for downscaling relative humidity, wind speed, and cloud cover can be found in Anandhi (2007). Results indicate that for the past period the mean statistic of the observed data is modeled fairly well by the downscaling models. It was observed that precipitation and maximum and minimum temperature are projected to increase in future for the A1B, A2, and B1 scenarios, whereas no trend is discerned with COMMIT. The projected increase in trend is high for the A2 scenario and is least for the B1 scenario.

6.1.4.3 STAGE 3: DISAGGREGATION

The SWAT hydrologic model chosen for generation of streamflow projections requires daily sequences of hydrometeorological variables as input. For this purpose, the monthly sequences of hydrometeorological variables obtained using SVM downscaling models in stage 2 are disaggregated to daily scale using the KNN technique. The technique uses the statistical relationship between the observed values of a variable at monthly and daily time scales to disaggregate a monthly sequence of the variable obtained from the downscaling model. The algorithm of disaggregation based on the KNN technique can be found in Anandhi (2007). Several variants of KNN models are available in the literature (Mehrotra *et al.*, 2006; Mehrotra and Sharma, 2007).

Monthly sequences of each of the hydrometeorological variables obtained from the SVM downscaling models are disaggregated to daily scale by the KNN technique. A trend was observed in future monthly projections of precipitation and maximum and minimum temperature. Therefore, for disaggregating these variables, a window of size three months was considered to explore the temporal relationship between a wide range of monthly and daily values to form the conditioning set $\mathbf{z}_\tau$ for each month τ. Future monthly projections of wind speed and relative humidity did not show a trend, hence a smaller window size (one month) was used to form the conditioning set for disaggregating these two variables.

Mean daily values of disaggregated hydrometeorological variables were compared with those observed in the region for the validation period (1994–2000). The results indicated that, in general, daily sequences of the variables obtained using the KNN disaggregation technique are consistent with those observed in the study region. However, the performance of the technique in disaggregating the various hydrometeorological variables varied. Among the five variables disaggregated, the performance of the KNN technique in disaggregating values of maximum and minimum temperatures is the best, followed by relative humidity, wind speed, and precipitation.

6.1.4.4 STAGE 4: GENERATION OF STREAMFLOW

In this stage, the disaggregated hydrometeorological variables at catchment scale were fed into the SWAT hydrologic model with an ArcView interface to predict streamflows at daily time scale. The hydrologic model was developed to capture the relationship between historical data of the hydrometeorological variables at catchment scale and historical streamflow data in the Malaprabha catchment. To apply the model for streamflow simulation using the curve number method, the observed data on climate variables, namely precipitation, maximum and minimum temperature, wind speed, and relative humidity, were used as inputs in addition to other processed inputs such as DEM, land use, and soil map. For this analysis, the processed inputs were considered to be the same for the past and the future time periods. The ArcView GIS interface of SWAT provides an easy-to-use graphical user interface for organizing all the required inputs. Delineating the sub-watershed boundaries, defining the HRUs, generating SWAT input files, creating agricultural management scenarios, executing SWAT simulations, and reading and charting of results were all carried out by the various tools available in the interface.

The daily sequences of hydrometeorological variables (i.e., precipitation, maximum and minimum temperatures, relative humidity, and wind speed) obtained for the period 1978–2100 using the KNN disaggregation technique were given as inputs to the calibrated and validated SWAT model, and the output was aggregated to obtain monthly streamflow sequences for each of the four emissions scenarios. In order to compare the models' capabilities in simulating the dynamics of monthly runoff series, the monthly runoff values computed using different models were analyzed and correlated with observed values using a linear regression equation.

Prediction of future streamflows for changed climates The hydrometeorological variables downscaled using predictors in CGCM3 data sets for the period 2001–2100 were input to the SWAT to generate future projections of streamflows. The mean annual, mean monsoon, and mean non-monsoon streamflows projected in the Malaprabha catchment for four emissions scenarios

Table 6.3 *Percentage change in mean annual, mean monsoon, and mean non-monsoon streamflows projected with respect to their historical counterparts for the different scenarios used (A1B, A2, B1, and COMMIT)*

	Percentage change in streamflow from 2001 to 2100			
Season	A1B	A2	B1	COMMIT
Annual	5 to 108	5 to150	12 to 60	−3 to 20
Monsoon	−19 to 57	−17 to 96	−16 to 21	−24 to −13
Non-monsoon	117 to 346	105 to 391	142 to 245	98 to 174

(A1B, A2, B1, and COMMIT) were analyzed for the period 2001–2100. For this purpose, the future projections of streamflow for each scenario were divided into five parts (2001–20, 2021–40, 2041–60, 2061–80 and 2081–2100). For each part, the percentage change in projected mean streamflow with respect to its historical counterpart was computed for the four emissions scenarios. The results are shown in Figure 6.10 and Table 6.3.

Comparison of the percentage change in future streamflows projected across different scenarios (Figure 6.10) indicates that the SWAT model has large differences between the four scenarios considered at annual, monsoon, and non-monsoon seasons. It can be observed that the increase in trend is greater after 2060 for the A1B and A2 scenarios. The projected increase in streamflow is high for the A2 scenario, whereas it is least for the B1 scenario. The scenario A2 has the highest concentration of equivalent carbon dioxide (CO_2), equal to 850 ppm, while that for the A1B, B1, and COMMIT scenarios is 720 ppm, 550 ppm, and ~370 ppm, respectively. A rise in the concentration of equivalent CO_2 in the atmosphere causes the Earth's average temperature to increase, which in turn causes an increase in evaporation, especially at lower latitudes. The evaporated water will eventually precipitate. Increase in precipitation results in increased streamflow. In the COMMIT scenario, where the emissions are held the same as in the year 2000, no significant trend in the projected future streamflow could be discerned.

Results of trend analysis are presented at annual scale in Table 6.4. It can be seen that on an annual scale, the streamflow is projected to increase during 2021–2100 for the A1B, A2, and B1 scenarios, and during 2061–2100 for the COMMIT scenario.

6.1.5 Summary of the case study

The differences between future streamflow projections in the Malaprabha Reservoir catchment, India, are investigated by obtaining them from CGCM3 data sets for four emissions scenarios: A1B, A2, B1, and COMMIT using the hydrologic model

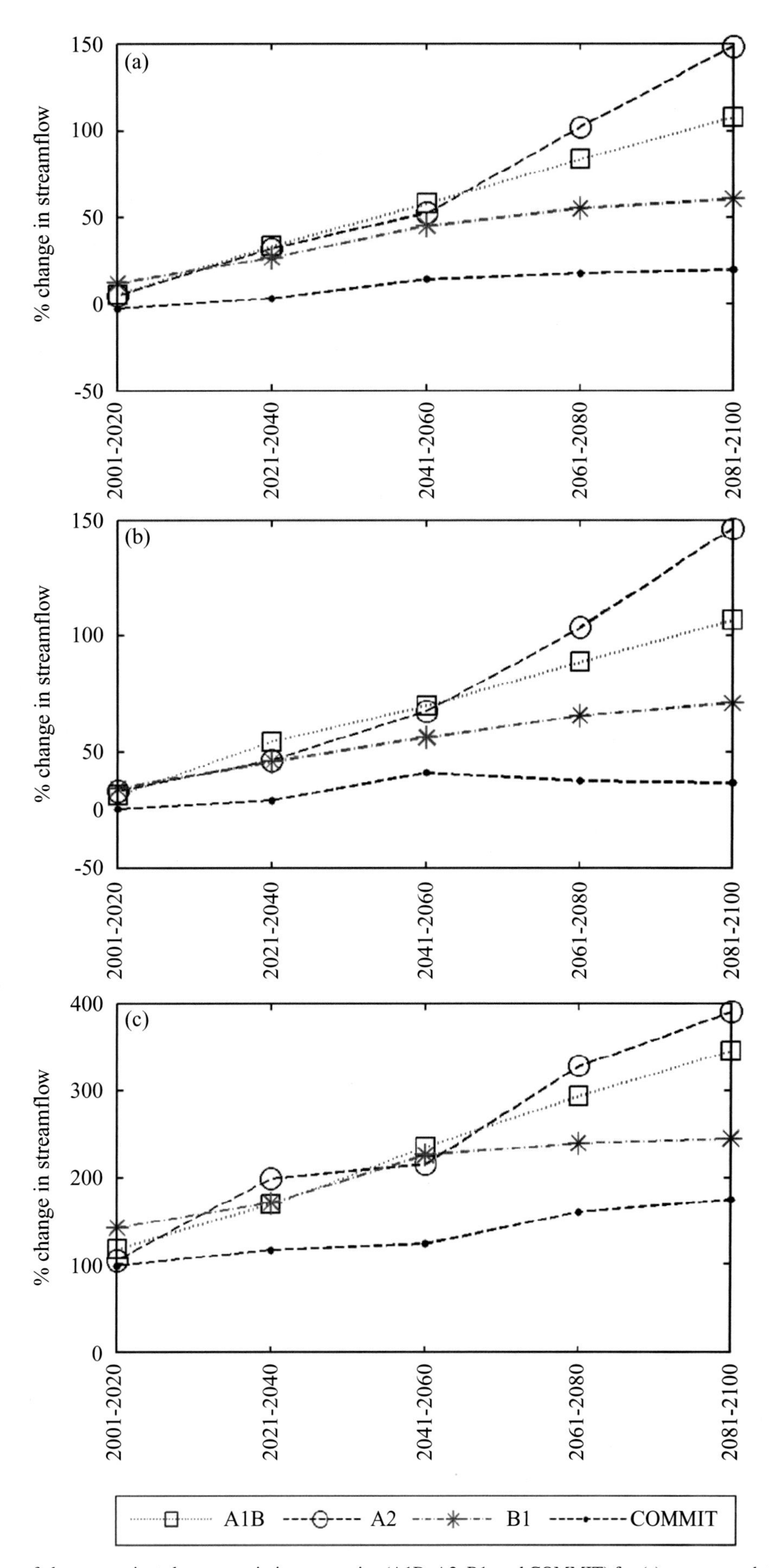

Figure 6.10 Comparison of changes projected across emissions scenarios (A1B, A2, B1, and COMMIT) for (a) mean annual, (b) mean monsoon, and (c) mean non-monsoon streamflows.

Table 6.4 *Results of trend analysis carried out for annual streamflows using null hypothesis considering 99% confidence level. Increasing trend is denoted by + and no change in trend is denoted by 0*

	Duration (in years)				
Scenario	2001–20	2021–40	2041–60	2061–80	2081–2100
SRES A1B	0	+	+	+	+
SRES A2	0	+	+	+	+
SRES B1	0	+	+	+	+
COMMIT	0	0	0	+	+

SWAT. On an annual scale, the streamflow in the Malaprabha catchment is projected to increase during 2021–2100 for the A1B, A2, and B1 scenarios, and during 2061–2100 for the COMMIT scenario by SWAT. Results indicate that differences in future streamflow projections pertinent to differences in climate change scenario and downscaling model are significant even for a chosen GCM. Hence, there is a need to devote effort towards developing new methodologies for comprehensive integrated analysis of uncertainties in projections of streamflows pertinent to a wider range of factors, such as differences in (i) reanalysis data sets, (ii) spatial domain of each predictor climate variable surrounding the catchment, (iii) GCMs, (iv) LSAV data available for different runs of a GCM, (v) climate change scenarios, (vi) downscaling strategies, and (vii) hydrologic models.

6.2 CASE STUDY: MAHANADI RIVER BASIN, INDIA

This section presents a case study of long-term climate change impacts on flood discharges and river flows in the Mahanadi River basin, India. Uncertainties associated with impacts on flows are addressed in the case study, with possibility-weighted probability distributions.

6.2.1 Mahanadi River basin

The Mahanadi River basin (Figure 6.11) is situated in eastern India between 19° 20′ N and 23° 35′ N latitude and 80° 30′ E and 86° 50′ E longitude. The Mahanadi is a rain-fed river flowing eastwards draining an area of 141,600 km^2 into the Bay of Bengal. Being situated in a coastal region, the river basin is quite likely to be affected by climate change. There has been a reported increase in hydrologic extremes in the recent past. The observed increase in temperature in this region is 1.1 °C/century, whereas the average increase for the whole of India is about 0.4 °C/century (Rao and Kumar, 1992; Rao, 1993, 1995). A major portion of the annual rainfall in this region is received during the summer monsoon (June to September). This basin is highly vulnerable to floods, and has suffered from frequent catastrophic flood disasters. The flood during the monsoon of 2001, inundating 38% of the basin area, was the worst ever flood recorded in this basin during the past century (Asokan and Dutta, 2008). This basin has also been predicted to be the worst affected by floods under climate change, out of all the major river basins in India (Gosain *et al.*, 2006).

6.2.2 Future flood peaks and water availability

Hydrologic impacts of climate change in the Mahanadi River basin with respect to water resources availability and floods are analyzed at a sub-catchment level using a physically distributed hydrologic model, taking as input the precipitation for future scenarios under climate change from the GCMs (Asokan and Dutta, 2008). The month of September is projected to bring the highest rainfall in this river basin for the years 2000–2100 under a future climate change scenario and peak flows for this wet month for years 2025, 2050, 2075, and 2100 are compared to that of the base year (2000) to compute percentage increase in peak discharges in future.

An example of a physically based distributed hydrologic model is the Institute of Industrial Science Distributed Hydrologic Model (IISDHM), which consists of five major flow components of the hydrologic cycle: interception and evapotranspiration, unsaturated zone, saturated zone, overland surface flow, and river network flow (Asokan and Dutta, 2008). Water availability is estimated from this distributed hydrologic model using future daily precipitation output of the CCCma GCM (CGCM2) as input. The major spatial data sets required include watershed boundary, topography, land use, soil, aquifer layers, river network, and cross-sections, while the temporal data sets required for this model include rainfall data, evapotranspiration and soil parameter data, and upstream boundary water level and discharge.

In order to get the spatial distribution of rainfall over the basin, a Thiessen polygon based simple downscaling technique is used. Thiessen polygons are derived from the locations of the ground-based rainfall gauging stations and different weighting factors are assigned to different polygons based on a magnitude–distance–elevation method. Large-scale CGCM2 precipitation for each grid over the basin is multiplied with the weighting factor. Thus, a non-uniform distribution of raw GCM data for the grid of the DHM is obtained. The most commonly used approach of statistical downscaling with transfer functions that involves identification of climate predictors is not adopted The CGCM2 grids laid over the Thiessen polygons and the rain gauge stations in the Mahanadi River basin are shown in Figure 6.12.

The DHM is calibrated against the observed daily discharge for the year 1998 at two gauging stations, one for an upstream sub-catchment and the other for a downstream sub-catchment. Model

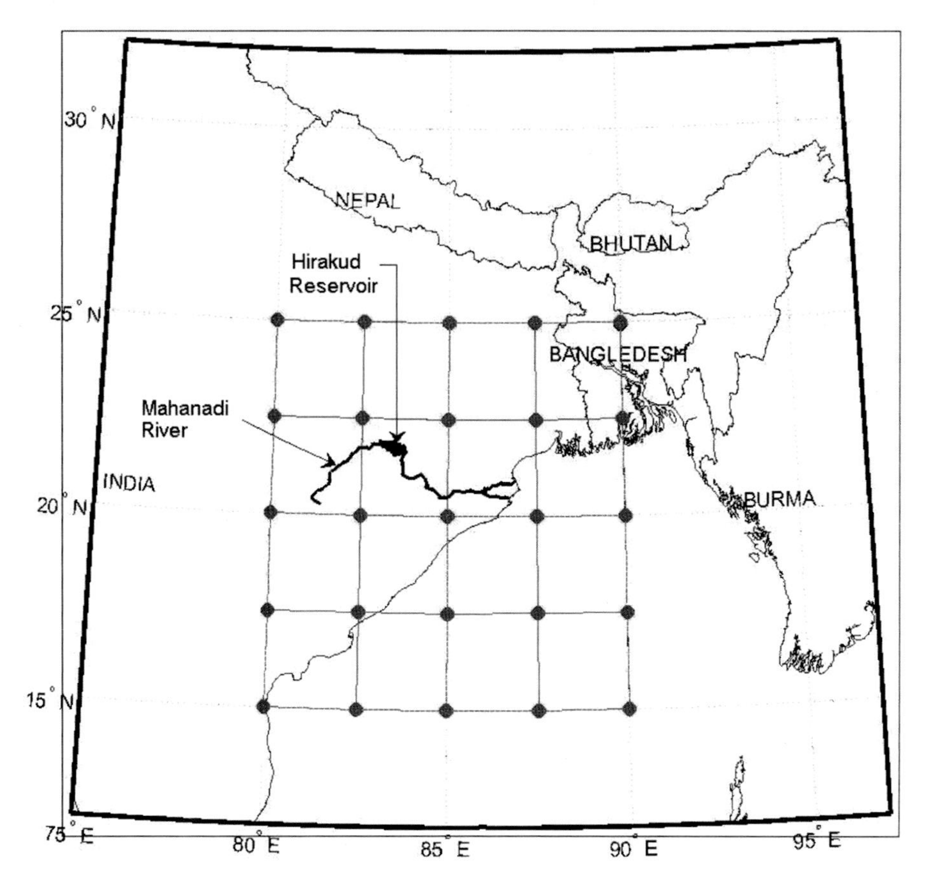

Figure 6.11 Mahanadi River basin with NCEP grids superposed (*Source*: Mujumdar and Ghosh, 2008).

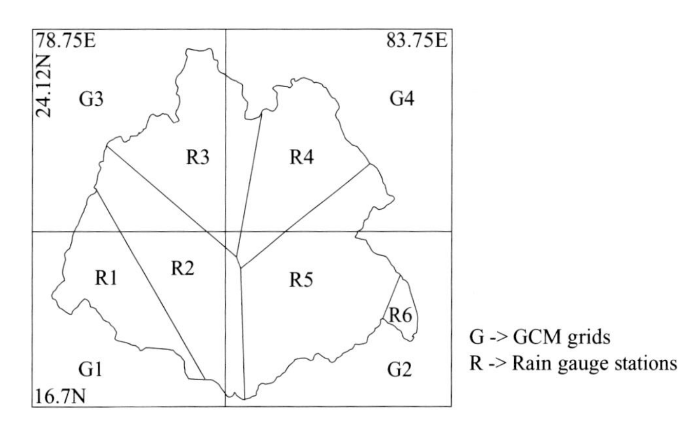

Figure 6.12 CGCM2 grids over the Mahanadi River basin with the Thiessen polygons (*Source*: Asokan and Dutta, 2008).

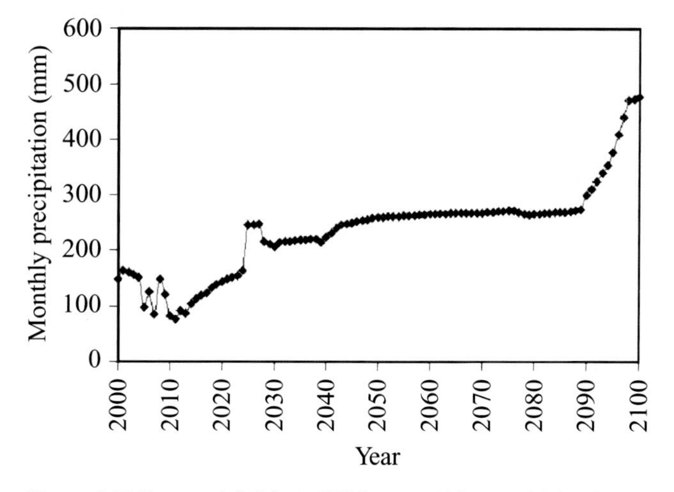

Figure 6.13 Future rainfall from GCM output (A2 scenario) for the month of September (*Source*: Asokan and Dutta, 2008).

validation is done with the corresponding data for the year 1996. The calibrated model is used for simulating daily flows for the year 2000 with the observed rainfall, and for the years 2025, 2050, 2075, and 2100 with the downscaled rainfall simulated with the A2 scenario from CGCM2. Figure 6.13 shows the future monthly precipitation for the monsoon month of September, as obtained from CGCM2. The increasing trend in monsoon precipitation illustrated in the figure hints at the possible intensification of floods during this month. With the simulated precipitation, future river discharges are obtained from the calibrated DHM, and future peak discharges under climate change are projected.

The percentage increases in future peak discharge in Mahanadi for the month of September as compared to the base year values are plotted in Figure 6.14. It is observed that the highest increase in peak runoff (38%) is likely to occur during September, within

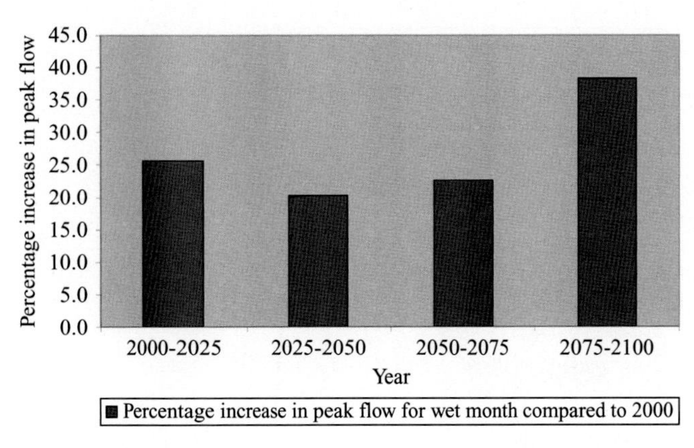

Figure 6.14 Percentage variation in peak monsoon discharge (*Source*: Raje and Mujumdar, 2010).

the period 2075–2100. This indicates future intensification of monsoon floods in the river basin under climate change conditions.

A climate change impact assessment study with respect to floods in the Mahanadi River basin was also carried out using a temporal water balance model, SWAT, driven by weather conditions predicted by the Hadley Center regional climate model HadRM2, with the IS92a scenario. (Gosain *et al.*, 2006). SWAT, as discussed in Section 6.3, is a distributed continuous model that simulates the hydrologic cycle components at daily time steps. In this combined HadRM2–SWAT approach, data on the terrain, land use, soil, and weather at desired locations of the basin are used to generate the hydrologic time series for the present/control (representing the period 1981–2000) and the future/greenhouse gases (GHG) (representing the period 2041–60) simulated weather data obtained from corresponding HadRM2 runs. The control runs are made with the emissions of 1990 held constant, and the GHG runs are made with the IS92a scenario. Land use is assumed to be unchanging for the future GHG climate scenario.

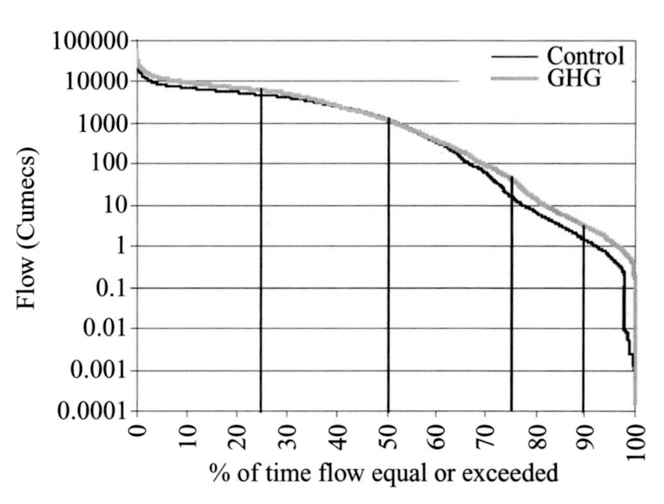

Figure 6.16 Flow duration curves for the control and GHG scenarios (*Source*: Gosain *et al.*, 2006).

Figure 6.15 shows the annual average precipitation, actual evapotranspiration, and water yield as simulated by this approach over the Mahanadi River basin. It can be observed that there is a likely increase in precipitation in the future under climate change conditions, and a corresponding increase in evapotranspiration and water yield. Figure 6.16 depicts the flow duration curves for the two scenarios. To study the impact of climate change on the dependability of the water yield, four arbitrarily selected levels of 25, 50, 75, and 90% are shown on the figure. It is found that the flow for all the dependability levels increases for the GHG scenario over the corresponding control flow magnitude, except for the 50% level of dependability, at which the flow marginally reduces. These conclusions are, however, with respect to simulations provided by one model with one scenario. As discussed in the next section, the use of different model–scenario combinations may lead to different projections – not only with respect to

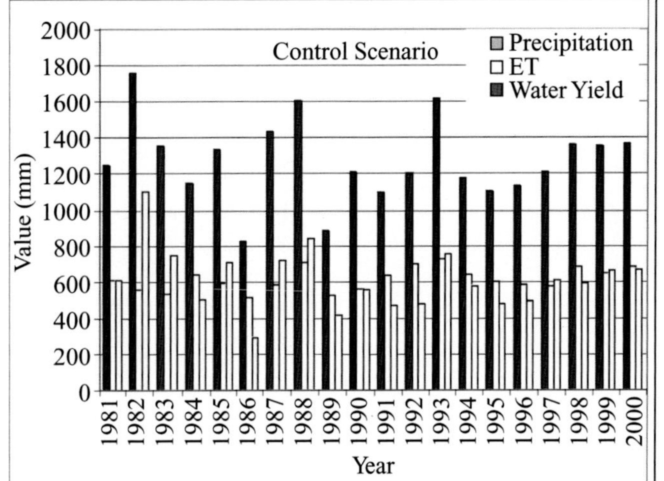

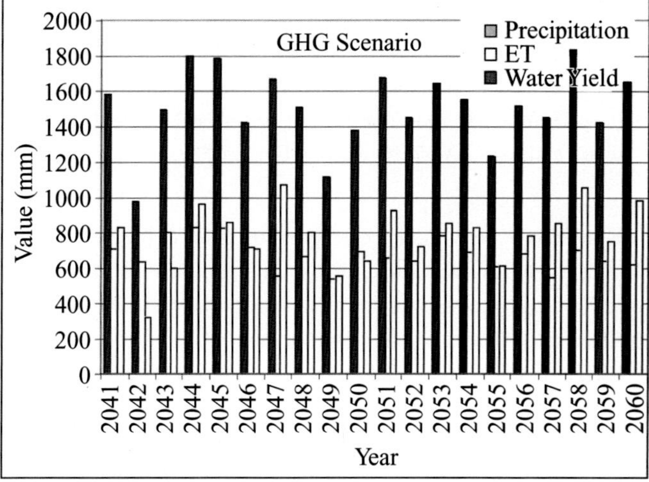

Figure 6.15 Annual components for control and GHG scenarios (*Source*: Gosain *et al.*, 2006).

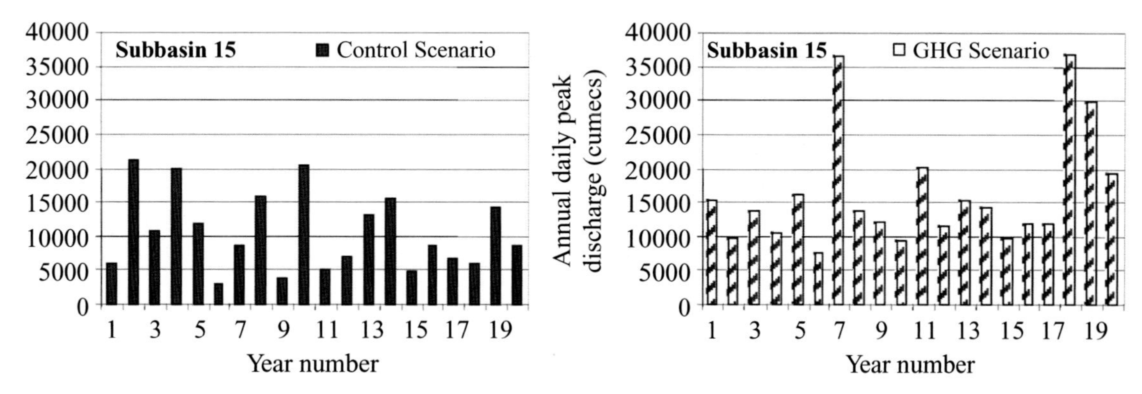

Figure 6.17 Annual maximum daily peak discharges for a sub-catchment (*Source*: Gosain *et al.*, 2006).

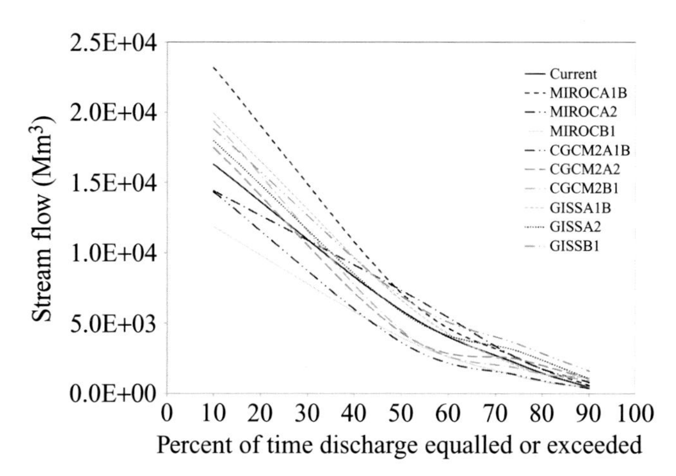

Figure 6.18 Range of projected future flow duration curves for monsoon inflows at Hirakud (2045–65).

the magnitude of the likely change but also with respect to the direction of the change itself.

A more detailed analysis of one sub-catchment (named Subcatchment 15) in the basin is presented in Figure 6.17. The annual maximum daily peak discharge is found to increase significantly as compared to the control scenarios and the extremely high flow value of 30,000 cumec is likely to be surpassed thrice in the future time range considered. These findings, emphasizing intensified flood risks in this already flood-prone river basin, have implications on the existing flood control infrastructure in the region.

Next, a more rigorous statistical downscaling method is used to obtain the flow duration curves, for the monsoon flows. Figure 6.18 shows the flow duration curves for monsoon streamflows in Mahanadi, projected using the relatively new conditional random field downscaling model for time slice 2045–65 for a range of GCM combinations with the IPCC Fourth Assessment Report scenarios. The GCMs used are CGCM2 (Meteorological Research Institute, Japan), MIROC3.2 medium resolution (Center for Climate System Research, Japan), and GISS model E20/Russell (NASA Goddard Institute for Space Studies, USA). It is seen that high flows increase in most scenarios for 2045–65, concurrent with the other studies. However, it is also worth noting that the number of scenarios showing an increase in high flows decreases by 2075–95.

6.2.3 Uncertainty modeling

As seen from Figure 6.18, different model–scenario combinations lead to different projections, implying significant uncertainty in the projections. The model and scenario uncertainties are addressed with a possibility theory (Mujumdar and Ghosh, 2008), to provide weighted probability distributions of the flows to the Hirakud Reservoir on the river.

The possibilistic approach for uncertainty modeling is useful in obtaining a weighted mean projection of future streamflows from the GCMs, which can directly be applied in water resources management systems. This approach is discussed here to quantify uncertainties for monsoon streamflow forecasting of the Mahanadi River, upstream from the Hirakud Dam (21.32° N, 83.4° E). The Hirakud Reservoir is a multipurpose project, created by constructing a dam across the river in Sambalpur district, Orissa state, India. There is no major control structure upstream of the Hirakud Reservoir and hence the inflow to the dam is considered as unregulated flow. Groundwater contribution is also known to be insignificant during the monsoon season.

In the possibilistic approach to model uncertainty in hydrologic impacts of climate change, the different GCMs and scenarios are assigned a possibility distribution, measured in terms of their ability to reflect hydrologic behavior under climate change based on their performance in the recent past (years 1991–2005) under climate forcing. The possibility theory is an uncertainty theory devoted to addressing partially inconsistent knowledge and linguistic information based on intuition. The intuition about the future hydrologic condition is derived based on the performance of GCMs with associated scenarios. This intuition is then used to obtain a possibility mass function, with possibility values assigned to the GCMs and scenarios. "Possibility assigned to a GCM" is interpreted here as the possibility with which the future

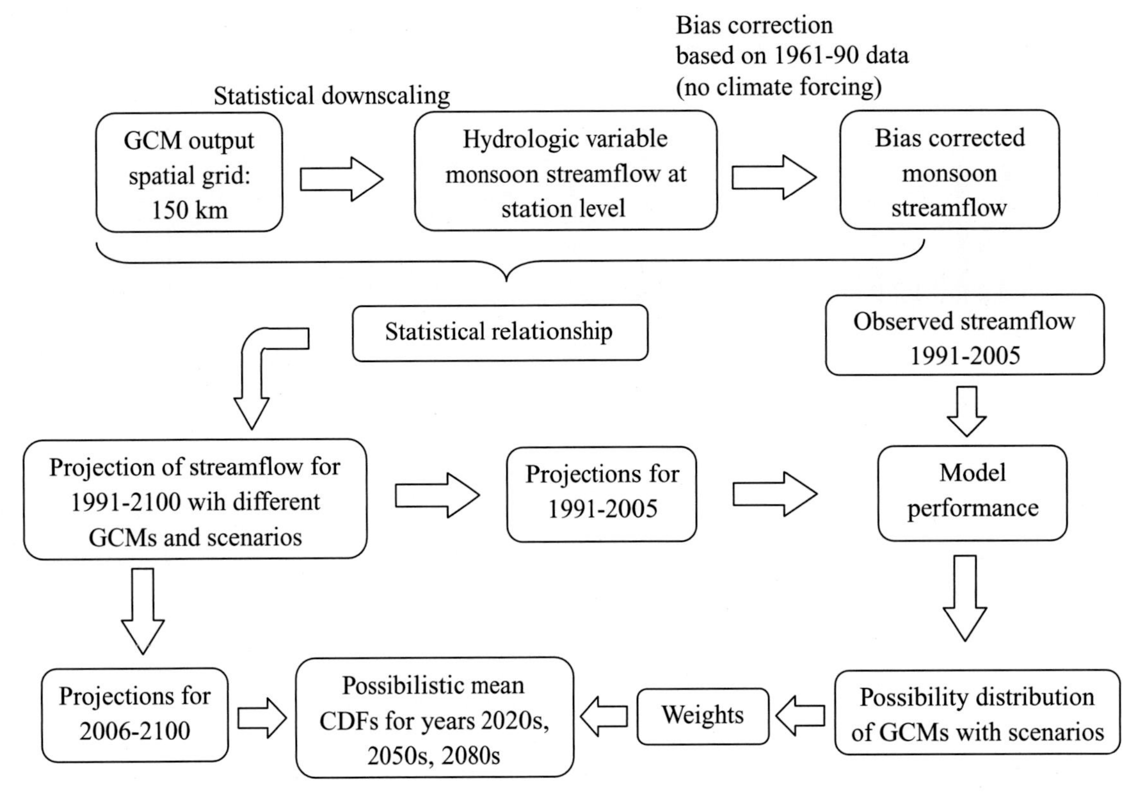

Figure 6.19 Overview of the possibilistic approach for modeling GCM and scenario uncertainty (*Source*: Mujumdar and Ghosh, 2008).

hydrologic variable of interest is best modeled by the downscaled output of the GCM. Similarly, "possibility assigned to a scenario" denotes the possibility with which the scenario best represents the climate forcing resulting in the change in the hydrologic variable.

The possibility values thus computed are used as weights in deriving a possibilistic mean CDF (weighted CDF) of future hydrologic variable for time slices 2020s (years 2006–35), 2050s (years 2036–65), and 2080s (years 2066–95). Figure 6.19 presents an overview of the possibility model. An advantage of the possibilistic approach over the more commonly used probabilistic approach is that it not only projects the hydrologic variable for the future considering GCM and scenario uncertainty, but also assigns possibilities to the GCMs and scenarios to determine how effectively the GCMs model climate change and which of the scenarios best represent the present situation under the threat of global-warming induced climate forcing.

Streamflows in the river are derived from GCM projections of large-scale climatological data by statistical downscaling and bias correction. The statistical downscaling model is based on principal component analysis, fuzzy clustering, and relevance vector machine. Surface air temperature at 2 m, mean sea level pressure, geopotential height at a pressure level of 500 hPa, and surface specific humidity are considered as the predictors for modeling Mahanadi streamflow in the monsoon season.

Three GCMs with two scenarios, A2 and B2, are considered: (i) CCSR/NIES coupled model developed by the Center for Climate System Research/National Institute for Environmental Studies, Japan; (ii) Hadley Climate Model 3 (HadCM3), developed by the Hadley Centre for Climate Prediction and Research, UK; and (iii) Coupled Global Climate Model 2 (CGCM2), developed by the Canadian Center for Climate Modeling and Analysis, Canada. This anaysis is done with the Third Assessment Report (TAR) scenarios of the IPCC. The simulation periods of the models CCSR/NIES, HadCM3, and CGCM2 are 1890–2100, 1950–2100 and 1900–2100, respectively (data publicly available at: http://www.mad.zmaw.de/IPCC_DDC/html/SRES_TAR/index.html).

The CDFs of the future streamflow derived with multiple GCMs and scenarios are presented in Figure 6.21. The figure shows that the uncertainty bandwidth for 1990–2005 for probabilities in the range 0.2–0.5 is high and becomes smaller at lower or higher probabilities. This points to higher disagreement between the simulations of GCMs for medium flow in the observed period. Figure 6.20 reveals significant dissimilarity among the projections of GCMs and scenarios. The possibility values computed for GCMs and scenarios are presented in Figure 6.21. The difference between the possibility values of two GCMs for a given scenario is found to be higher than that between the possibility values for two scenarios of a given GCM, which denotes that

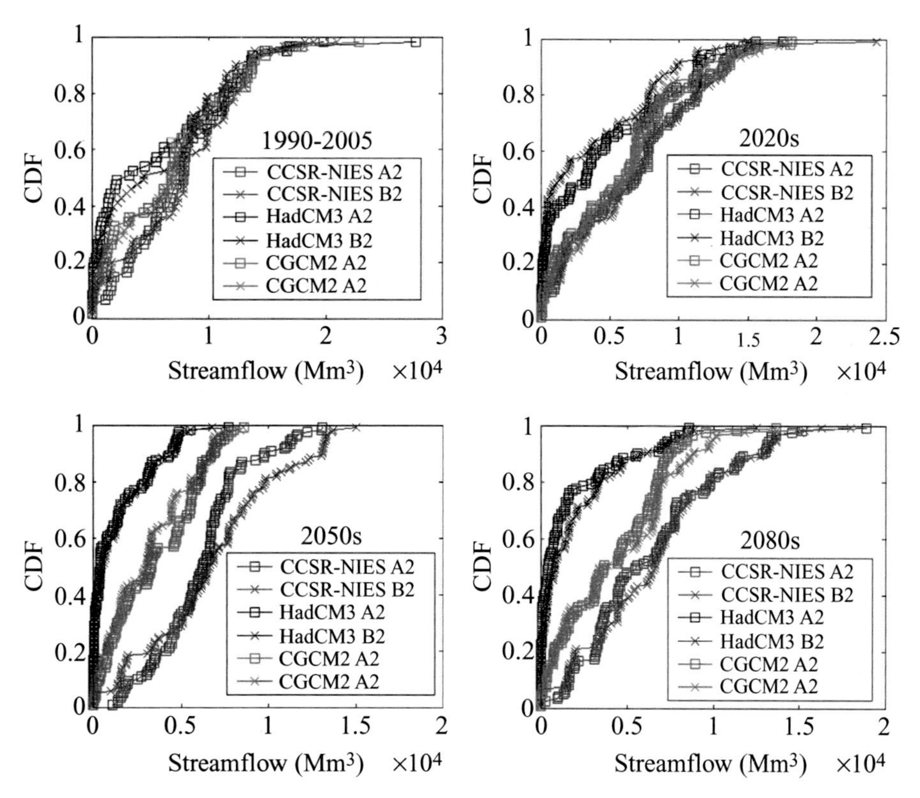

Figure 6.20 Predicted streamflow with different GCMs and scenarios (*Source*: Mujumdar and Ghosh, 2008). See also color plates.

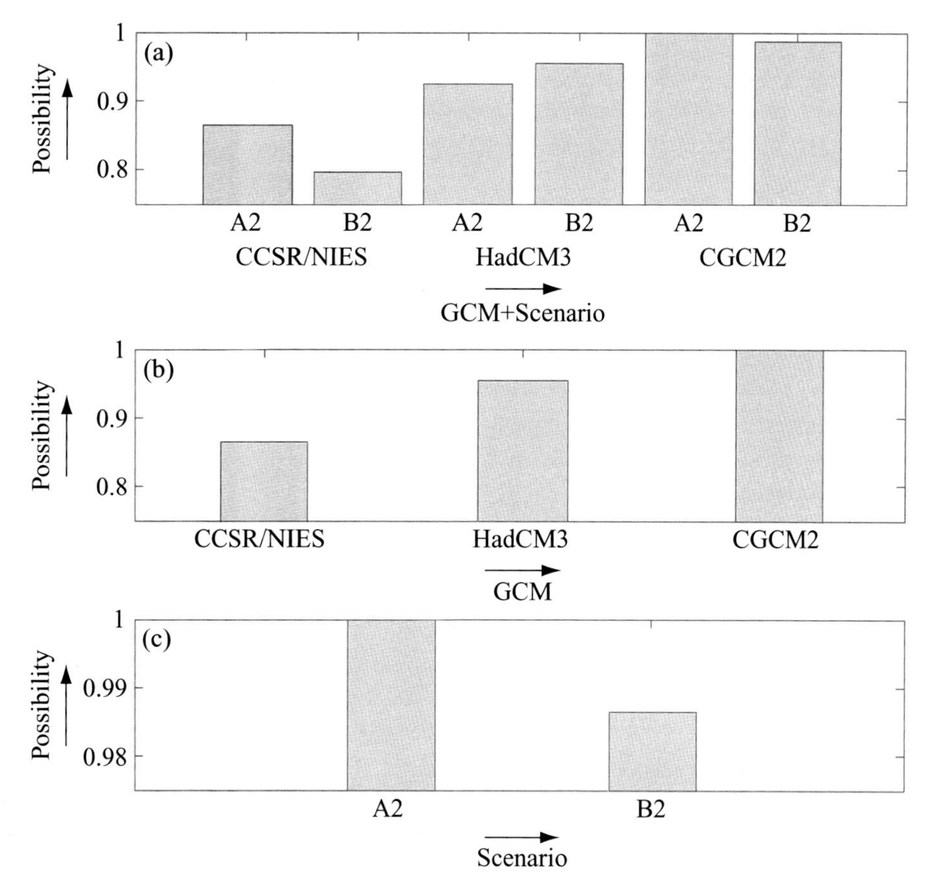

Figure 6.21 Possibility values assigned to GCMs and scenarios (*Source*: Mujumdar and Ghosh, 2008).

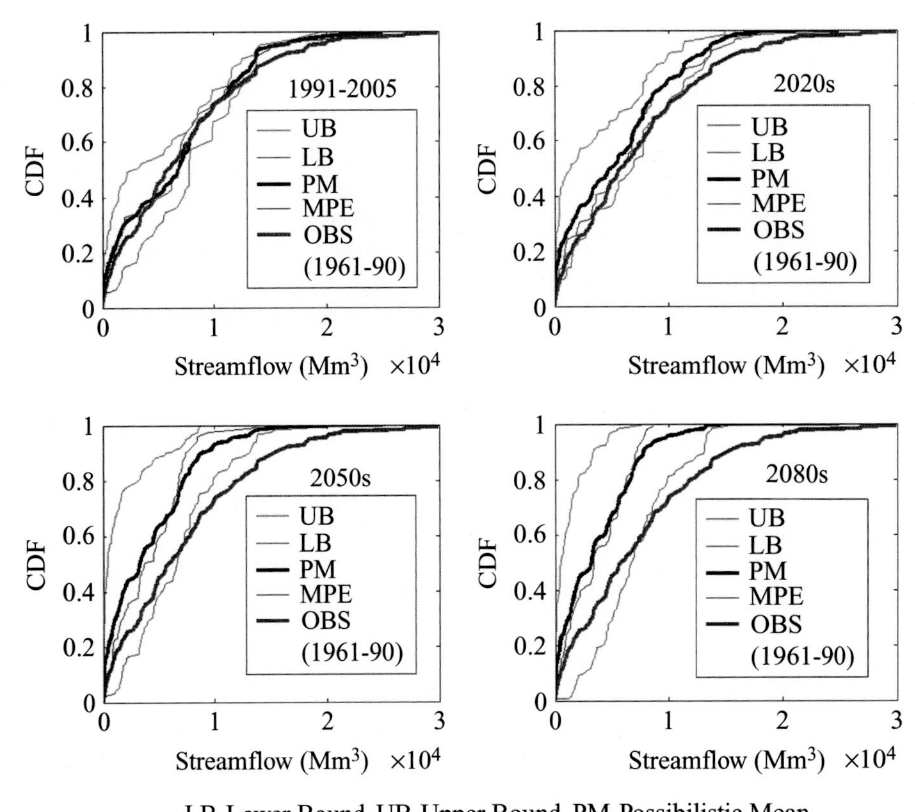

Figure 6.22 Upper bound (UB), lower bound (LB), and possibilistic mean CDF (*Source*: Mujumdar and Ghosh, 2008). See also color plates.

the uncertainty due to selection of GCM is greater than scenario uncertainty. A large difference is not observed between the possibilities for the two scenarios considered. This is because of the fact that the signal of climate forcing is not very pronounced in the initial time period (1991–2005) and therefore the results obtained by modeling climate forcing by GCMs are not significantly different from each other. A GCM/scenario with a possibility 1 does not imply that the particular GCM/scenario perfectly projects climate change, but in this case it points to an ignorance of the existence of any better GCMs or scenarios in modeling climate change impact on streamflow at the river-basin scale. Differences between possibility values assigned to GCMs are expected to grow with the passage of time, which emphasizes the importance of possibilistic models in the future.

Possibilistic mean CDF derived with the possibilities assigned to GCMs and scenarios, and the most possible CDF are presented in Figure 6.22 for 1961–90, 2020s, 2050s, and 2080s. For each of the discrete streamflow values at equal intervals, maximum and minimum CDF values are obtained from the CDFs generated using the projections with three GCMs and two scenarios. The results present a decreasing trend in the monsoon flows of the Mahanadi River at the Hirakud Dam.

It is interesting to note that, on the one hand, the projections indicate an increase in the flood peaks, while on the other hand, they indicate a decrease in the monsoon seasonal flows that determine primarily water availability for irrigation, hydropower, and municipal and industrial water supply. The rule curves for Hirakud Reservoir operation may need to be derived keeping these projections in view. An initial study towards such an exercise is presented in Raje and Mujumdar (2010c).

6.3 FUTURE PERSPECTIVES

The greatest challenge that climate change has posed to hydrologists is the non-stationarity induced by it. Most hydrologic models are built on the assumption of stationarity. The calibration and validation of models based on historical data implies that the future will be similar to the observed past – in a statistical sense. However, hydrologists are now confronted with the fact that the future will not be similar to the past. Additionally, relatively rapid changes are occurring in many river basins due to significant change in the land use patterns (e.g., Camorani *et al.*, 2005; Clarke, 2007). This is especially true in most developing countries, where rapid changes in land use are likely to introduce abrupt changes in hydrologic processes. Hydrologic models must therefore account for the non-stationarity introduced not only by climate change but also by rapid land use changes. Detection of abrupt or gradual changes in the hydrologic cycle component is of crucial importance, being the basis of planning for flood control (Kundzwicz, 2004). Global change has rendered the concept of stationarity in hydrology defunct, with changes in water cycle components resulting from a range of human activities, such as manmade reservoirs, land use changes, deforestation, and urbanization, that increase peak runoff rates and reduce the time-to-peak contributing to flooding. Milly *et al.*, (2008) have brought out the issues of non-stationarity in the context of water management. They state, "Stationarity – the idea that natural systems fluctuate within an unchanging envelope of variability – is a fundamental concept that permeates training and practice in water resources engineering," and further, "In view of the magnitude and ubiquity of the hydroclimatic change apparently now under way, however, we assert that stationarity is dead and should no longer serve as a central, default assumption in water resource risk assessment and planning." The latter remark is particularly important in assessing the flood risks. The short-duration high-intensity rainfall events causing floods are likely to be changing – and, in most cases, increasing – in many regions of the world, and it is essential to quantify such recent changes for use in flood frequency analysis, which conventionally uses extreme value distributions fitted with historical data with a tacit assumption of stationarity of extreme events. Cordery *et al.* (2004) discuss hydrologic models in a non-stationary environment. They infer that if some variables included in the model are non-stationary, then even with a small change in such variables there could be large effects on the model's ability to satisfactorily reproduce the reality. In conventional modeling for hydrologic impact assessment with the transfer function approach of downscaling discussed in Chapter 3, one or more predictor variables may be non-stationary, and it is critical to identify and include such non-stationarity in the impact models. An extra degree of uncertainty may need to be added where the assumption of stationarity in the variables is suspect.

Another challenge in modeling for floods under climate change is quantifying and constraining uncertainties in the projections. Unlike normal, smoothened processes, such as monthly or seasonal flows, peak flood discharges cannot readily be directly related to climate change through the conventional GCM–scenario-downscaling approach. The short time duration of flood events contributes to an additional source of uncertainty in the impact assessment. The GCMs' ability to simulate the climate at small time intervals of a day or less is extremely limited, and stochastic weather generator techniques (such as the k-nearest neighbor weather generator, discussed in Chapter 3) have been often used to overcome this limitation. Assessing the likely change in the IDF relationships, along with the associated uncertainties, due to climate change is an important research problem in this context. Hydrologic designs must then incorporate the modified rainfall intensities and return periods.

From a modeling point of view, integration of land surface–subsurface water interactions in the models to gain a clearer physical understanding is necessary. This calls for catchment/system-scale models of large rivers that link key components of the hydrologic cycle. Modeling of complex interactions of key hydrologic processes at catchment scales is still a challenging task, much more so in the context of floods. Hydrologic processes of precipitation, evaporation/evapotranspiration, infiltration, overland flow, flow through the vadoze zone, groundwater dynamics, and baseflow contribution to river flow need integration through river-basin simulation models with the capability to accept inputs from climate models that simulate future projections under prescribed emissions scenarios. It is also of interest to study the hydrometeorological feedbacks to the climate models, providing linkages between climate models, land surface models that include a surface hydrology component, soil moisture models, vadoze zone models and groundwater models. Such an exercise will be extremely data-intensive. Space technology products that are continually improving in their accuracy of estimations are of immense use in this context.

Most catchments, especially in the developing world, are poorly gauged, and this poses a great challenge for understanding floods in these regions. The IAHS decade on "Prediction in Ungauged Basins" has contributed immensely to the science of hydrology of floods. The ungauged basins provide an immense opportunity for the hydrologists to provide new understanding of the processes and to develop novel methodologies of predictions in ungauged basins, thus making hydrology at once exciting as both basic and applied science. In his commentary, Sivapalan (2003) writes, "Hydrology should be an exciting field of study; instead, it appears to be fragmented, deeply rooted in empiricism, and struggling to realize its full potential," and laments that this is essentially due to tensions between expectations of hydrology as a natural earth science and as an applied engineering science. It is important to develop hydrology as a basic science, while at the same time strengthening its professional engineering flavor to provide implementable solutions to the water-related problems that humanity faces. Questions posed by climate change to the hydrologists provide an immense opportunity to achieve such a balance.

6.4 SUMMARY

In this chapter, case studies of the Malaprabha Reservoir catchment and the Mahanadi River basin in India are presented, synthesizing the models and methodologies discussed earlier in the book. Methods of statistical downscaling techniques, identification of predictors, use of RS and GIS, and quantification of uncertainties are discussed through the case studies. In the case study of the Malaprabha Reservoir catchment presented, emphasis is on detailed discussion of statistical downscaling and use of a hydrologic model for streamflow projections. Possible increases in flood peak discharges due to climate change and quantification of uncertainties are dealt with in the case study of the Mahanadi River basin. Future perspectives on hydrologic modeling in the context of climate change are provided.

References

Abbott, M. B., Bathurst, J. C., Cunge, J. A., OConnell, P. E., and Rasmussen, J. (1986a). An introduction to the European Hydrologic System–Système Hydrologique Européen, SHE, 1: History and philosophy of a physically-based, distributed modeling system. *J. Hydrol.*, **87**, 45–59.

Abbott, M. B., Bathurst, J. C., Cunge, J. A., OConnell, P. E., and Rasmussen, J. (1986b). An introduction to the European Hydrologic System–Système Hydrologique Européen, SHE, 2: Structure of a physically-based, distributed modeling system. *J. Hydrol.*, **87**, 61–77.

Ackerman, C. T. (2002). *HEC-GeoRAS 3.1.* US Army Corps of Engineers, Institute for Water Resources. Hydrologic Engineering Centre, Davis, Calif.

Adler, R. F., and Negri, A. J. (1988). Satellite infrared technique to estimate tropical convective and stratiform rainfall. *J. Appl. Meteorol.*, **27**, 30–51.

Allison, E. W., Brown, R. J., Press, H., and Gairns, J. (1989). Monitoring drought affected vegetation with AVHRR. In *Quantitative Remote Sensing: An Economic Tool for the Nineties, Proceedings IGRASS'89*, IEEE, 1961–1965.

American Society of Photogrammetry (1975). *Manual of Remote Sensing*, Falls Church, Va.

Anandhi, A. (2007). Impact assessment of climate change on hydrometeorology of Indian river basin for IPCC SRES scenarios. PhD thesis, Indian Institute of Science, India.

Anandhi, A., Srinivas, V. V., Nanjundiah, R. S., and Kumar, D. N. (2008). Downscaling precipitation to river basin in India for IPCC SRES scenarios using support vector machine. *Int. J. Climatol.*, **28**, 401–420.

Anandhi, A., Srinivas, V. V., Kumar, D. N., and Nanjundiah, R. S. (2009). Role of predictors in downscaling surface temperature to river basin in India for IPCC SRES scenarios using support vector machine. *Int. J. Climatol.*, **29**, 583–603.

Anderson, M., and Kavvas, M. L. (2002). A global hydrology model. In *Mathematical Models of Watershed Hydrology*, V. P. Singh and D. K. Frevert, eds., Littleton, Colo.: Water Resources Publications.

Andrews, W. H., Riley, J. P., and Masteller, M. B. (1978). *Mathematical Modeling of a Sociological and Hydrological System.* ISSR Research Monograph, Utah Water Research Laboratory, Utah State University, Logan, Utah.

Arnell, N. W. (1992). Factors controlling the effects of climate change on river flow regimes in a humid temperate environment. *J. Hydrol.*, **132**, 321–342.

Arnell, N. W. (1999a). A simple water balance model for the simulation of streamflow over a large geographic domain. *J. Hydrol.*, **217**, 314–335.

Arnell, N. W. (1999b). Climate change and global water resources. *Global Environ. Change*, **9**, S31–S49.

Arnold, J. G., Srinivasan, R., Muttiah, R. S., and Williams, J. R. (1998). Large area hydrologic modeling and assessment. Part I: Model development. *J. Am. Water Resour. Assoc.*, **34**(1), 73–89.

Aron, G., and Lakatos, D. F., (1980). *Penn State Urban Runoff Model: Users Manual.* Institute for Research on Land and Water Resources, Pennsylvania State University, University Park, Pa.

Arora, V. K., and Boer, G. J. (2001). Effects of simulated climate change on the hydrology of major river basins. *J. Geophys. Res.*, **106**, 3335–3348.

ASCE (2000a). Task Committee on Application of Artificial Neural Networks in Hydrology. Artificial neural networks in hydrology, I: Preliminary concepts. *J. Hydrol. Eng.*, **5**, 115–123.

ASCE (2000b). Task Committee on Application of Artificial Neural Networks in Hydrology. Artificial neural networks in hydrology, II: Hydrologic applications. *J. Hydrol. Eng.*, **5**, 124–137.

Asokan, S. M., and Dutta D. (2008). Analysis of water resources in the Mahanadi River Basin, India under projected climate conditions. *Hydrol Process.*, **22**, 3589–3603.

Baguis, P., Boukhris, O., Ntegeka, V., *et al.* (2008). *Climate Change Impact on Hydrological Extremes Along Rivers and Urban Drainage Systems. I. Literature Review.* Belgian Science Policy: SSD Research Programme, Technical report CCI-HYDR project by K. U. Leuven, Hydraulics Section and Royal Meteorological Institute of Belgium, May 2008.

Baker, D. B., Richards, R. P., Loftus, T. T., and Kramer, J. W. (2004). A new flashiness index: characteristics and applications to Midwestern rivers and streams. *J. Am. Water Res. Assoc.*, **40**(2), 503–522.

Balakrishnan, P. (1987). *Issues in Water Resources Development and Management and the Role of Remote Sensing.* Report ISRO-NNRMS-TR-07–86, Indian Space Research Organization, Bangalore, India.

Bardossy, A., and Disse, M. (1993). Fuzzy rule-based models for infiltration. *Water Resour. Res.*, **29**(2): 373–382.

Bardossy, A., and Plate, E. J. (1992). Space-time model for daily rainfall using atmospheric circulation patterns. *Water Resour. Res.*, **28**(5), 1247–1259.

Bardossy, A., Duckstein, L., and Bogardi, I. (1995). Fuzzy rule-based classification of atmospheric circulation patterns. *Int. J. Climatol.*, **15**(10), 1087–1097.

Barret, E. C., and Martin, D. W. (1981). *The Use of Satellite Data in Rainfall Monitoring.* New York: Academic Press.

Bastiaanssen, W. G. M. (1998). *Remote Sensing in Water Resources Management: The State of the Art.* Colombo, Sri Lanka: International Water Management Institute, 118.

Bathurst, J. C., Wicks, J. M., and O'Connell, P. E. (1995). The SHE/SHESED basin scale water flow and sediment transport modeling system. In *Computer Models of Watershed Hydrology*, V. P. Singh, ed., Littleton, Colo.: Water Resources Publications, 563–594.

Beasley, D. B., Monke, E. J., and Huggins, L. F. (1977). *ANSWERS: A Model for Watershed Planning.* Purdue Agricultural Experimental Station Paper No. 7038, Purdue University, West Lafayette, Ind.

Becker, A., and Pfutzner, B. (1987). EGMO-system approach and subroutines for river basin modeling. *Acta Hydrophysica*, **31**, 125–141.

Bell, V. A., Kay, A. L., Jones, R. G., and Moore, R. J. (2007). Use of a grid-based hydrological model and regional climate model outputs to assess changing flood risk. *Int. J. Climatol.*, **27**, 1657–1671.

Benestad, R. E. (2001). A comparison between two empirical downscaling strategies. *Int. J. Climatol.*, **21**, 1645–1668.

Bergstrom, S. (1976). *Development and Application of a Conceptual Runoff Model for Scandinavian Countries.* SMHI Report No. 7, Norrkoping, Sweden.

Bergstrom, S. (1992). *The HBV Model: Its Structure and Applications.* SMHI Report No. 4, Norrkoping, Sweden.

Bergstrom, S. (1995). The HBV model. In *Computer Models of Watershed Hydrology*, V. P. Singh, ed., Littleton, Colo.: Water Resources Publications.

Beven, K. J. (1995). TOPMODEL. In *Computer Models of Watershed Hydrology*, V. P. Singh, ed., Littleton, Colo.: Water Resources Publications.

Beven, K. J., and Kirkby, M. J. (1976). *Toward a Simple Physically Based Variable Contributing Area of Catchment Hydrology.* Working Paper No. 154, School of Geography, University of Leeds, UK.

Beven, K. J., and Kirkby, M. J. (1979). A physically-based variable contributing area model of basin hydrology. *Hydrol. Sci. Bull.*, **24**(1), 43–69.

Beven, K. J., Calver, A., and Morris, E. (1987). *The Institute of Hydrology Distributed Model*. Institute of Hydrology Report No. 98, Wallingford, UK.

Bhanumurthy, V., Manjusree, P., Srinivasa Rao, G. (2010). Flood Disaster Management. In *Remote Sensing Applications*, P. S. Roy, R. S. Dwivedi, and D. Vijayan, eds., Hyderabad, India: National Remote Sensing Center.

Bhatt, C. M., Rao, G. S., Manjushree, P., and Bhanumurthy, V. (2010). Space based disaster management of 2008 Kosi floods, North Bihar, India. *J. Indian Soc. Remote Sensing*, **38**, 99–108.

Bicknell, B. R., Imhoff, J. L., Kittle, J. L., Donigian, A. S., and Johanson, R. C. (1993). *Hydrologic Simulation Program – Fortran: Users Manual for Release 10*. US EPA Environmental Research Laboratory, Athens, Ga.

Bishop, C. M. (1995). *Neural Networks for Pattern Recognition*. New York: Oxford University Press.

Blanchard, B. J. (1974). *Third ERTS 1 Symposium*, NASA SP **351**, 1-B, 1089–1098.

Bo, Z., Islam, S., and Eltahir, E. A. B. (1994). Aggregation-disaggregation properties of a stochastic rainfall model. *Water Resour. Res.*, **30**, 3423–3435.

Bobba, A. G., Bukata, R. P., and Jerome, J. H. (1992). Digitally processed satellite data as a tool in detecting potential groundwater flow systems. *J. Hydrol.* **131**, 25–62.

Booij, M. J. (2005). Impact of climate change on river flooding assessed with different spatial model resolutions. *J. Hydrol.*, **303**, 176–198.

Bonnici, A. M. (2006). Teaching and learning resources for web GIS and related technologies. Web GIS Development, www.webGISdev.com (accessed April 15, 2012).

Borah, D. K., and Bera, M. (2000). *Hydrologic Modeling of the Court Creek Watershed*. Contract Report No. 2000–04, Illinois State Water Survey, Champaign, Ill.

Borah, D. K., Bera, M., Shaw, S., and Keefer, L. (1999). *Dynamic Modeling and Monitoring of Water, Sediment, Nutrients and Pesticides in Agricultural Watersheds During Storm Events*. Contract Report No. 655, Illinois State Water Survey, Champaign, Ill.

Boughton, M. E. (1966). A mathematical model for relating runoff to rainfall with daily data. *Civ. Eng. Trans. Inst. Eng. Aust.*, **CE8**(1), 83–97.

Bouraoui, F., Braud, I., and Dillaha, T. A. (2002). ANSWERS: A non-point source pollution model for water, sediment and nutrient losses. In *Mathematical Models of Small Watershed Hydrology and Applications*, V. P. Singh and D. K. Frevert, eds., Littleton, Colo.: Water Resources Publications.

Boyd, M. J., Pilgrim, D. H., and Cordery, I. (1979). *A Watershed Bounded Network Model for Flood Estimation: Computer Programs and User Guide*. Water Research Laboratory Report No. 154, University of New South Wales, Sydney, Australia.

Boyd, M. J., Rigby, E. H., and van Drie, R. (1996). WBNM: a comprehensive flood model for natural and urban catchments. *Proceedings 7th International Conference on Urban Drainage*, Institution of Engineers, Sydney, Australia, 329–334.

Braun, J., and Sambridge, M. (1997). Modelling landscape evolution on geological time scales: A new method based on irregular spatial discretization. *Basin Res.*, **9**, 27–52.

Buchanan, M. D. (1979). Effective utilisation of colour in multidimensional data presentation. *Proc. Soc. Photo-Opt. Eng.*, **199**, 9–19.

Burnash, R. J. C. (1975). The NWS river forecast systemcatchment modeling. In *Computer Models of Watershed Hydrology*, V. P. Singh, ed., Littleton, Colo.: Water Resources Publications.

Burnash, R. J. C., Ferral, R. L., and McGuire, R. A. (1973). *A Generalized Streamflow Simulation System: Conceptual Modeling for Digital Computers*. Report, US Department of Commerce, National Weather Service, Silver Springs, Md., and State of California, Department of Water Resources, Sacramento, Calif.

Burrough, P. A. (1986). *Principles of Geographical Information Systems for Land Resources Assessment*. Oxford, UK: Oxford University Press.

Bussieres, N., Louie P. Y. T., and Hogg, W. (1989). *Implementation of an Algorithm to Estimate Regional Evapotranspiration Using Satellite Data*. Canadian Climate Centre, Downsview.

Calver, A., and Wood, W. L. (1995). The Institute of Hydrology distributed model. In *Computer Models of Watershed Hydrohydrology*, V. P. Singh, ed., Littleton, Colo.: Water Resources Publications.

Camorani, G., Castellarin, A., and Brath, A. (2005). Effects of land-use changes on the hydrologic response of reclamation system. *Phys. Chem. Earth*, **30**, 561–574.

Campbell, J.-B. (1996). *Introduction to Remote Sensing*. New York: The Guilford Press, 622.

Carlson, D. H., and Thurow, T. L. (1992). *SPUR-91: Workbook and User Guide*. Texas Agricultural Experimental Station MP-1743, College Station, Tex.

Carlson, D. H., Thurow, T. L., and Wight, J. R. (1995). SPUR-91: Simulation of production and utilization of rangelands. In *Computer Models of Watershed Hydrology*, V. P. Singh, ed., Littleton, Colo.: Water Resources Publications.

Carroll, T., and Baglio, J. (1989). Techniques for near real-time snow cover mapping using AVHRR satellite data. Poster paper at 57th Western Snow Conference, NOAA/NWS, Minneapolis, Minn.

Chandra, S., Sharma, K. P., and Kashyap, O. (1984). Watershed studies using simulation models for Upper Yamuna catchment. In *Application of Remote Sensing Methods to Hydrology*, University of Roorkee, India.

Chang, A. T. C., Foster, J. L., Hall, D. K., Rango, A., and Hartline, B. K. (1982). Snow water equivalent estimation by microwave radiometry. *Cold Regions Res Technol.*, **5**, 259–267.

Charles, S. P., Bates, B. C., Whetton, P. H., and Hughes, J. P. (1999), Validation of downscaling models for changed climate conditions: case study of southwestern Australia. *Climate Res.*, **12**, 1–14.

Charles, S. P., Bates, B. C., Smith, I. N., and Hughes, J. P. (2004). Statistical downscaling of daily precipitation from observed and modelled atmospheric fields. *Hydrol. Proc.*, **18**, 1373–1394.

Chaudhry, M. H. (1987). *Open-Channel Flow*. Englewood Cliffs, N.J.: Prentice-Hall, Inc.

Chaudhry M. H. (1993). *Open-Channel Flow*. Englewood Cliffs, N.J.: Prentice-Hall, Inc.

Chen, D. M., and Stow, D. (2002). The effect of training strategies on supervised classification at different spatial resolutions. *Photogramm. Eng. Remote Sensing*, **68**(11), 1155–1161.

Chen, S.-H., Lin, Y.-H., Chang, L.-C., and Chang, F.-J. (2006). The strategy of building a flood forecast model by neuro-fuzzy network. *Hydrol. Process.* **20**, 1525–1540.

Chiew, F. H. S., and McMahon, T. A. (1994). Application of the daily rainfall-runoff model MODHYDROLOG to twenty eight Australian catchments. *J. Hydrol.*, **153**, 383–416.

Chiew, F. H. S., Peel, M. C., and Western, A. W. (2002). Application and testing of the simple rainfall-runoff model SIMHYD. In *Mathematical Models of Small Watershed Hydrology and Applications*, V. P. Singh and D. K. Frevert, eds., Littleton, Colo.: Water Resources Publications.

Choi, J. Y., Engel, B. A., Pandey, S., and Harbor, J. (2001). Web-based decision support system for evaluation of hydrological impact of urban sprawl. *Proceedings ASAE Annual Meeting*, Paper No. 012026, American Society for Agricultural Engineers, St. Joseph, Mich.

Chow, V. T. (1959). *Open Channel Hydraulics*. New York: McGraw Hill.

Chow, V. T., Maidment, D. R., and Mays, L. W. (1988). *Applied Hydrology*. New York: McGraw Hill.

Christensen, N. S., Wood, A. W., Voisin, N., Lettenmaier, D. P., and Palmer, R. N. (2004). The effects of climate change on the hydrology and water resources of the Colorado river basin. *Climatic Change*, **62**, 337–363.

Chu, H.-J. (2009). The Muskingum flood routing model using a neuro-fuzzy approach. *KSCE J. Civil Eng.*, **13**(5), 371–376.

Claps, P., and Laio, F. (2003). Can continuous streamflow data support flood frequency analysis? An alternative to the partial duration series approach. *Water Resour. Res.*, **39**(8), 1216, doi:10.1029/2002WR001868.

Clarke, R. T. (2007). Hydrological prediction in a non-stationary world. *Hydrol. Earth Syst. Sci.*, **11**(1), 408–414.

Conway, D., and Jones, P. D. (1998). The use of weather types and air flow indices for GCM downscaling. *J. Hydrol.*, **212–213**, 348–361.

Conway, D., Wilby, R. L., and Jones, P. D. (1996). Precipitation and air flow indices over the British Isles. *Clim. Res.*, **7**(2), 169–183.

Cordery, I., Mehrotra, R., Sharma, A., and Nazemosadat, M. J. (2004). Hydrological models in a non stationary environment. In *Hydrology: Science & Practice for the 21st Century, Vol. I*, H. Wheater, and C. Kirby, eds., London: British Hydrological Society, 103–107.

Crane, R. G., and Hewitson, B. C. (1998). Doubled CO2 precipitation changes for the Susquehanna basin: Down-scaling from the genesis general circulation model. *Int. J. Climatol.*, **18**(1), 65–76.

Crawford, N. H., and Linsley, R. K. (1966). *Digital Simulation in Hydrology: Stanford Watershed Model IV*. Technical Report No. 39, Stanford University, Palo Alto, Calif.

Croley, T. E. (1982). *Great Lake Basins Runoff Modeling*. NOAA Technical Memo No. EER GLERL-39, National Technical Information Service, Springfield, Va.

Croley, T. E. (1983). *Lake Ontario Basin Runoff Modeling*. NOAA Technical Memo No. ERL GLERL-43, Great Lakes Environmental Research Laboratory, Ann Arbor, Mich.
Cunge, J. A., Holly, F. M. Jr., and Verwey, A. (1980). *Practical Aspects of Computational River Hydraulics*. London: Pitman.
Dankers, R., Christensen, O. B., Feyen, L., Kalas, M., and de Roo, A. (2007). Evaluation of very high-resolution climate model data for simulating flood hazards in the Upper Danube Basin. *J. Hydrol.*, **347**(3–4), 319–331.
Dawdy, D. R., and O'Donnell, T. (1965). Mathematical models of catchment behavior. *J. Hydraul. Div., Am. Soc. Civ. Eng.*, **91**(HY4), 123–127.
Dawdy, D. R., Litchy, R. W., and Bergmann, J. M. (1970). *Rainfall–Runoff Simulation Model for Estimation of Flood Peaks for Small Drainage Basins*. USGS Open File Report, Washington, D.C.
Dawdy, D. R., Schaake, J. C., and Alley, W. M. (1978). *Users Guide for Distributed Routing Rainfall-Runoff Model*. USGS Water Resources Investigation Report No. 78–90, Gulf Coast Hydroscience Center, NSTL, Miss.
Deepashree Raje (2009). Hydrologic impacts of climate change: quantification of uncertainties. PhD thesis, Indian Institute of Science, India.
Dempster, A. P. (1967). Upper and lower probabilities induced by a multivalued mapping. *Ann. Statist.*, **28**, 325–339.
Dessai, S. X. R. (2005). Robust adaptation decisions amid climate change uncertainties. PhD thesis, Tyndall Centre, University of East Anglia, UK, 283.
Di Luzio, M., Srinivasan, R., Arnold, J. G., and Neitsch, S. L. (2002). *Soil and Water Assessment Tool. ArcView GIS Interface Manual: Version 2000*. GSWRL Report 02–03, BRC Report 02–07, Texas Water Resources Institute TR-193, College Station, Tex., 346.
Dibike, Y. B., and Coulibaly, P. (2005). Hydrologic impact of climate change in the Saguenay watershed: Comparison of downscaling methods and hydrologic models. *J. Hydrol.*, **307**, 145–163.
Doan, J. H. (2000). *Geospatial Hydrologic Modelling Extension HEC-GeoHMS*. US Army Corps of Engineers, Institute for Water Resources, Hydrologic Engineering Center, Davis, Calif.
Döll, P., Kaspar, F., and Lehner, B. (2003). A global hydrological model for deriving water availability indicators: model tuning and validation. *J. Hydrol.*, **270**, 105–134.
Donald, J. R., Seglenieks, F. R., Soulis, E. D., Kouwen, N., and Mullins, D. W. (1992). Mapping of partial snow cover during melt season using c-band SAR imagery. *Proceedings of 15th Canadian Symposium on Remote Sensing, Toronto, Canada*,170–175.
Donald, J. R., Soulis, E. D., Thompson, N., and Malla, S. B. (1990). Using GOES visible data to extend snow course in southern Ontario. In *Application of Remote Sensing in Hydrology*, G. W. Kite and A. Wankiewicz, eds., Proceedings Symposium No. 5, NHRI, Saskatoon, Canada, 69–78.
Donigian, A. S., Beyerlein, D. C., Davis, H. H., and Crawford, N. H. (1977). *Agricultural Runoff Management (ARM) Model Version II: Refinement and Testing*. Report No. EPA-600/3–77-098, US EPA Environmental Research Laboratory, Athens, Ga.
Doty, B., and Kinter, J. I. (1993). The Grid Analysis and Display System (GrADS): a desktop tool for earth science visualization. *American Geophysical Union 1993 Fall Meeting*, San Fransico, Calif., 6–10 December.
Dousset, B. (1989). AVHRR derived cloudiness and surface temperature patterns over the Los Angeles area and their relationship to land use. In *Quantitative Remote Sensing: An Economic Tool for the Nineties, Proceedings IGRASS'89*, IEEE, 2132–2137.
Dugdale, G., Hardy, S., and Milford, J. R. (1991). Daily catchment rainfall estimated from Meteosat. *Hydrol. Process.*, **5**, 261–270.
Dunn, S. M. (1998). Large scale hydrological modelling using small scale processes. In *Hydrology in a Changing Environment, Vol. 1*, C. Kirby and H. S. Wheater, eds., Chichester, UK: Wiley & Sons.
Ehlers, M., Greenlee, D., Smith, T., and Star, J. (1991). Integration of remote sensing and GIS: Data and data access. *Photogramm. Eng. Remote Sensing*, **57**(6), 669–675.
English, P., Richardson, P., Glover, M., Cresswell, H., and Gallant, J. (2004). *Interpreting Airborne Geophysics as an Adjunct to Hydrogeological Investigations for Salinity Management: Honeysuckle Creek Catchment, Victoria*. 18/04, CSIRO Land and Water, Australia.
ESA (2002). *SMOS Mission Objectives and Scientific Requirements*. http://esamultimedia.esa.int/docs/SMOS_MRD_V5.pdf.
ESA (2004). *Soil Moisture Retrieval by a Future Space-Borne Earth Observation Mission*. Report 14662/00/NL, University of Reading, Reading, UK.
Everett, D. (2001). Global precipitation measurement, satellites, orbits, and coverage. In *Proceedings of IEEE Geoscience and Remote Sensing Symposium, IGARSS '02*.
Ewen, J., Parkin, G., and O'Connell, P. E. (2000). SHETRAN: Distributed river basin flow and transport modeling system. *J. Hydrol. Eng.*, **5**(3), 250–258.
Fairchild, J., and Leymarie, P. (1991). Drainage networks from grid digital elevation models. *Water Resour. Res.*, **3**, 709–717.
Feldman, A. D. (1981). HEC models for water resources system simulation: Theory and experience. *Adv. Hydrosci.*, **12**, 297–423.
Follansbee, W. A. (1973). *Estimation of Average Daily Rainfall from Satellite Cloud Photographs*. NOAA Technical Memo NESS 44, NOAA, Washington, D.C..
Fortin, J. P., Turcotte, R., Massicotte, S., *et al.* (2001a). A distributed watershed model compatible with remote sensing and GIS data. I: Description of model. *J. Hydrol. Eng.*, **6**(2), 91–99.
Fortin, J. P., Turcotte, R., Massicotte, S., *et al.* (2001b). A distributed watershed model compatible with remote sensing and GIS data. II: Application to Chaudiere watershed. *J. Hydrol. Eng.*, **6**(2), 100–108.
Fowler, H. J., Blenkinsop, S., and Tebaldi, C. (2007). Linking climate change modelling to impacts studies: Recent advances in downscaling techniques for hydrological modelling. *Int. J. Climatol.*, **27**(12), 1547–1578.
Freeman, J. A., and Skapura, D. M. (1991). *Neural Networks: Algorithms, Applications and Programming Techniques*. Reading, Mass.: Addison-Wesley Publishing Co.
Frei, C., Schär, C., Lüthi, D., and Davies, H. C. (1998). Heavy precipitation processes in a warmer climate. *Geophys. Res. Lett.*, **25**(9), 1431–1434.
Frere, M. H., Onstad, C. A., and Holtan, H. N. (1975). *ACTMO, an Agricultural Chemical Transport Model*. Report No. ARS-H-3, USDA, Washington, D.C.
Gangopadhyay, S., Clark, M., and Rajagopalan, B. (2005). Statistical downscaling using K-nearest neighbors. *Water Resour. Res.*, **41**, W02024, doi:10.1029/2004WR003444.
Gao, H., Tang, Q., Shi, X., *et al.* (2010). Water budget record from variable infiltration capacity (VIC) model. In *Algorithm Theoretical Basis Document for Terrestrial Water Cycle Data Records* (in review). (Available at http://www.hydro.washington.edu/Lettenmaier/Models/VIC/Documentation/References.shtml, accessed 17 December 2011.)
Gelfand, A. E., Zhu, L., and Carlin, B. P. (2000). On the change of support problem for spatio-temporal data. *Biostatistics*, **2**(1), 31–45, doi: 10.1093/biostatistics/2.1.31.
Georgakakos, K. P., Sperfslage, J. A., Tsintikidis, D., *et al.* (1999). *Design and Tests of an integrated Hydrometeorological Forecast System for Operational Estimation and Prediction of Rainfall and Streamflow in the Mountainous Panama Canal Watershed*. HRC Technical Report No. 2, Hydrologic Research Center, San Diego, Calif.
Ghosh, S., and Mujumdar, P. P. (2006). Future rainfall scenario over Orissa with GCM projections by statistical downscaling. *Curr. Sci.*, **90**(3), 396–404.
Ghosh, S., and Mujumdar, P. P. (2007). Nonparametric methods for modeling GCM and scenario uncertainty in drought assessment. *Water Resour. Res.*, **43**(7), W07405, doi:10.1029/2006WR005351.
Ghosh, S., and Mujumdar, P. P. (2008). Statistical downscaling of GCM simulations to streamflow using relevance vector machine. *Adv. Water Resour.*, **31**, 132–146.
Ghosh, S., and Mujumdar, P. P. (2009). Climate change impact assessment: Uncertainty modeling with imprecise probability. *J. Geophys. Res.*, **114**, D18113, doi:10.1029/2008JD011648.
Gibson, P. J., and Power, C.H. (2000). *Introductory Remote Sensing: Digital Image Processing and Applications*. London: Routledge.
Giorgi, F., and Mearns, L. O. (2003). Probability of regional climate change based on the Reliability Ensemble Averaging (REA) method. *Geophys. Res. Lett.*, **30**(12).
Goel, N. K., Kurothe, R. S., Mathur, B. S., and Vogel, R. M. (2000). A derived flood frequency distribution for correlated rainfall intensity and duration. *J. Hydrol.*, **228**, 56–67.
Goodess, C., and Palutikof, J. (1998). Development of daily rainfall scenarios for southeast Spain using a circulation-type approach to downscaling. *Int. J. Climatol.*, **18**(10), 1051–1083.
Gopakumar, R., and Mujumdar, P. P. (2009). A fuzzy dynamic flood routing model for natural channels. *Hydrol. Process.*, **23**, 1753–1767.
Gosain, A. K., Rao S., and Basuray, D. (2006). Climate change impact assessment on hydrology of Indian river basins. *Curr. Sci.*, **90**(3), 346–354.

Grayson, R. B., Bloschl, G., and Moore, I. D. (1995). Distributed parameter hydrologic modeling using vector elevation data: THALES and TAPEC-C. In *Computer Models of Watershed Hydrology*, V. P. Singh, ed., Littleton Colo.: Water Resources Publications.

Griensven, A.V. (2005). AVSWAT-X SWAT-2005 Advanced Workshop. In *SWAT 2005 3rd International Conference*, Zurich, Switzerland.

Guo, S., and Wang, G. (1994). *Water balance in the semi-arid regions: Yellow River*, No. 12 (in Chinese).

Guthrie, J., Dartiguenave, C., and Ries, K. (2009). Web services in the US Geological Survey StreamStats Web Application. *International Conference on Advanced Geographic Information Systems and Web Services*, 60–63.

Haferman, J. L., Krajewski, W. F., Smith, T. F., and Sanchez, A. (1994). Three dimensional aspects of radiative transfer in remote sensing of precipitation: Application to the 1986-COHMEX storm. *J. Appl. Meteorol.*, **33**, 1609–1622.

Hall, D. K. (1996). Remote sensing applications to hydrology: Imaging radar. *J. Hydrol. Sci.*, **41**(4), 609–624.

Hallada, W. A. (1984). Mapping bathymetry with Landsat 4 thematic mapper, preliminary findings. In *Proceedings of 9th Canadian Symposium on Remote Sensing, Ottawa, Canada*, 277–285.

Hawk, K. L., and Eagleson, P. (1992). *Climatology of Station Storm Rainfall in the Continental U.S.: Parameters of the Bartlett-Lewis and Poisson Rectangular Pulses Models*. Technical Report 336, Massachusetts Institute of Technology, Department of Civil Engineering, Cambridge, Mass.

Hay, L. E., McCabe, G. J., Wolock, D. M., and Ayers, M. A. (1991). Simulation of precipitation by weather type analysis. *Water Resour. Res.*, **27**, 493–501.

Haykin, S. (1994). *Neural Computing: A Comprehensive Foundation*. New York: Macmillan.

HEC (Hydrologic Engineering Center) (1981). *HEC-1 Flood Hydrograph Package: Users Manual*. US Army Corps of Engineers, Davis, Calif.

HEC (Hydrologic Engineering Center) (2000). *Hydrologic Modeling System HEC-HMS: Users Manual, Version 2*. US Army Corps of Engineers, Davis, Calif.

Henderson, F. M. (1966). *Open Channel Flow*. New York: Macmillan.

Hingray B., and Haha, M. B. (2005). Statistical performances of various deterministic and stochastic models for rainfall series disaggregation. *Atmos. Res.*, **77**, 152–175.

Hjelmfelt, A. T. (1986). Estimating peak runoff from field-size watersheds. *Bull. Am. Water Res. Assoc.*, **22**(2), 267–274.

Holtan, H. N., and Lopez, N. C. (1971). *USDAHL-70 Model of Watershed Hydrology*. USDA-ARS Technical Bulletin No. 1435, Agricultural Research Station, Beltsville, Md.

Holtan, H. N., Stilner, G. J., Henson, W. H., and Lopez, N. C. (1974). *USDAHL-74 Model of Watershed Hydrology*. USDA-ARS Plant Physiology Research Report No. 4, Agricultural Research Station, Beltsville, Md.

Horton, R. E. (1939). Analysis of runoff-plot experiment with varying infiltration capacity. *Trans. Am. Geophys. Union*, **20**, 693–711.

Hosking, J. R. M., and Wallis, J. R. (1987). Parameter and quantile estimation for the generalized Pareto distribution. *Technometrics*, **29**(3), 339–349. http://www.webgisdev.com/webpro/web1/wpro1_files/frame.htm

Huber, W. C. (1995). EPA storm water management model SWMM. In *Computer Models of Watershed Hydrology*, V. P. Singh, ed., Littleton, Colo.: Water Resources Publications.

Huber, W. C., and Dickinson, R. E. (1988). *Storm Water Management Model: Users Manual, Version 4*. Report No. EPA/600/3-88/001a, US EPA, Environmental Research Laboratory, Athens, Ga.

Huete, A. R., Liu, H. Q., Batchily, K., and van Leeuwen, W. (1997). A comparison of vegetation indices over a global set of TM images for EOS-MODIS. *Remote Sensing Environ.*, **59**, 440–451.

Huggins, L. F., and Monke, E. J. (1970). *Mathematical Simulation of Hydrologic Events of Ungaged Watersheds*. Technical Report No. 14, Water Resources Research Center, Purdue University, West Lafayette, Ind.

Hughes, J. P., and Guttorp, P. (1994). A class of stochastic-models for relating synoptic atmospheric patterns to regional hydrologic phenomena. *Water Resour. Res.*, **30**(5), 1535–1546.

Hughes, J. P., Guttorp, P., and Charles, S. P. (1999). A non-homogeneous hidden Markov model for precipitation occurrence. *J. R. Statist. Soc. C*, **48**, 15–30.

IPCC (2007). *Climate Change 2007:The Physical Science Basis*, Contribution of Working Group I to the Fourth Assessment Report of the Intergovernmental Panel on Climate Change, S. Solomon *et al.*, eds., Cambridge University Press, Cambridge, UK.

ISRO (Indian Space Research Organization) (2009). Megha-Tropiques Announcement of Opportunity, http://www.isro.org/news/pdf/MEGHA-TROPIQUES.pdf (accessed on 26/12/2011).

Javanmard, S., Yatagai, A., Nodzu, M. I., BodaghJamali, J., and Kawamoto, H. (2010). Comparing high-resolution gridded precipitation data with satellite rainfall estimates of TRMM 3B42 over Iran. *Adv. Geosci.*, **25**, 119–125.

Jeniffer, K., Su, Z., Woldai, T., and Maathuis, B. (2010). Estimation of spatial–temporal rainfall distribution using remote sensing techniques: A case study of Makanya catchment, Tanzania. *Int. J. Appl. Earth Obs. Geoinform.*, **12S**, S90–S99.

Jenson, S. K., and Domingue, J. O. (1988). Extracting topographic structure from digital elevation data for geographic information system analysis. *Photogramm. Eng. Remote Sensing.*, **54**(11), 1593–1600.

Jetten, V., Govers, G., and Hessel, R. (2003). Erosion models: Quality of spatial predictions. *Hydrol. Process.*, **17**, 887–900.

Jiang, T., Chen, D. Y., Xu, Y. C., *et al.* (2007). Comparison of hydrological impacts of climate change simulated by six hydrological models in the Dongjiang Basin, South China. *J. Hydrol.*, **336**, 316–333.

Johnson, F., and Sharma, A. (2009). Measurement of GCM skill in predicting variables relevant for hydroclimatological assessments. *J. Climate*, **22**, 4373–4382, doi: 10.1175/2009JCLI2681.1.

Jones, N. L., Wright, S. G., and Maidment, D. R. (1990). Watershed delineation with triangle-based terrain models. *J. Hydraul. Eng.*, **16**(10), 1232– 1251.

Jones, P. D., Hulme, M., and Briffa, K. R. (1993). A comparison of Lamb circulation types with an objective classification scheme. *Int. J. Climatol.*, **13**(6), 655–63.

Julien, P. Y., and Saghafian, B. (1991). *CASC2D Users Manual*. Department of Civil Engineering Report, Colorado State University, Fort Collins, Colo.

Kalnay, E., Kanamitsu, M., Kistler, R., *et al.* (1996). The NCEP/NCAR 40-year reanalysis project. *Bull. Am. Meteorol. Soc.*, **77**, 437–471.

Kalyanaraman, S. (1999). Remote sensing data from IRS Satellites: Past, present and future. *J. Indian Soc. Remote Sensing*, **27**(2), 59–70.

Kaur, R., and Rabindranathan, S. (1999) Ground validation of an algorithm for estimating surface suspended sediment concentrations from multi-spectral reflectance data. *J. Indian Soc. Remote Sensing*, **27**(4), 235–251.

Kavvas, M. L., *et al.* (1998). A regional scale land surface parameterization based on areally-averaged hydrological conservation equation. *Hydrol. Sci. J.*, **43**(4), 611–631.

Kay, A. L., Davies, H. N., Bell, V. A., and Jones, R. G. (2009). Comparison of uncertainty sources for climate change impacts: Flood frequency in England. *Climatic Change*, **92**(1–2), 41–63.

Khatibi, R., Ghorbani, M. A., Kashani, M. H., and Kisi, O. (2011). Comparison of three artificial intelligence techniques for discharge routing. *J. Hydrol.*, **403**, 201–212.

Kilsby, C. G., Jones, P. D., Burton, A., *et al.* (2007). A daily weather generator for use in climate change studies. *Environ. Modell. Software*, **22**(12), 1705–1719.

Kirpich, Z. P. (1940). Time of concentration of small agricultural watersheds. *Civil Eng.*, **10**(6), 362.

Kite, G. W. (1989). Using NOAA data for hydrological modelling. In *Quantitative Remote Sensing: An Economic Tool for the Nineties, Proceedings IGRASS'89*, IEEE. 553–558.

Kite, G. W. (1995). The SLURP model. In *Computer Models of Watershed Hydrology*, V. P. Singh, ed., Littleton, Colo.: Water Resources Publications.

Kite, G. W., and Kouwen, N. (1992). Watershed modelling using land classification *Water Resour. Res.*, **28**(12), 3193–3200.

Kite, G. W., and Pietroniro, A. (1996). Remote sensing applications in hydrological modelling. *J. Hydrol. Sci.*, **41**(4), 563–592.

Kite, G. W., Dalton, A., and Dion, K. (1994). Simulation of streamflow in a macroscale watershed using general circulation model data. *Water Resour. Res.*, **30**, 1547–1599.

Kleinen, T., and Petschel-Held, G. (2007). Integrated assessment of changes in flooding probabilities due to climate change. *Climatic Change*, **81**, 283–312, doi:10.1007/s10584-006-9159-6.

Knisel, W. G., and Williams, J. R. (1995). Hydrology components of CREAMS and GLEAMS models. *Computer Models of Watershed Hydrology*, V. P. Singh, ed., Littleton, Colo.: Water Resources Publications.

Knisel, W. G., Leonard, R. A., Davis, F. M., and Nicks, A. D. (1993). *GLEAMS Version 2.10, Part III, Users Manual*. Conservation Research Report, USDA, Washington, D.C.

Kohavi, R., and Provost, F. (1998). Glossary of terms. *Machine Learning*, **30**(2/3), 271–274.

Kokkonen, T., Koivusalo, H., Karvonen, T., and Lepisto, A. (1999). A semidistributed approach to rainfall-runoff modeling-aggregating responses from hydrologically similar areas. In *MODSIM99*, L. Oxley and F. Scrimgeour, eds., The Modelling and Simulation Society of Australia and New Zealand, Hamilton, New Zealand, 75–80.

Kosko, B. (1996). *Neural Networks and Fuzzy Systems*. Prentice-Hall of India (original edition: Prentice-Hall Inc., Englewood Cliffs, N.J., 1992).

Kouwen, N. (2000). *WATFLOOD/SPL: Hydrological Model and Flood Forecasting System*. Department of Civil Engineering, University of Waterloo, Waterloo, Ont.

Kouwen, N., Soulis, E. D., Pietroniro, A., Donald, J., and Harrington, R. (1993). Grouped response units for distributed hydrologic modeling. *J. Water Resour. Plann. Manage.*, **119**(3), 289–305.

Krishna Prasad, V., Yogesh Kant, and Badarinath, K. V. S. (1999). Vegetation discrimination using IRS-P3 WiFS temporal dataset: A case study from Rampa forests, Eastern Ghats, A.P. *J. Indian Soc. Remote Sensing*, **27**(3), 149–154.

Kumar, V. S., Haefner, H., and Seidel, K. (1991). Satellite snow cover mapping and snowmelt runoff modelling in Beas basin. *20th General Assembly of the International Union of Geodesy and Geophysics*, Vienna, Austria, August 11–24, 1991, 101–109.

Kumar, D. N., Lall, U., and Peterson, M. R. (2000). Multi-site disaggregation of monthly to daily streamflow. *Water Resour. Res.*, **36**(7), 1823–1833, doi: 10.1029/2000WR900049.

Kummerow, C., Barnes, W., Kozu, T., Shiue, J., Simpson, J. (1998). The Tropical Rainfall Measuring Mission (TRMM) sensor package. *J. Atmos. Oceanic Technol.*, **15**, 809–817.

Kundzwicz, Z. W. (2004). Searching for change in hydrologic data, Editorial. *Hydrol. Sci. J.*, **49**(1), 3–6.

Lafferty, J., McCallum, A., and Pereira, F. (2001). Conditional random fields: Probabilistic models for segmenting and labeling sequence data. In *Proceedings 18th International Conference on Machine Learning*, C. E. Brodley and A. P. Danyluk, eds., San Francisco, Calif.: Morgan Kaufmann, 282–289.

Lahmer, W., Becker, A., Muller-Wohlfelt, D.-I., and Pfutzner, B. (1999). A GIS-based approach for regional hydrological modeling. In *Regionalization in Hydrology*, B. Diekkruger, M. J. Kirkby, and U. Schroder, eds., IAHS Publication No. 254, International Association of Hydrological Sciences, 33–43.

Lall, U., and Sharma, A. (1996). A nearest neighbour bootstrap for time series resampling. *Water Resour. Res.*, **32**(3), 679–693.

Lamb, H. H. (1972). *British Isles Weather Types and a Register of Daily Sequence of Circulation Patterns, 1861 –1971*. Geophysical Memoir, 116, HMSO, London, 85.

Lane, W. L. (1982). Corrected parameter estimates for disaggregation schemes. In *Statistical Analysis of Rainfall and Runoff*, V. P. Singh, ed., Littleton, Colo.: Water Resources Publications, 505–530.

Lang, M., Ouarda, T. B. M. J., and Bobée, B. (1999). Towards operational guidelines for over-threshold modelling. *J. Hydrol.*, **225**, 103– 117.

Laurenson, E. M. (1964). A catchment storage model for runoff routing. *J. Hydrol.*, **2**, 141–163.

Laurenson, E. M., and Mein, R. G. (1993). *RORB Version 4 Runoff Routing Program: Users Manual*. Monash University, Department of Civil Engineering, Monash, Victoria, Australia.

Laurenson, E. M., and Mein, R. G. (1995). RORB: Hydrograph synthesis by runoff routing. *Computer Models of Watershed Hydrology*, V. P. Singh, ed., Littleton, Colo.: Water Resources Publications.

Ledoux, E., Girard, G., de Marsily, G., and Deschenes, J. (1989). Spatially distributed modeling: Conceptual approach, coupling surface water and ground water. In *Unsaturated Flow Hydrologic Modeling: Theory And Practice*, H. J. Morel-Seytoux, ed., NATO ASI Series S 275, Boston: Kluwer Academic, 435–454.

Leonard J. A., Karmer, M., and Ungar, L. H. (1992). Using radial basis functions to approximate a function and its error bounds. *IEEE Trans. Neural Networks*, **3**(4), 624–627.

Levizzani, V., and Amorati, R. (2002). A review of satellite-based rainfall estimation methods. A look back and a perspective. In *Proceedings of the 2000 EUMETSAT Meteorological Satellite Data User's Conference*, 29 May–2 June 2000, Bologna, Italy, 344–353.

Liang, X., and Xie, Z. (2001). A new surface runoff parameterization with subgrid-scale soil heterogeneity for land surface models. *Adv. Water Resour.*, **24**(9–10), 1173–1193.

Liang, X., Lettenmaier, D. P., Wood, E. F., and Burges, S. J. (1994). A simple hydrologically based model of land surface water and energy fluxes for GSMs. *J. Geophys. Res.*, **99**(D7), 14415–14428.

Lillesand, T. M. and Kiefer, R. W. (2002). *Remote Sensing and Image Interpretation*. New York: John Wiley & Sons.

Loeve, M. (1955). *Probability Theory*. Princeton, N.J.: van Nostrand Company.

Ma, X., and Cheng, W. (1998). A modeling of hydrological processes in a large low plain area including lakes and ponds. *J. Jpn. Soc. Hydrol. Water Resour.*, **9**, 320–329.

Ma, X., Fukushima, Y., Hashimoto, T., Hiyama, T., and Nakashima, T. (1999). Application of a simple SVAT model in a mountain catchment model under temperate humid climate. *J. Jpn. Soc. Hydrol. Water Resour.*, **12**, 285–294.

Madsen, H., Rasmussen, P. F., and Rosbjerg, D. (1997). Comparison of annual maximum series and partial duration series methods for modeling extreme hydrologic events. 1. At-site modelling. *Water Resour. Res.*, **33**(4), 747–757.

Maidment, D. R. (1993). *Handbook of Hydrology*. New York: McGraw Hill.

Mamdani, E. H., and Assilian, S. (1975). An experiment in linguistic synthesis with a fuzzy logic controller. *Int. J. Man-Machine Stud.*, **7**: 1–13.

Mandelbrot, B. B. (1974). Intermittent turbulence in self-similar cascades: Divergence of high moments and dimension of the carrier. *J. Fluid Mech.*, **62**, 331–358.

Margulis, S. A., and Entekhabi, D. (2001). Temporal disaggregation of satellite-derived monthly precipitation estimates and the resulting propagation of error in partitioning of water at the land surface. *Hydrol. Earth System Sci.*, **5**(1), 27–38.

Martz, L. W., and de Jong, E. (1988). CATCH: a FORTRAN program for measuring catchment area from digital elevation models. *Comput. Geosci.*, **14**, **5**, 627–640.

MATLAB (1995). *Fuzzy Logic Toolbox*. The MathWorks Inc.

Maurer, E. P., and Duffy, P. B. (2005). Uncertainty in projections of streamflow changes due to climate change in California. *Geophys. Res. Lett.*, **32**(3), L03704.

Mays, L. W. (1996). *Water Resources Handbook*. New York: McGraw-Hill.

McCabe, G. J., and Wolock, D. M. (2002). A step increase in streamflow in the conterminous United States. *Geophys. Res. Lett.*, **29**(24), 38.1–38.4.

McCuen, R. H., and Snyder, W. M. (1986). *Hydrologic Modelling: Statistical Methods and Applications*. Upper Saddle River, N.J.: Prentice-Hall.

McKee, T. B., Doesken, N. J., and Kleist, J. (1993). The relationship of drought frequency and duration to time scale. In *Proceedings of the Eighth Conference on Applied Climatology, American Meteorological Society*, 179–184.

Meenu, R., Rehana, S., and Mujumdar, P. P. (2012). Assessment of hydrologic impacts of climate change in Tunga-Bhadra river basin, India with HEC-HMS and SDSM. *Hydrol. Process.*, accepted, DOI: 10.1002/hyp.9220.

Mehrotra, R., and Sharma, A. (2006). A nonparametric stochastic downscaling framework for daily rainfall at multiple locations. *J. Geophys. Res.*, **111**, D15101, doi:10.1029/2005JD006637.

Mehrotra, R., and Sharma, A. (2007). Preserving low-frequency variability in generated daily rainfall sequences. *J. Hydrology*, **345**, 102–120.

Mehrotra, R., and Sharma, A. (2009). Evaluating spatio-temporal representations in daily rainfall sequences from three stochastic multi-site weather generation approaches. *Adv. Water Resour.*, **32**(6), 948–962.

Mehrotra, R., Srikanthan, R., and Sharma, A. (2006). A comparison of three stochastic multi-site precipitation occurrence generators. *J. Hydrol.*, **331**, 280–292.

Meijerink, A. M. J. (1996). Remote sensing applications to hydrology: groundwater. *J. Hydrol. Sci.*, **41**(4), 549–562.

Mejia, J. M., and Rousselle, J. (1976). Disaggregation models in hydrology revisited. *Water Resour. Res.*, **12**(2), 185–186.

Menabde, M., Seed, A., and Pegram, G. (1999). A simple scaling model for extreme rainfall. *Water Resour. Res.*, **35**(1), 335–339.

Metcalf and Eddy, Inc., University of Florida, and Water Resources Engineers, Inc. (1971). *Storm Water Management Model, Vol. 1: Final Report*. EPA Report No. 11024DOC07/71 (NITS PB-203289), EPA, Washington, D.C.

Mezghani, A., and Hingray, B. (2009). A combined downscaling-disaggregation weather generator for stochastic generation of multisite hourly weather variables over complex terrain: Development and multi-scale validation for the Upper Rhone River basin. *J. Hydrol.*, **377** (3–4), 245–260.

Milly, P. C. D., Wetherald, R. T., Dunne, K. A., and Delworth, T. L. (2002). Increasing risk of great floods in a changing climate. *Nature*, **415**, 514–517.

Milly, P. C. D., Betancourt, J., Falkenmark, M., *et al.* (2008). Stationarity is dead: Whither water management. *Science*, **319**, 573–574.

Minville, M., Brissette, F., and Leconte, R. (2008). Uncertainty of the impact of climate change on the hydrology of a nordic watershed. *J. Hydrol.*, **358**(1–2), 70–83.

Mishra, S. K., and Singh, V. P. (2003). *Soil Conservation Service Curve Number (SCS-CN) Methodology*. Water Science and Technology Library, Vol. 42, Dordrecht, the Netherlands: Kluwer Academic Publishers, 513.

Mitasova, H., Hofierka, J., Zlocha, M., and Iverson, L. (2007). Modelling topographic potential for erosion and deposition using GIS. *Int. J. Geogr. Inf. Syst.*, 629–641.

Moik, H. (1980). *Digital Processing of Remotely Sensed Images*. NASA SP no. 431, Washington, D.C.

Montgomery, D., and Dietrich, W. (1988). Where do channels begin? *Nature*, **336**, 232–234.

Moore, I. D., Grayson, R. B., and Ladson, A. R. (1991). Digital terrain modelling: A review of hydrological, geomorphological and biological applications. *Hydrol. Process.*, **5**, 3–30.

Moore, R. J. (1985). The probability distributed principle and run-off production at point and basin scales. *Hydrol. Sci. J.*, **30**(2), 273–295.

Morin, G., Paquet, P., and Sochanski, W. (1995). *Le Modèle de Simulation de Quantité et de Qualité CEQUEAU, Manuel de Référence*. INRS Eau Rapport de Recherche No. 433, Sainte-Foy, Que.

Morin, G., Sochanski, W., and Paquet, P. (1998). *Le Modèle de Simulation de Quantité et de Qualité CEQUEAU-ONU, Manuel de Référence*. Organisation des Nations-Unies et INRS Eau Rapport de Recherche No. 519, Sainte-Foy, Que.

Morisette, J. T., Privette, J. L., Justice, C. O. (2002). A framework for the validation of MODIS Land products. *Remote Sensing Environ.*, **83**, 77–96.

Morris, M. D. (1991). Factorial sampling plans for preliminary computational experiments. *Technometrics*, **33**, 161–174.

Mujumdar, P. P., and Ghosh, S. (2008). Modeling GCM and scenario uncertainty using a possibilistic approach: Application to the Mahanadi River, India. *Water Resour. Res.*, **44**, W06407.

Murty, B. S., Panday, S., and Huyakorn, P. S. (2003). Sub-timing in fluid flow and transport simulations. *Adv. Water Resour.*, **26**, 477–489.

Naden, P. S. (1992). Analysis and use of peaks-over-threshold data in flood estimation. In *Floods and Flood Management*, A. J. Saul, ed., Dordrecht: Kluwer Academic, 131–143.

Nagesh Kumar, D. (2003). Strengths and weakness of ANN and its potential application in hydrology. *Proceedings Workshop of Artificial Neural Networks in Hydraulic Engineering*, Indian Society for Hydraulics, Pune, 1–8.

NAS (National Academy of Sciences) (2006). *Assessment of the Benefits of Extending the Tropical Rainfall Measuring Mission: A Perspective from the Research and Operations Communities*. Interim Report, http://www.nap.edu/catalog/11195.html (accessed December 26, 2011).

NASDA (2001). *TRMM Data Users Handbook*. National Space Development Agency of Japan, http://www.eorc.jaxa.jp/TRMM/document/text/handbook_e.pdf (accessed December 26, 2011).

Nash, J. E., and Sutcliffe, J. V. (1970). River flow forecasting through conceptual models. Part I: A discussion of principles. *J. Hydrol.*, **10**, 282–290.

Natale, L., and Todini, E. (1976a). A stable estimator for large models. 1: Theoretical development and Monte Carlo experiments. *Water Resour. Res.*, **12**(4), 667–671.

Natale, L., and Todini, E. (1976b). A stable estimator for large models. 2: Real world hydrologic applications. *Water Resour. Res.*, **12**(4), 672– 675.

Natale, L., and Todini, E. (1977). A constrained parameter estimation technique for linear models in hydrology. In *Mathematical Models of Surface Water Hydrology*, T. A. Ciriani, U. Maione, and J. R. Wallis, eds., London: Wiley, 109–147.

Needs Assessment Report (2010). *Bihar Kosi Flood – 2008*. Prepared by Government of Bihar, World Bank, and Global Facility for Disaster Reduction and Recovery.

Neitsch, S., Arnold, J., Kiniry, J., and Williams, J. (2000). *Soil and Water Assessment Tool User's Manual Version 2000*. Blackland Research Center, Texas Agricultural Experiment Station, Temple, TX.

Neitsch, S., Arnold, J., Kiniry, J., and Williams, J. (2001). *Soil and Water Assessment Tool: Theoretical Documentation*. Blackland Research Center, Texas Agricultural Experiment Station, Temple, Tex.

Nijssen, B., Lettenmaier, D. P., Lohmann, D., and Wood, E. F. (2001). Predicting the discharge of global rivers. *J. Climate*, **14**(15), 3307–3323.

Nohara, D., Kitoh, A., Hosaka, M., and Oki, T. (2006). Impact of climate change on river discharge projected by multimodel ensemble. *J. Hydrometeorol.*, **7**, 1076–1089.

Noilhan, J., and Mahfouf, J. F. (1996). The ISBA land surface parameterization scheme, global planet. *Climate Change*, **13**, 145–159.

O'Callaghan, J. F., and Mark, D. M. (1984). The extraction of drainage networks from digital elevation data. *Computer Vision, Graphics and Image Processing*, **28**, 323–344.

Ogden, F. L. (1998). *CASC2D Version 1.18 Reference Manual*. Department of Civil and Environmental Engineering Report U-37, CT1665–1679, University of Connecticut, Storrs, Conn.

Oliver, C., and Quegan, S. (2004). *Understanding Synthetic Aperture Radar Images*. Raleigh, N.C.: SciTech Publishing.

O'Loughlin, E. M. (1986). Prediction of surface saturation zones in natural catchments by topographic analysis. *Water Resour. Res.* **22**(5), 794–804.

Olsson, J. (1998). Evaluation of a cascade model for temporal rainfall disaggregation. *Hydrol. Earth Syst. Sci.*, **2**, 19–30.

Ormsbee, L. E. (1989). Rainfall disaggregation model for continuous hydrologic modelling. *J. Hydraul. Eng.*, **115**(4), 507– 525.

Ozga-Zielinska, M., and Brzezinski, J. (1994). *Applied Hydrology*. Warsaw, Poland: Wydawnictawa Naukowe, PWN (in Polish).

Pandey, S., Gunn, R., Lim, K. J., Engel, B., and Harbor, J. (2000). Developing a web-enabled tool to assess long-term hydrologic impacts of land-use change: information technology issues and a case study. *URISA J.* **12**(4), 5–17.

Panigrahi, D. P., and Mujumdar, P. P. (2000). Reservoir operation modelling with fuzzy logic. *Water Resour. Manage.*, **14**, 89–109.

Papadakis, I., Napiorkowski, J., and Schultz, G. A. (1993). Monthly runoff generation by nonlinear model using multi spectral and multi temporal satellite imagery. *Adv. Space Res.*, **13**(5), 181–186.

Peck, E. L., McQuivey, R., Keefer, T. N., Johnson, E. R., and Erekson, J. (1981). *Review of Hydrologic Models for Evaluating Use of Remote Sensing Capabilities*. NASA CR 166674, Goddard Space Flight Center, Greenbelt, Md.

Perumal, M., Moramarco, T., Sahoo, B., and Barbetta, S. (2007). A methodology for discharge estimation and rating curve development at ungauged river sites. *Water Resour. Res.*, **43**, W02412, doi:10.1029/2005WR004609.

Perumal, M., Moramarco, T., Sahoo, B., and Barbetta, S. (2010). On the practical applicability of the VPMS routing method for rating curve development at ungauged river sites. *Water Resour. Res.*, **46**, W03522, doi:10.1029/2009WR008103.

Petty, G. W., and Krajewski, W. F. (1996). Satellite estimation of precipitation over land. *J. Hydrol. Sci.*, **41**(4), 433–452.

Peuker, T., and Douglas, D. (1975). Detection of surface-specific points by local parallel processing of digital terrain elevation data. *Computer Graphics and Image Processing*, **4**, 375–387.

Pietroniro, A., Soulis, E. D., Kouwen, N., Rotunno, O., and Mullins, D.W. (1993). Using wide swath C-band SAR imagery for basin soil moisture mapping. *Canadian J. Remote Sensing*, Special issue, January, 77–82.

Pietroniro, A., Wishart, W., and Solomon, S. I. (1989). Use of remote sensing data for investigating water resources in Africa. In *Quantitative Remote Sensing: An Economic Tool for the Nineties, Proc. IGRASS'89*, IEEE, 2169–2172.

Pilling, C., and Jones, J. A. A. (1999). High resolution equilibrium and transient climate change scenario implications for British runoff. *Hydrol. Process.*, **13**, 2877–2895.

Plate, E. J. (2009). Classification of hydrological models for flood management. *Hydrol. Earth Syst. Sci.*, **13**, 1939–1951.

Potter, J. W., and McMahon, T. A. (1976). *The Monash Model: User Manual for Daily Program HYDROLOG*. Resource Report 2/76, Department of Civil Engineering, Monash University, Monash, Victoria, Australia.

Price, J. C. (1980). The potential of remotely sensed thermal infrared data to infer surface soil moisture and evaporation. *Water Resour. Res.*, **16**, 787–795.

Prudhomme, C., and Davies, H. (2009). Assessing uncertainties in climate change impact analyses on the river flow regimes in the UK. Part 2: Future climate. *Climatic Change*, **93**(1–2), 197–222.

Prudhomme, C., Reynard, N., and Crooks, S. (2002). Downscaling of global climate models for flood frequency analysis: Where are we now? *Hydrol. Process.*, **16**(6), 1137–1150.

Prudhomme, C., Jakob, D., and Svensson, C. (2003). Uncertainty and climate change impact on the flood regime of small UK catchments. *J. Hydrol.*, **277**, 1–23.

Quick, M. C. (1995). The UBC watershed model. In *Computer Models of Watershed Hydrology*, V. P. Singh, ed., Littleton, Colo.: Water Resources Publications.

Quick, M. C., and Pipes, A. (1977). UBC watershed model. *Hydrol. Sci. Bull.*, **XXI**(1/3), 285–295.

Ragan, R. M., and Jackson, T.J. (1980). Runoff synthesis using Landsat and SCS model. *J. Hydraulics Division, ASCE*, **106**(HY5), 667–678.

Rajagopalan, B., and Lall, U. (1999). A k-nearest-neighbor simulator for daily precipitation and other variables. *Water Resour. Res.*, **35**(10), 3089–3101.

Raje, D., and Mujumdar, P. P. (2009). A conditional random field based downscaling method for assessment of climate change impact on multisite daily precipitation in the Mahanadi basin. *Water Resour. Res.*, **45**(10), W10404, doi: 10.1029 / 2008WR007487.

Raje, D., and Mujumdar, P. P. (2010a), Constraining uncertainty in regional hydrologic impacts of climate change: Nonstationarity in downscaling. *Water Resour. Res.*, **46**, W07543, doi:10.1029/2009WR008425.

Raje, D., and Mujumdar, P. P. (2010b). Hydrologic drought prediction under climate change: Uncertainty modeling with Dempster–Shafer and Bayesian approaches. *Adv. Water Resour.*, **33**(9), 1176–1186, doi: 10.1016 / j.advwatres.2010.08.001.

Raje, D., and Mujumdar, P. P. (2010c). Reservoir performance under uncertainty in hydrologic impacts of climate change. *Adv. Water Resour.*, **33**, 312–326.

Rango, A. (1995). The snowmelt runoff model (SRM). In *Computer Models of Watershed Hydrology*, V. P. Singh, ed., Littleton, Colo.: Water Resources Publications.

Rango, A. (1996). Space-borne remote sensing for snow hydrology applications. *J. Hydrol. Sci.*, **41**(4), 477–494.

Rao, P. G. (1993). Climatic changes and trends over a major river basin in India. *Climate Res.*, **2**, 215–223.

Rao, P. G. (1995). Effect of climate change on streamflows in the Mahanadi river basin, India. *Water Int.*, **20**, 205– 212.

Rao, P. G., and Kumar, K. K. (1992). Climatic shifts over Mahanadi river basin. *Current Sci.*, **63**, 192–196.

Rawls, W. J., and Brakensiek, D. L. (1983). A procedure to predict Green and Ampt infiltration parameters. *Proceedings of the American Society of Agricultural Engineers Conference on Advances in Infiltration*, ASAE, St. Joseph, MI, 102–112.

Rawls, W. J., Brakensiek, D. L., and Soni, B. (1983). Agricultural management effects on soil water process, Part I: Soil water retention and Green–Ampt infiltration parameters. *Trans. ASAE*, 1747–1752.

Refsgaard, J. C., and Storm, B. (1995). MIKE SHE. In *Computer Models of Watershed Hydrology*, V. P. Singh, ed., Littleton, Colo.: Water Resources Publications.

Richardson, C. W., and Wright, D. A. (1984). *WGEN: A Model for Generating Daily Weather Variables*. ARS-8, US Department of Agriculture, Agricultural Research Service, Washington, D.C.

Rigby, E. H., Boyd, M. J., and vanDrie, R. (1999). Experiences in developing the hydrology model: WBNM2000. *Proceedings 8th International Conference on Urban Drainage*, Institution of Engineers, **3**, 1374–1381.

Ritchie, J. C., and Schiebe, F. R. (1986). Monitoring suspended sediment with remote sensing techniques. In *Hydrologic Application of Space Technology, IAHS*, **160**, 233–243.

Rockwood, D. M. (1982). Theory and practice of the SSARR model as related to analyzing and forecasting the response of hydrologic systems. In *Applied Modeling in Catchment Hydrology*, V. P. Singh, ed., Littleton, Colo.: Water Resources Publications, 87–106.

Rodriguez-Iturbe, I., Gupta, V. K., and Waymire, E. (1984). Scale considerations in the modelling of temporal rainfall. *Water Resour. Res.*, **20**, 1611–1619.

Rodriguez-Iturbe, I., Cox, D. R., and Isham, V. (1987). Some models for rainfall based on stochastic point processes. *Proc. R. Soc. London A*, **410**, 269–288.

Ross, T. J. (1997). *Fuzzy Logic with Engineering Applications*. Electrical Engineering Series, New York: McGraw-Hill.

Roubens, M. (1982). Fuzzy clustering algorithms and their cluster validity. *Eur. J. Oper. Res.*, **10**, 294–301.

Sabbins, F. F. Jr. (1986). *Remote Sensing: Principles and Interpretation*. New York: W. H. Freeman & Co.

Salmonson, V. V. (1983). Water resources assessment. In *Manual of Remote Sensing*, J. Colwell, ed., Bethesda, Md.: American Society of Photogrametry and Remote Sensing, 1497–1570.

Salvucci, G., and Song, C. (2000). Derived distributions of storm depth and frequency conditioned on monthly total precipitation: Adding value to historical and satellite-derived estimates of monthly precipitation. *J. Hydrometeorol.*, **1**, 113–120.

Scaefer, M. G., and Barker, B. L., (1999). *Stochastic Modeling of Extreme Floods for A. R. Bowman Dam*. MGS Engineering Consultants Report, Olympia, Wash.

Schmidli, J., Frei, C., and Vidale, P. L. (2006). Downscaling from GCM precipitation: A benchmark for dynamical and statistical downscaling methods. *Int. J. Climatol.*, **26**, 679–689.

Schmugge, T. J., Kustas, W. P., Ritchie, J. C., Jackson, T. J., and Rango, A. (2002). Remote sensing in hydrology. *Adv. Water Resour.*, **25**(8–12), 1367–1385.

Schultz, G. A. (1988). Remote sensing in hydrology. *J. Hydrol.*, **100**, 239–265.

Schultz, G. A. (1996). Remote sensing applications to hydrology: Runoff. *J. Hydrol. Sci.*, **41**(4), 453–476.

Sellers, P. J., Meeson, B. W., Hall, F. G., *et al.* (1995). Remote sensing of the land surface for studies of global change: Models – algorithms – experiments. *Remote Sensing Environ.*, **51**(3), 3–26.

Sentz, K., and Ferson, S. (2002). *Combination of Evidence in Dempster–Shafer Theory*. Sandia National Laboratories, 2002–4015.

Shafer, G. (1976). *A Mathematical Theory of Evidence*. Princeton, N.J.: Princeton University Press.

Short, N. M. (1999). *Remote Sensing Tutorial: Online Handbook*. NASA Goddard Space Flight Center, Greenbelt, Md.

Showalter, P. S. (2001). Remote sensing's use in disaster research: A review. *Disaster Prevent. Manage.*, **10**(1), 21–29, doi: 10.1108/09653560110381796.

Shu, C., and Ouarda, T. B. M. J. (2008). Regional flood frequency analysis at ungauged sites using the adaptive neuro-fuzzy inference system. *J. Hydrol.*, **349**, 31–43.

Simonovic, S. P. (2012). *Floods in a Changing Climate: Risk Management*. Cambridge, UK: Cambridge University Press.

Simonovic, S. P., and Li, L. (2004). Sensitivity of the Red River basin flood protection system to climate variability and change. *Water Resour. Manage.*, **18** (2), 89–110.

Simpson, J., Adler, R. F., and North, G. R. (1988). A proposed tropical rainfall measuring mission TRMM satellite. *Bull. Am. Meteorol. Soc.*, **3**(69), 278–295.

Singh, V. P. (1992). *Elementary Hydrology*. Upper Saddle River, N.J.: Prentice Hall.

Singh, V., and Woolhiser, D. (2002). Mathematical modeling of watershed hydrology. *J. Hydrol. Eng.*, **7**(4), 270–292.

Sittner, W. T., Scauss, C. E., and Munro, J. C. (1969). Continuous hydrograph synthesis with an API-type hydrologic model. *Water Resour. Res.*, **5**(5), 1007–1022.

Sivapalan, M. (2003). Prediction in ungauged basins: A grand challenge for theoretical hydrology. *Hydrol. Process.*, **17**, 3163–3170.

Sivapalan, M., Ruprecht, J. K., and Viney, N. R. (1996a). Water and salt balance modeling to predict the effects of land use changes in forested catchments: 1. Small catchment water balance model. *Hydrol. Process.*, **10**, 393–411.

Sivapalan, M., Viney, N. R., and Jeevaraj, C. G. (1996b). Water and salt balance modeling to predict the effects of land use changes in forested catchments: 3. The large catchment model. *Hydrol. Process.*, **10**, 429–4446.

Sivapalan, M., Viney, N. R., and Ruprecht, J. K. (1996c). Water and salt balance modeling to predict the effects of land use changes in forested catchments: 2. Coupled model of water and salt balances. *Hydrol. Process.*, **10**, 413–428.

Slingsby, A. (2003). An object-oriented approach to hydrological modelling using triangular irregular networks. *Proceedings of GISRUK03*, City University, London.

Slough, K., and Kite, G. W. (1992). Remote sensing estimates of snow water equivalent for hydrologic modelling. *Candian Water Resour. J.*, **17**(4), 323–330.

Smith, R. E., Goodrich, D. C., Woolhiser, D. A., and Unkrich, C. L. (1995). KINEROS – A kinematic runoff and erosion model. In *Computer Models of Watershed Hydrology*, V. P. Singh, ed., Littleton, Colo.: Water Resources Publications.

Soil Conservation Service (SCS) (1965). *Computer Model for Project Formulation Hydrology*. Technical Release No. 20, USDA, Washington, D.C.

Soil Conservation Service (SCS) (1969). Hydrology. In *SCS National Engineering Handbook*, USDA, Washington, D.C., Section-4.

Speers, D. D. (1995). SSARR model. *Computer Models of Watershed Hydrology*, V. P. Singh, ed., Littleton, Colo.: Water Resources Publications.

Stephenson, D. (1989). A modular model for simulating continuous or event runoff. *IAHS Publication*, **181**, 83–91.

Stephenson, D., and Randell, B. (1999). *Streamflow Prediction Model for the Caledon Catchment.* ESKOM Report No. RES/RR/00171, Cleveland, South Africa.

Stern, R. D., and Coe, R. (1984). A model-fitting analysis of daily rainfall data. *J. R. Statist. Soc. A*, **147**(1), 1–34.

Strupczewski, W. G., Singh, V. P., and Feluch, W. (2001). Non-stationary approach to at-site flood frequency modelling. I: Maximum likelihood estimation. *J. Hydrol.*, **248**, 123–142.

Subimal Ghosh (2007). Hydrologic impacts of climate change: Uncertainty modeling. PhD thesis, Indian Institute of Science, India.

Sugawara, M. (1995). Tank model. In *Computer Models of Watershed Hydrology*, V. P. Singh, ed., Littleton, Colo.: Water Resources Publications.

Sugawara, M., *et al.* (1974). Tank model and its application to Bird Creek, Wollombi Brook, Bikin River, Kitsu River, Sanga River and Nam Mune. Research Note, National Research Center for Disaster Prevention, No. 11, Kyoto, Japan, 1–64.

Swain, P. H., and Davis, S.M. (1978). *Remote Sensing: The Quantitative Approach.* New York: McGraw-Hil.

Tarboton, D. G. (1997). A new method for the determination of flow directions and upslope areas in grid digital elevation models. *Water Resour. Res.*, **33** (2), 309–319.

Tarboton, D. G. (2002). *Terrain Analysis Using Digital Elevation Models (TauDEM).* Utah State University, Logan, Utah.

Tatli, H., Dalfes, H. N., and Mentes, S. (2005). Surface air temperature variability over Turkey and its connection to large-scale upper air circulation via multivariate techniques. *Int. J. Climatol.*, **25**, 161–180.

Tennessee Valley Authority (TVA) (1972). *A Continuous Daily-Streamflow Model: Upper Bear Creek Experimental Project.* Research Paper No. 8, Knoxville, Tenn.

Tisseuil, C., Vrac, M., Lek, S., and Wade, A. J. (2010). Statistical downscaling of river flows. *J. Hydrol.*, **385**, 279–291.

Todini, E. (1988a). *Il modello afflussi deflussi del flume Arno. Relazione Generale dello studio per conto della Regione Toscana.* Technical Report, University of Bologna, Italy (in Italian).

Todini, E. (1988b). Rainfall runoff modelling: Past, present and future. *J. Hydrol.*, **100**, 341–352.

Todini, E. (1995). New trends in modeling soil processes from hillslopes to GCM scales. In *The Role of Water and Hydrological Cycle in Global Change*, H. R. Oliver and S. A. Oliver, eds., NATO Advanced Study Institute, Series 1: Global, Dordrecht, the Netherlands: Kluwer Academic.

Todini, E. (1996). The ARNO rainfall-runoff model. *J. Hydrol.*, **175**, 339–382.

Trigo, R. M., and Palutikof, J. P. (1999). Simulation of daily temperatures for climate change scenarios over Portugal: a neural network model approach. *Climate Res.*, **13**(1), 45–59.

Tripathi, S., Srinivas, V., and Nanjundiah, R. (2006). Downscaling of precipitation for climate change scenarios: A support vector machine approach. *J. Hydrol.*, **330**(3–4), 621–640.

Turcotte, R., Fortin, J.-P., Rousseau, A. N., Massicotte, S., and Villeneuve, J.-P. (2001). Determination of the drainage structure of a watershed using a digital elevation model and a digital river and lake network. *J. Hydrol.*, **240**, 225–242.

US Army Corps of Engineers (1987). *SSARR Users Manual.* North Pacific Division, Portland, Ore.

US Army Corps of Engineers (2002). *HEC-RAS Users Manual.* Hydrologic Engineering Center, Davis, Calif.

US Army Corps of Engineers (2008). *HEC HMS User Manual, Version 3.2.* Hydrologic Engineering Center, Davis, Calif.

US Department of Agriculture (1972). *Soil Conservation Service, National Engineering Handbook, Section 4, Hydrology.* Washington, D.C.: US Government Printing Office.

USDA (1980). *CREAMS: A Field Scale Model for Chemicals, Runoff and Erosion from Agricultural Management Systems.* W. G. Knisel, ed., Conservation Research Report No. 26, Washington, D.C.

Valencia, D. R., and Schaake, J. C. (1973). Disaggregation process in stochastic hydrology. *Water Resour. Res.*, **9**(3), 580–585.

Vandenberg, A. (1989). A physical model of vertical integration, drain discharge and surface runoff from layered soils. NHRI Paper No. 42. *IWD Tech. Bull.*, 161, National Hydrologic Research Institute, Saskatoon, Sask.

Viessman, W. Jr., Lewis, G. L., and Knapp, J. W. (1989). *Introduction to Hydrology.* New York: Harper Collins.

Vörösmarty, C. J., Moore, B., Grace, A. L., *et al.* (1989). Continental scale models of water balance and fluvial transport: An application to South America. *Global Biogeochem. Cycles.*, **3**, 241–265.

Vörösmarty, C. J., Federer, C. A., and Schloss, A. (1998). Potential evaporation functions compared on US watersheds: Implications for global-scale water balance and terrestrial ecosystem modelling. *J. Hydrol.*, **207**, 147–169.

Vörösmarty, C. J., Gutowski, W. J., Person, M., Chen, T.-C., and Case, D. (1993). Linked atmosphere–hydrology models at the macroscale. In *Macroscale Modeling of the Hydrosphere*, W. B. Wilkinson, ed., IAHS Publication No. 214,. 3–27.

Walker, A. E., Gray, S. A., Goodison, B. E., and O'Neill, R. A. (1990). Analysis of MOS-1 microwave scanning radiometer data for Canadian Prairie snow cover. In *Application of Remote Sensing in Hydrology. Proceedings Symposium No. 5, NHRI, Saskatoon, Canada*, G. W. Kite and A. Wankiewicz, eds., 319–330.

Wang, Y., Leung, L. R., McGregor, J. L., *et al.* (2004). Regional climate modeling: Progress, challenges, and prospects. *J. Meteorol. Soc. Japan*, **82**, 1599–1628.

Wankiewicz, A. (1989). Microwave satellite forecasting of snowmelt runoff. In *Quantitative Remote Sensing: An Economic Tool for the Nineties, Proceedings IGRASS'89*, IEEE, 1235–1238.

Wetherald, R. T. (2009). Changes of variability in response to increasing greenhouse gases. Part II: Hydrology. *J. Climate*, **22**, 6089–6103.

Whiting, J. (1990). Determination of characteristics for hydrologic modeling using remote sensing. In *Application of Remote Sensing in Hydrology. Proceedings Symposium No. 5, NHRI, Saskatoon, Canada*, G. W. Kite and A. Wankiewicz, eds., 79–91.

Wight, J. R., and Skiles, J. W., eds. (1987). *SPUR-Simulation of Production and Utilization of Rangelands: Documentation and User Guide.* Report No. ARS-63, USDA, Washington, D.C.

Wigley, T. M. L., Jones, P. D., Briffa, K. R., and Smith, G. (1990). Obtaining sub-grid-scale information from coarse-resolution general-circulation model output. *J. Geophys. Res. Atmos.*, **95**(D2), 1943–1953.

Wigmosta, M. S., Vail, L. W., and Lettenmaier, D. P. (1994). A distributed hydrology–vegetation model for complex terrain. *Water Resour. Res.*, **30**(6), 1665–1679.

Wilby, R. L. (2005). Uncertainty in water resource model parameters used for climate change impact assessment. *Hydrol. Process.*, **19**(16), 3201–3219.

Wilby, R. L., and Wigley, T. M. L. (2000). Precipitation predictors for downscaling: Observed and General Circulation Model relationships. *Int. J. Climatol.*, **20**(6), 641–661.

Wilby, R. L., Beven, K. J., and Reynard, N. S. (2008). Climate change and fluvial flood risk in the UK: More of the same? *Hydrol. Process.*, **22**, 2511–2523.

Wilks, D. S. (1992). Adapting stochastic weather generation algorithms for climate change studies. *Climatic Change*, **22**(1), 67–84.

Wilks, D. S. (2010). Use of stochastic weather generators for precipitation downscaling. *Climate Change*, **1**(6), 898–907.

Wilks, D. S., and Wilby, R. L. (1999). The weather generation game: A review of stochastic weather models. *Prog. Phys. Geogr.*, **23**(3), 329–357.

Williams, J. R. (1995a). The EPIC model. In *Computer Models of Watershed Hydrology*, V. P. Singh, ed., Littleton, Colo.: Water Resources Publications.

Williams, J. R., (1995b). SWRRB – A watershed scale model for soil and water resources management. In *Computer Models of Watershed Hydrology*, V. P. Singh, ed., Littleton, Colo.: Water Resources Publications.

Williams, J. R., Jones, C. A., and Dyke, P. T. (1984). The EPIC model and its application. *Proceedings ICRISAT-IBSNAT-SYSS Symposium on Minimum Data Sets for Agrotechnology Transfer*, 111–121.

Williams, J. R., Nicks, A. D., and Arnold, J. G. (1985). Simulator for water resources in rural basins. *J. Hydraul. Eng.*, **111**(6), 970–986.

Willmott, C. J., Rowe, C. M., and Philpot, W. D. (1985). Small-scale climate maps: A sensitivity analysis of some common assumptions associated with grid-point interpolation and contouring. *Am. Cartog.*, **12**, 5–16.

WMO (World Meteorological Organization) (1988). *Concept of the Global Energy and Water Experiments (GEWEX).* Report WCRP 5, Geneva, Switzerland.

Woolhiser, D. A., Smith, R. E., and Goodrich, D. C. (1990). *KINEROS – A Kinematic Runoff and Erosion Model: Documentation and User Manual.* Report No. ARS-77, USDA, Washington, D.C.

Wu, C. L., and Chau, K.W. (2006). Evaluation of several algorithms in forecasting flood. *IEA/AIE 2006, LNAI*, **4031**, 111 – 116.

Xu, C. Y. (1999a). Climate change and hydrologic models: A review of existing gaps and recent research developments. *Water Resour. Manage.*, **13**(5), 369–382.

Xu, C. Y. (1999b). Operational testing of a water balance model for predicting climate change impacts. *Agric. Forest Meteorol.*, **98–99**(1–4), 295–304.
Xu, C. Y. (2000). Modelling the effects of climate change on water resources in Central Sweden. *Water Resour. Manage.*, **14**, 177–189.
Xu, C. Y., Widén, E., Halldin, S. (2005). Modelling hydrological consequences of climate change: Progress and challenges. *Adv. Atmos. Sci.*, **22**, 789–797.
Yang, D., Herath, S., and Musiake, K. (1998). Development of a geomorphology-based hydrological model for large catchments. *Ann. J. Hydraul. Eng.*, **42**, 169–174.
Yates, D. N. (1997). Approaches to continental scale runoff for integrated assessment models. *J. Hydrol.*, **291**, 289–310.
Yates, D., Gangopadhyay, S., Rajagopalan, B., and Strzepek, K. (2003). A technique for generating regional climate scenarios using a nearest neighbor algorithm. *Water Resour. Res.*, **39**(7), 1199, doi:10.1029/2002WR001769.
Yoo, D. H. (2002). Numerical model of surface runoff, infiltration, river discharge and groundwater flow – SIRG. In *Mathematical Models of Small Watershed Hydrology and Applications*, V. P. Singh and D. K. Frevert, eds., Littleton, Colo.: Water Resources Publications.
Young, M. D. B., and Gowing, J. W. (1996). *PARCHED-THIRST Model User Guide*. Report, University of Newcastle upon Tyne.
Young, R. A., Onstad, C. A., and Bosch, D. D. (1995). AGNPS: An agricultural nonpoint source model. In *Computer Models of Watershed Hydrology*, V. P. Singh, ed., Littleton, Colo.: Water Resources Publications.
Young, R. A., Onstad, C. A., Bosch, D. D., and Anderson, W. P. (1989). AGNPS: A nonpoint source pollution model for evaluating agricultural watershed. *J. Soil Water Conservat.*, **44**, 168–173.
Yu, Z. (1996). Development of a physically-based distributed-parameter watershed (basin-scale hydrologic model) and its application to Big Darby Creek watershed. PhD dissertation, Ohio State University, Columbus, Ohio.
Yu, Z., and Schwartz, F. W. (1998). Application of integrated basinscale hydrologic model to simulate surface water and groundwater interactions in Big Darby Creek watershed, Ohio. *J. Am. Water Resour. Assoc.*, **34**, 409–425.
Yu, Z., Barron, E. J., Yarnal, B., *et al.* (1999). Simulating the river-basin response to atmospheric forcing by linking a mesoscale meteorological model and a hydrologic model system. *J. Hydrol.*, **218**, 72–91.
Yu, Z., Pollard, D., and Cheng, L. (2006). On continental-scale hydrologic simulations with a coupled hydrologic model. *J. Hydrol.*, **331**, 110–124.
Zadeh, L. A. (1965). *Fuzzy Sets. Information and Control*, **8**, 338–353.
Zhao, R. J., and Liu, X. R. (1995). The Xinjiang model. In *Computer Models of Watershed Hydrology*, V. P. Singh, ed., Littleton, Colo.: Water Resources Publications.
Zhao, R. J., Zhuang, Y.-L., Fang, L. R., Liu, X. R., and Zhang, Q. S. (1980). The Xinanjiang model. In *Proceedings Oxford Symposium on Hydrological Forecasting*, IAHS Publication No. 129, International Association of Hydrological Sciences, Wallingford, UK, 351–356.
Zimmermann, H. J. (1996). *Fuzzy Set Theory and Its Applications*. New Delhi, India: Allied Publishers (original edition: Kluwer Academic Publishers, 1991).

Index

Printed in the United States
By Bookmasters